Readings in Macroeconomics

Readings in Macroeconomics

Tim Jenkinson

OXFORD UNIVERSITY PRESS
1996

Oxford University Press, Walton Street, Oxford OX2 6DP

Oxford New York
Athens Auckland Bangkok Bogota Bombay
Buenos Airs Calcutta Cape Town Dar es Salaam
Delhi Florence Hong Kong Istanbul Karachi
Kuala Lumpur Madras Madrid Melbourne
Mexico City Nairobi Paris Singapore
Taipei Tokyo Toronto
and associated companies in
Berlin Ibadan

Oxford is a trade mark of Oxford University Press

Published in the United States by
Oxford University Press Inc., New York

© Oxford University Press, 1996

British Library Cataloguing in Publication Data
Data available

Library of Congress Cataloging in Publication Data
Readings in macroeconomics / [edited by] Tim Jenkinson.
Includes bibliographical references.
1. Macroeconomics. I. Jenkinson, Tim, 1961-
HB172.5.R4 1996 339—dc20 95-51467
ISBN 0-19-829065-9
ISBN 0-19-829064-0 (Pbk)

Printed in Great Britain
on acid-free paper by
Bookcraft (Bath) Ltd., Midsomer Norton, Avon

Preface

Demand often results in supply. The idea of producing a book collecting together many of the important and widely cited papers published in the *Oxford Review of Economic Policy* initially came from my students. They complained to me about the difficulty of finding copies of the *Review* in libraries when everyone taking a course was after the same issue. They told me how good it would be to have a selection of the most useful papers readily available. They even told me they might buy such a collection. And so the idea to produce this reader on macroeconomics, along with the companion volume on microeconomics, was born.

The first volume of the *Oxford Review of Economic Policy* was published in 1985. Two features have continued to distinguish the *Review* from other economics journals: the thematic approach, whereby each issue is focused on a particular subject, with leading authorities commissioned to contribute, and an insistence that articles should be written in a non-technical style. From the start the editors have stressed that articles should be accessible to students, policy-makers, journalists—in fact, anyone with an interest in economic issues. There has also been an insistence that policy implications should be discussed fully, rather than relegated to an occasional footnote. As much of economics has become more technical, specialized, and mathematically sophisticated, the role played by the *Review* in cutting through the algebra and stating ideas in words has become increasingly valuable. As a result, the *Review* has established a strong and growing readership amongst students, academics, journalists, teachers, and economists in the private and public sectors, as well as an impressive list of contributing authors.

These volumes of collected papers draw together some of the most important and widely cited articles published in the first ten volumes of the *Oxford Review of Economic Policy*. I hope that the books will fulfil a number of aims. First, they will ensure that the original articles remain easily accessible. Second, they will provide a resource to those teaching core macro- and microeconomics courses, and may be used as a ready-made 'reading pack' by lecturers (hence avoiding the trouble and cost of obtaining copyright clearance on individual articles). Third, they will enable students and others interested in economic policy to acquire a selection of the most useful papers at a low cost and reduce the need to spend hours slaving over the photocopier.

The production of these volumes also presented a useful opportunity to update or revise the original articles in the light of developments since they were first published. Many of the authors have taken advantage of this opportunity and updated and revised their papers to take into account recent advances in the literature, policy developments, and more recent data. This periodic spring-cleaning should ensure that even the papers published in early volumes of the *Review* stay relevant and useful. Some papers, especially those published relatively recently, have not been changed.

By far the most difficult task in putting together these volumes was to decide which papers to include. Over the last ten years a large number of high quality and widely cited papers have been published in the *Review*—many more than could be included in two books of a reasonable length. In making the selection I was concerned to cover all the main areas of macro- and microeconomics, which inevitably meant that some of the less 'mainstream' areas were neglected altogether. This is not to deny the significance and importance of these areas. Indeed, it is hard to think of an economic policy area more important than, say, the economics of transition to market economies, to which the *Review* devoted two issues in 1991–2, even though this specific topic is excluded from these more general volumes. A slightly different problem was that some issues contained many important articles, but only one or two could be included. There are thus many interesting and important papers not included in this selection.

An important criterion used in selecting papers was that the volume should contain a balance between those papers that surveyed a particular theoretical literature and those that were more focused on applied policy issues. It was also decided to go for depth of coverage rather than breadth. Consequently, the volume contains a variety of articles on each of the five selected areas:

- Fiscal Policy
- Exchange Rates and Monetary Policy
- Consumption and the Balance of Payments
- Inflation and Unemployment
- Growth and Productivity

Whilst this has resulted in the exclusion of several other areas—such as business cycles and international policy coordination—all the selected subject areas have been the focus of extensive academic research and policy debate in recent years, and would be covered by intermediate or advanced courses on macroeconomics.

Finally, I should like to thank a number of people who helped turn the idea to produce these volumes into reality in little more than six months. An important role was played by Andrew Balls who, rather than spend the weeks immediately after his final examinations relaxing on the beach, instead read large numbers of articles and advised me on the content of the volumes, the required revisions to papers, and also on student preferences. Most of the editorial work on the volumes took place while I was on leave at Dartmouth College, USA. Dartmouth provided me with a haven from normal administrative chores, and an excellent e-mail link that kept me in touch with those who actually produced the books. Among these, Tracy Mawson and Jenni Scott at OUP supervised the production process with great efficiency and speed, and Alison Gomm shouldered much of the responsibility for dealing with revisions, liaising with authors, and much more with her characteristic effectiveness. Over the years that I have been managing editor of the *Oxford Review of Economic Policy* I have accumulated a tremendous debt of gratitude to Alison, whose contribution has been far greater than that of a conventional production editor.

Last and certainly not least, the success of the *Oxford Review of Economic Policy* owes much to the efforts and vision of its editors—Chris Allsopp, Andrea Boltho, Dieter Helm, Gerry Holtham, Colin Mayer, Ken Mayhew, and Derek Morris—greatly assisted by an active and supportive editorial board. Their editorial input was critical in the commissioning and production of the original papers.

Tim Jenkinson
January 1996

Contents

List of Figures, Tables, and Boxes

List of Figures

List of Tables

List of Boxes

PART I

FISCAL POLICY

The role of fiscal policy in the 1990s

CHRISTOPHER ALLSOPP

New College, Oxford[1]

I. Introduction

European and British economic policy-makers face a number of extremely difficult policy choices. There is a widespread feeling that economic policy is in a muddle: that the conventional framework of the 1980s, which emphasized monetary control and labour-market flexibility, is no longer adequate for the problems of the 1990s.

In Europe, this is most obvious in the growing concern over unemployment which, at about 11 per cent, has now reached a level exceeding even that of the Great Depression. Most developed countries, moreover, face —or, at least, think they face—a fiscal crisis. The fact that problems of high public borrowing and rising national debt stocks are self-evidently in part a result of recession, slow growth, and (in Europe) persistent unemployment, does not make the situation any more comfortable: conventional responses of the fiscal kind, whether macroeconomic or in the form of more directed labour-market measures, seem to be blocked off. And, though the situation varies considerably between countries, there are limits to the use of monetary policy imposed by worries over inflation or by exchange-rate objectives (such as those which were embodied in the disintegrating exchange-rate mechanism (ERM)). Even without formal exchange-rate commitments or

objectives, increasing international integration limits the freedom of manoeuvre for any one country.

With unemployment moving to the top of the European policy agenda, it is natural that attention should focus on supply-side or structural responses to the problem. Such an assignment of instruments fits in with the policy presumptions of the 1980s, with macroeconomic policy (especially monetary and exchange-rate policy) assigned to the control of inflation whilst industrial and labour-market policy is supposed to look after the real side. But real-side problems— unemployment and, for the UK, poor productivity and competitiveness—look large in relation to the instruments that can be deployed. Moreover, supply-oriented policies, too, are increasingly constrained by international integration and, within the European Union (EU), by institutional developments such as those called for in the Maastricht Treaty.

The issues that arise during transition to larger political/economic units are difficult and contentious enough, even in tranquil times. In the 1990s they are doubly difficult. The European economy has had to face the shock of German economic and monetary union (GEMU) and the policy responses to it. The ERM has effectively been broken and the Maastricht timetable of moves towards European monetary union (EMU) postponed—at the very least. (Even if it had remained on track, the UK was not committed to the social chapter or to monetary union within the timetable.) Above all, the persistence of the longer-term structural problems, of unemployment and competitiveness, intensify other difficulties and raise the spectre of competitive rather than co-operative responses by individual countries. The main positive feature is

First Published in *Oxford Review of Economic Policy*, vol. 9, no. 3 (1993). This version has been updated and revised to incorporate recent events.

[1] The author would like to thank Christopher Bliss, Andrea Boltho, Tim Jenkinson, Ken Mayhew, and John Walker for helpful discussions during the course of the preparation of this paper. The usual disclaimers apply.

that inflationary pressure, world-wide and in most of Europe, is low and interest rates are falling.

Much of this paper is concerned with the policy issues as they appear for the UK. These cannot, however, be looked at in isolation. The UK having been forcibly ejected from the ERM of the European monetary system (EMS) in 1992, then pursued, somewhat by accident, a reflationary recovery strategy based on sharply lowered interest rates, exchange-rate depreciation, and fiscal stimulus.[2] The fiscal deficit rose to about 8 per cent of gross domestic product (GDP). A combination of tax rises (and announced future tax rises) introduced in the November 1993 Budget and economic recovery markedly lowered expected deficits and debt, so that the UK became one of the few European countries expected to meet the Maastricht fiscal criteria for 1997. The fiscal improvement was, however, heavily dependent on sustained recovery, and the underlying issue of how to combine fiscal consolidation with recovery remains.

But this is not the only nor even the main issue facing UK policy-makers. The ERM had, in effect, replaced the medium-term financial strategy (MTFS) as the anchor or framework for macroeconomic policy. It was supposed to provide a stable and credible framework for the longer-run control of inflation. That framework no longer exists and policy is being run on the basis of short-term expediency, with interest rates as the main instrument for affecting both growth and inflation. A new framework within which monetary, fiscal, and exchange-rate policies can be co-ordinated is urgently needed.

That is not just a problem for the UK. Other countries, too, need to redefine their policies within Europe. And the European Union as a collective, needs an agreed framework within which national policies can be co-ordinated. That is not easy, given that the previous framework is widely seen to have failed in important respects—as demonstrated, especially, by persistent and rising unemployment—and that its macroeconomic backbone, the developing EMS, fell apart at the first real hurdle. Even more than in the UK, the problem of combining 'fiscal consolidation' with improved growth and employment appears difficult to solve.

The main theme of this paper is that the underlying framework of policy, in the UK and in Europe, needs to change. Macroeconomic policy, as practised in the

1980s, has failed to produce a stable environment: it has interacted unfavourably with shocks (such as oil-price changes, the European investment boom in the late 1980s, and, in the 1990s, GEMU). A different assignment of instruments is needed, with, in particular, a more active role for fiscal policy in offsetting shocks.

The lack of such a framework is the most important reason for the break-up of the ERM. Fixed exchange rates tie monetary policy. Without capital controls to give some autonomy to domestic monetary policies (they were swept away as part of the single-market programme) and without any other policies to manage the economy, it is only a matter of time before some shock makes policy incredible, leading to uncontainable speculation. The obvious candidate for an additional instrument is fiscal policy. (That was well understood in the Bretton Woods era.)

But the use of fiscal policy for stabilization purposes, or to offset particular shocks, is not simple. The dynamics may be very difficult. About the very worst thing that governments could do, given the problems, is to tackle problems of debt and deficits head on, targeting debt or deficits directly: that is a recipe for even more instability, and for not actually meeting the objectives. And short-term mistakes, because of hysteresis effects, can turn into long-term problems—such as persistent high unemployment.

The difficulties with fiscal policy account for the obvious attraction of alternative ways of getting out of recession, such as lowering interest rates and devaluing the currency. But there are two big dangers in this. The first is that this may be a beggar-thy-neighbour strategy, competitive devaluation on the 1930s' model. Alternatively, if generalized, the devaluation gain is lost and the system is relying on interest-rate declines. This may be appropriate, or it may not. It relates to the second problem, which is that monetary and exchange-rate policy is being assigned to the short-term stabilization of expenditure, and this looks like the opposite of a sensible longer-term strategy against inflation, which needs stable exchange rates or some other medium-term anchor against price and wage pressures.

This takes the argument back to fiscal policy. The difficulties, especially the unfavourable dynamics, with fiscal policies have to be faced. One thing that would help would be to *reframe* the problems of deficits and debt as being about the *private sector*. Public deficits are highly endogenous. Their cure requires changes in the private sector. That is where the focus of policy action needs to be.

After some recovery in 1994, the European economy slowed again in 1995. The slowdown suggests a fall in interest rates. Despite this, there are serious risks that

[2] It is interesting to contrast this response in September 1992 to that in France in August 1993. Unlike in the UK, the forcing out of the franc did not lead to a U-turn on policy: French interest rates were kept (uncomfortably) high and depreciation limited to about 3 per cent. In essentials, policy was not changed, despite the failure of its institutional base.

Table 1.1. Debt Ratios and Deficits

	Net public debt % of nominal GNP/GDP				General government financial balances % of nominal GNP/GDP			
	1980	1985	1990	1993[a]	1980	1985	1990	1993[a]
Europe	26.3	39.5	40.9	50.6	−3.3	−4.4	−3.4	−6.4
Germany	14.3	21.8	22.8	27.8	−2.9	−1.2	−2.0	−4.1
France	14.3	22.9	25.0	35.2	0.0	−2.9	−1.5	−5.7
Italy	53.6	81.5	98.9	111.6	−8.6	−12.6	−10.9	−9.5
UK	47.5	45.8	28.5	42.6	−3.4	−2.9	−1.3	−8.3
Belgium	68.9	112.2	119.7	128.5	−9.2	−8.8	−5.7	−6.6
Netherlands	25.0	42.0	57.0	60.6	−4.1	−4.6	−4.9	−3.6
Spain	7.8	27.6	31.3	40.0	−2.5	−6.9	−3.9	−5.4
Sweden	−13.6	15.6	−5.4	16.2	−4.0	−3.8	4.2	−13.0

Note: [a] OECD forecasts.
Source: OECD.

recovery will be delayed or very slow. (Slow growth in Europe is also a major risk for the UK.) This suggests that Europe is likely to need a co-ordinated set of macroeconomic policy measures including fiscal expansion.

This, however, is not to deny that structural problems exist, nor to claim that macroeconomic management by itself will be enough. Rather, the argument is that continued recession and slow growth, combined with the perceived fiscal crisis, would rule out needed supply-side measures as well. A moderate macro-economic expansion, combined with measures on the supply side, would have the best chance of dealing, over time, with structural problems, and the best chance of *improving* the fiscal balance in the medium term.

The sections below deal, respectively, with fiscal policy, with monetary and exchange-rate issues, and with supply-side policies. Section V concludes.

II. Fiscal policy

The conventional wisdom of the 1980s played down the role of fiscal policy. In Europe, there has been, ever since the first oil shock, a general objective of lowering deficits and national debt levels. In the Organization for Economic Co-operation and Development (OECD) area as a whole, national debt has nearly doubled from about 35 per cent of GDP in the early 1970s to about 67 per cent in 1993, and it is continuing to rise rapidly. (Table 1.1 shows some data for selected European countries.) In the UK, the picture is different—debt has fallen during most of the period since the war, but rose extremely rapidly in the first half of the 1990s, both absolutely and as a share of GDP. (See Figure 1.2, later in this section.) At a more mundane level, Luxembourg was the *only* EU country which, in the early 1990s, met the fiscal convergence criteria laid down in the Maastricht treaty (government deficits limited to 3 per cent of GDP and debt ratios limited to 60 per cent).

The role of fiscal policy within the overall framework of policy was made most clear in the case of the MTFS adopted in 1980 in the UK. (The underlying justification was, however, never spelled out convincingly—see Allsopp, 1985.) Broadly, however, it was claimed that fiscal policy was important as an adjunct to monetary policy and it was frequently argued, though without much justification, that control of the public-sector borrowing requirement (PSBR) was necessary in order to control 'money'. The direct effects of fiscal policy on private-sector spending were played down: there was an objective of restraining public expenditure; stabilization policy of a fiscal kind was regarded as inappropriate and ineffective and, in the medium or long term, low public borrowing (and stable or falling national debt) was seen as a component in the provision of a stable nominal financial framework. In the later 1980s, as policy became more pragmatic, fiscal effects seemed to be given more weight, but serious errors were made (e.g. with hindsight there are few defenders of the 1988 budget, which should have been more restrictive).

Less formal but similar views conditioned policy in many other European countries. In practical terms, the most important aspects were a general objective of lowering public borrowing and debt levels, and an unwillingness to use fiscal policy to manage the economy or to offset short-term or long-term shocks.

These views, it is argued below, have been damaging

and need to change. This section starts by outlining some theoretical considerations relating to the proper role of fiscal policy before going on to illustrate the issues in three important cases. These are first, the credit boom and its aftermath in the UK, second, the fiscal effects of GEMU, and, finally, the longer-term problem of European debt.

1. Some theoretical considerations

There is no dispute amongst economists that public expenditure, whether real public expenditure, such as spending on roads, or transfers, such as income support, raises microeconomic issues of resource allocation. One line of argument that might follow from such a focus is that expenditure programmes, as well as the tax and benefit system, should be designed with medium-term resource-allocation and welfare objectives in mind—see Flemming (1993) on arguments for a medium-term approach.

Even if this view is taken, however, there are still questions of a more macroeconomic kind that need to be addressed. One concerns the appropriate size of the public sector's surplus or deficit position in the medium term, which in turn involves questions about the significance of different levels of public-sector debt. Another concerns stabilization policy. Should components of expenditure or tax be varied in response to shocks or, for example, counter-cyclically? This is not just a question of discretionary policy but also involves the degree of built-in or automatic stabilization. Moreover, the counterpart to tax and expenditure variations will be variations in the budget balance. What is the appropriate response to these and how should they be financed?

Keynesian stabilization policy, of course, is normally taken as involving acceptance of the automatic stabilizers and of discretionary policy as well, implying a tolerant attitude, at least in the short term, to consequences for the budget balance. This framework was rejected with the earlier and more rigid versions of the MTFS in the UK. Variations in the deficit or surplus position can, however, be justified on other grounds: for example, by the desirability of smoothing tax rates in the face of shocks or cyclical developments to avoid distortions and welfare losses.

The key question is whether (and if so why and how) changes in the fiscal position affect the *private* economy. Most of the issues can be conceptualized in terms of a simple choice between tax and bond finance—i.e. by posing the question of what difference would be made if a given bit of public expenditure were financed by current taxes as opposed to running a budget deficit and financing it by borrowing in the form of government bonds.

Ricardian equivalence

One of the most influential stories about fiscal policy which conditioned attitudes during the 1980s was the neo-Ricardian equivalence proposition which had been extended and popularized by Barro (1974). It is a proposition that needs to be taken seriously—even by those who believe in fiscal stabilization.

Ricardian equivalence suggests that the answer to the above question on tax versus bond finance is that it should make no difference: this aspect of fiscal policy has no effect. Ministers should not worry about taxes versus deficits. (They should worry about expenditure because that would have real resource effects.) Raising taxes to cut borrowing would not threaten to abort a recovery or lower the growth rate. By the same token, not doing that and continuing to run a large budget deficit would not matter either.

One way of looking at Ricardian equivalence is the proposition that national debt in the form of government bonds, the cumulation of bond-financed deficits, should (if distributional and other 'small' effects are ignored or assumed away) have no macroeconomic effects. Basically, the present value of future tax payments balances the value of the bonds: bonds are not net wealth. Alternatively, interest payments on national debt are just a transfer payment, which is how they are treated in the national accounts. (Barro's 1974 contribution was to deal rigorously with a standard objection—that individuals who do not live forever would apply a higher discount rate (discounting distant tax payments altogether) so that the present value of tax payments would not sum to the current value of the bonds.)

There are, of course, lots of eminently sensible objections to the pure proposition (see, for example, the critical discussion in Tobin, 1980). But it is difficult to escape the basic thrust. Consider, for example, a government policy of handing out to the private sector a quantity of government gilt-edged stock (a bond drop by analogy with Friedman's famous 'helicopter money' drop). Would such a policy make a country richer as a direct consequence of people having more wealth? Surely not.

But, crucially, to accept the proposition that domestic national debts are not net wealth in the sense that a rise in the capital stock would be, does *not* mean that fiscal policy has no effects on private-sector behaviour. A bond handout (a fiscal deficit financed by bonds) would be likely to affect spending behaviour, and hence real activity and/or inflation, in an economy where a significant portion of economic agents (firms or households) were rationed and subject to *cash flow and liquidity* constraints. (Such constraints are, of course quite explicitly assumed away in Barro's analysis.)

Consider the switch from financing an amount of expenditure by current taxes to a deficit financed by bonds. For the private sector as a whole, private after-tax incomes would rise in the current period and there would be a rise in the holdings of bonds. On average for the private sector, the Ricardian argument depends on the increment in income being balanced in present value terms by future tax payments to pay the additional interest on the bonds. A representative (average) agent is simply unaffected by a contract which supplies money in the present that has to be paid back with interest in the future. But the 'contract' is a kind of *credit* of a particularly favourable sort. So the proposition will only stand if there are no credit constraints—i.e. if there are no cash flow or liquidity effects which limit access to a perfect capital market. Millionaires, it may be supposed, are not often limited by credit constraints or liquidity problems. Other agents—individuals and firms—are.

This, then, is the argument. A bond-financed fiscal stimulus works like a reduction in credit and liquidity constraints in imperfectly functioning markets. Such an analogy warns us that fiscal policy may lose its effectiveness as financial markets are liberalized and develop: in the extreme, the Ricardian paradigm—or something approaching it—could actually apply.[3]

If a fiscal stimulus is analogous to credit liberalization, there is a practical implication of importance when considering UK fiscal policy. The credit boom in Britain in the late 1980s was very *like* an expansionary fiscal impulse, though it arose for quite different reasons. The policy implications are developed below, but one is immediate and can be stated. The appropriate *offset* to a credit explosion would appear to be a fiscal tightening and not a rise in interest rates. And the appropriate offset to debt deflation is a budget deficit.

Some longer-term considerations

It has been argued above that variations in national debt through deficit finance will affect private-sector spending decisions in a way analogous to releasing credit constraints. This raises the question as to what

longer-term effects might be anticipated. Do high levels of national debt, reflecting sustained deficits, matter? Governments clearly think they do, but seem confused about why.

On an extreme Ricardian view, levels of (domestic) national debt do not matter. Even a small step towards reality suggests, however, that high levels are better avoided. If the taxes levied to pay the interest are distortionary, then there is a welfare loss, though, in practice, this loss is likely to be small.[4]

Most modern treatments of the problem accordingly focus on sustainability: primary deficits over the medium or longer term have to be consistent with convergence of the debt ratio. Alternatively, explosive tracks for debt need to be avoided. One of the most influential stories, due to Sargent and Wallace (1981), is that financial-market anticipations would lead to the immediate monetization of deficits expected to be explosive, which would be inflationary.

If debt stocks have the kind of money-like 'liquidity-constraint breaking' effects referred to above, a link via anticipated monetization is not needed; high levels of debt (and especially increases in debt) could be *directly* inflationary. Inflation would not, however, be the only possible effect. The main alternative is that national debt could crowd out the real capital stock.

A sketch of the argument would go as follows. National debt can be seen as a substitute for claims on the real capital stock in satisfying the private sector's demand for savings and assets to hold. High national debt, as well as other things that affect the demand for assets (such as state pension schemes) would thus, in long-run equilibrium, go with a lower capital stock. Thus, one concern of governments could be that they want more capital accumulation and that national debt somehow prevents this. The problem with this line of argument is that the causality could equally—and often more plausibly—go the other way, from deficient investment in relation to savings, leading to deficits and debt.

Returning to the policy problem, there seems to be agreement that explosive tracks for debt ratios need to be avoided, That, however, is consistent with wide variations in the level of debt, and certainly does not rule out quite long periods of deficits and rising debt,

---

[3] Note, however, that fundamental problems of assymetric information mean that there will always be rationing and capital markets can never be perfect. See Stiglitz and Weiss, 1981.

[4] If sustainability is looked at in terms of the maintenance of a particular *ratio* of debt to GDP, the relevant discount rate is the real rate of interest *minus* the real growth rate, and the continuing tax burden for realistic debt levels would be quite low. For example, if the debt ratio were 100 per cent, real interest rates 3 per cent, and growth 2 per cent, the tax burden would be 1 per cent of GDP. The welfare loss would be a fraction of that. The burden would be negative if growth exceeded real interest rates.

so long as deficits can plausibly be expected to adjust in the future, for example by policy action. Beyond that, high levels of debt are likely to have 'wealth effects', affecting savings behaviour of individuals and firms, which means that, starting from a position of full potential output, an increase in debt could be seen as generating inflation and/or crowding out of the private capital stock.[5]

Fiscal policy objectives and reaction functions

It should be clear from the above that irresponsible fiscal policies are possible (which is not the case in the pure Ricardian position) and should be avoided. The two dangers are inflation and some form of crowding out of the capital stock. It is easy to think of many examples of irresponsible fiscal policies leading to inflation. Moreover, there is an influential class of models where the rate of inflation is a function of the deficit: the equilibrating mechanism working through the inflation tax. Equally, if the tendency for excess demand and inflation were fought through high and perhaps rising interest rates, crowding out would result. Or, in an open economy with a credibly fixed exchange rate, excess fiscal expansion would be a cause of balance of payments deficits, and a reduction in the net wealth of the economy or region.

The key to all these examples, however, is *excess*. It is an excess of nominal spending, actual or potential, which, depending on circumstances and assumed policy changes, leads to the adverse outcomes. The problem is that it is a *non sequitur* to deduce from the possibilities outlined above any particular rule for public borrowing and debt. The appropriate position in the longer run will depend on a host of factors, including pension arrangements, the extent of financial development, and so on. The basic difficulty, as should be clear, is that the appropriate position of the public sector depends, not on the public sector, but on the *private* sector.

This can be illustrated by returning again to the Ricardian paradigm. In the formal model as used, for example, by Barro (1974), the private sector functions perfectly, interest rates are at their appropriate level, and the money supply pins the price level. Taxes are non-distortionary. The amount of national debt does not matter. Moving towards reality, however, stoch-

astic shocks affecting the private sector (or public expenditure) would affect the deficit. Targeting the deficit would involve varying taxes with welfare losses. The assumed characteristics of the private sector justify tax smoothing.

With wealth effects from changes in national debt, a focus on the private sector is even more important. There is, then, another way of looking at the policy issue. That is to allow the system to *reveal* the appropriate policy whilst bearing in mind the potential longer-term effects. The longer-term wealth effects may be stabilizing if they are allowed to feed through to policy responses of an appropriate kind.

A simple example illustrates. If deficits are run in the short term to offset recessionary forces, this will lead to a rise in debt stocks, the counterpart of which is a rise in assets in the private sector. If liquidity and wealth effects are present, there is an *integral* feedback raising spending. Policy, including, of course, any effect from automatic stabilizers, needs to adjust as the wealth effects come through, the public deficit falls and the debt track stabilizes or reverses. The accent, as has frequently been argued, is on 'balancing the economy, not the budget'. But the debt position is stable so long as wealth effects are sufficiently powerful and policy can indeed adjust as appropriate.

The practical situation may be more complex and more difficult. Nevertheless, there is still a case for concentrating on the *private sector*'s position. If public debt is rising on a worrying trend—as it is in the OECD countries—one policy reaction is to change the public-sector position: to raise taxes or cut spending. If this is done, what is happening is that the private sector has to sort out its own problems which are no longer being offset by the public sector. The process in real economies could be painful, and possibly not convergent. The alternative line would be to focus on the features of the economy that lead to (are the counterpart of) the public-debt trend. In a closed economy that implies a build-up of assets in excess of investment (private savings exceed private investment). That problem has to be addressed. For example, if the problem is deficiency of investment, then that should be the focus. If policy is successful, then the public-debt trend will also be stabilized so long as budgetary policy adjusts in an appropriate way. Again, a direct attack on the symptom—the public position—may simply fail or be counterproductive. The point is, yet again, to manage the economy not the budget. The budget position or the track of national debt may, however, be an important indicator that all is not well and that policy, directed towards the private sector, does indeed need to change.

---

[5] In formal approaches the longer-term effect of debt expansion is usually taken to be crowding out rather than inflation. This is because of the assumption of a fixed money supply. For policy purposes this is an unrealistic assumption. Crowding out would most plausibly occur owing to a policy reaction by the authorities, raising interest rates. In practice, governments usually seem more worried about inflation than crowding out. For a discussion of crowding out in this context, see Dornbusch, 1986, Ch. 7.

Dynamics

The usual way in which the dynamics of debt are treated is to focus on the government's budget constraint which relates, *by identity*, the government deficit to its financing counterpart. (Thus it is not really a constraint on behaviour except to the extent that even governments must obey accounting identities!) Ignoring money finance and seigniorage, the deficit, D, must be equal to the change in the bond stock, ΔB (the stock of bonds = B). But the deficit must be equal to the primary deficit $(G - T)$ plus interest flows, rB (G is government expenditure, T stands for taxes, and r is the interest rate):

$$\Delta B = D = (G - T) + rB.$$

This formula can be manipulated in various ways and it is easy to generate explosive paths, implying insolvency (see discussion in Buiter and Currie, this volume). For present purposes, with Y standing for GDP, r^* for the real rate of interest and g for the real rate of growth, it is useful to express the identity as:

$$\Delta(B/Y) = (G - T)/Y + B/Y(r^* - g).$$

This relates the growth in the debt/GDP ratio (the bond stock to GDP ratio) positively to the primary deficit as a proportion of GDP and positively to the existing bond stock and the excess of the real interest rate over the real rate of growth of the economy. Note that it is the real rate of interest minus the growth rate that matters here since we are concerned with ratios to GDP: this term picks up the burden of the existing level of debt.[6]

Now it is clear that, in the long run, if it is wished to stabilise the debt ratio, a primary surplus in the long run *must* be generated which is at least sufficient to cover the (growth adjusted) interest burden of that future debt stock. But starting, so to speak, from here, with budget deficits and a rising debt ratio, future surpluses must be high enough to cover not just the burden of existing debt, but the (adjusted) interest on the cumulated deficits (with interest). This looks like a very good argument for acting to contain rising debt ratios as soon as possible: the longer the deficit correction is postponed, the bigger the eventual correction needed.

---

[6] The derivation of the formula is as follows.

$$\Delta(B/Y) = B/Y(\Delta B/B - \Delta Y/Y).$$

Substituting for B from the government's budget constraint in the text yields,

$$\Delta(B/Y) = (G - T)/Y + B/Y(r - \Delta Y/Y).$$

The formula in the text follows from noting that $\Delta Y/Y$ is the growth of nominal GDP and is the sum of the rate of growth, g, and the rate of inflation. The real interest rate r^* is simply the nominal interest rate, r, minus the rate of inflation.

This implication, that deficits now must be balanced by larger surpluses in the future, is entirely valid. But it is not the whole story. The deficits or surpluses themselves are endogenous. Obviously, taxes and expenditure depend upon assumptions made about growth. (Hence the very real worry that slow growth will worsen the fiscal position). Beyond that, however, growth itself may depend on deficits and debt. That will alter the dynamics.

Above, wealth effects were described as stabilizing, due to integral feedback. In terms of the formulae above, what is being suggested is that, as debt rises, the economy is stimulated and deficits fall, turning, if the process goes far enough, into surpluses. This is, in practice, one of the reasons why economies recover: cumulative deficits in recession generate liquidity and wealth effects, raising consumption and investment. Part of the effect on the future budgetary position is through the automatic stabilizers. Part may arise as a 'discretionary' response, as the authorities raise taxes or cut expenditures to avoid excess spending. Policy is an essential part of the feedback and, of course, may be anticipated by financial markets or investors. (An important practical example is the UK in the early 1990s, where public deficits could be seen as allowing the private sector to get out of debt.)

But there is an important class of problems, well known in engineering, where the dynamics may be much more difficult to manage. These are problems where the initial control impulse has to be in the 'wrong' direction, apparently taking the system away from, rather than towards, the objective.

Consider a fiscal expansion designed to engender a recovery from recession. The medium-term objective would be to raise investment and consumption and for them both to grow at some sustainable rate. We may assume that if that objective were achieved, public borrowing would be acceptable and (it may be assumed) lower than in the recession.

The practical fiscal-policy alternatives are to cut taxes to stimulate consumption or to raise public expenditure. Private investment, which tends to lag the cycle, is hard to stimulate directly. Either consumption stimulated by lower taxes, or public spending could start the real recovery, but, in either case, the public deficit would worsen.

There are two dangers and a way to avoid them. The first danger is that the stimulus is insufficient, and the economy remains in recession (with, incidentally, high deficits). The second is that as the recovery develops, investment revives, and when that happens, real demand is excessive and an inflationary boom ensues. The ideal policy, and the only one that works, is to start

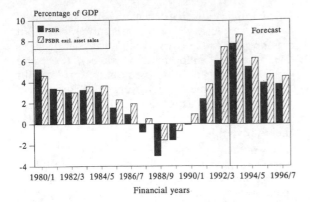

Figure 1.1. UK Public-sector Borrowing Requirement, 1980/1–1996/7

Source: Oxford Economic Forecasting model, 1993.

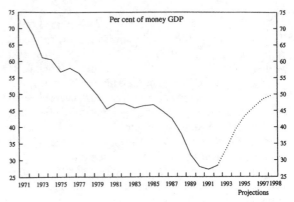

Figure 1.2. Net Public-sector Debt Ratio[a]

Note: [a] Net public-sector debt at end-March as a percentage of money GDP in four quarters centred on end-March.

Source: HM Treasury (1993), *Financial Statement and Budget Report, 1993–94.*

the recovery with tax cuts and expenditure increases (accepting the deficit increase) and, as investment comes through, to raise taxes or to cut expenditure to make room for it. The timing in this kind of system is not easy to get right.

There are physical or engineering analogies to this kind of problem. One is reversing a car. Another is the steering of a boat—say a tanker. Not only does the control impetus from the rudder seem to be in the wrong direction (which is why people have to learn to steer boats or back cars), but some of the state variables do actually move in the wrong direction before converging to their objective. Thus, if a tanker needs to change its course to the right, the bridge at the back starts, alarmingly, by moving to the left.[7] Likewise, in the economic example, deficits rise before falling and the point of maximum displacement is the point where things start to move in the right direction.

These analogies are not pursued further here, except to note one point. It is very important to discover what *kind* of system one is trying to control. It is dangerous for car drivers (or politicians) to assume they are driving forwards if they are in fact driving backwards.

Finally, in terms of the debt-dynamics formula above, it is notable that what is being described in those examples is the extreme case where the objective of stabilizing or pulling down debt in the longer term may *only* be achievable by making borrowing worse in the short term. That would be the more plausible if, in practice, an active fiscal policy would favourably affect growth in the longer term: that is, if it affected supply potential as well.

2. Fiscal policy in the UK

The deterioration in the UK's fiscal position in the early 1990s, after the surpluses of the late 1980s, shown in Figure 1.1 is extraordinarily marked, amounting to more than 10 per cent of GDP. The chart for debt ratios, Figure 1.2, is even more revealing, though no less worrying. What it shows is that the situation in the 1990s was, by peacetime standards, effectively unique. For the first time since the war, the debt trend moved sharply upwards. Alternatively, the real fiscal 'stimulus' being applied became strongly positive, having been negative through most of the 1970s and 1980s.[8]

In broad terms the reasons for the rapid fall in the national debt ratio in the late 1980s and the marked change in trend since are well known. The consumer borrowing boom, as well as rising business investment, meant that private-sector financial accumulation was unprecedentedly negative: the counterpart showed up as a public-sector surplus and as a balance-of-payments deficit (Figure 1.3). The biggest swing was in personal savings: the savings ratio fell to about 5 per cent in late 1988 before rising again to about 12 per cent in 1992 (Figure 1.4). There could not be a clearer example of a large change in *private*-sector financial behaviour (induced by financial liberalization, of course) showing up as a reduction in public borrowing and public debt. As some argued at the time, public irresponsibility was

---

[7] In technical terms the system to be controlled is 'nonminimum-phase'. I am indebted to Professor D. Clark not just for the terminology, but for an exposition of the control problems involved in this and other systems.

[8] Raw data for government borrowing and the PSBR can be very misleading, especially in times of inflation. In particular, the large deficits of the mid-1970s did not lead to a rising debt ratio because of the inflation tax, which is not included in the official accounts. See Miller (1985).

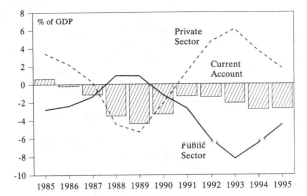

Figure 1.3. UK Sectoral Balances, 1985–95

Source: Oxford Economic Forecasting model, 1993.

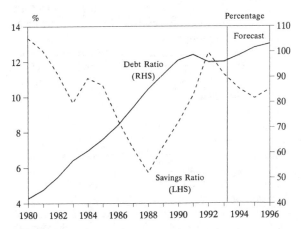

Figure 1.4. UK Personal Debt and Savings Ratio, 1980–96

Source: Oxford Economic Forecasting model, 1993.

replaced by private irresponsibility—though, to be fair, the public position was not, according to the debt figures, irresponsible before the boom.[9]

Whilst the cause of the public surplus in the late 1980s was the fall in savings, the mechanism was the boom itself and the automatic effects on revenues and expenditure. Thus the public sector was acting as a stabilizer, but to an inadequate degree given that the boom involved a major rise in the external deficit as well as excessive domestic demand. (If the economy had been 'closed' the credit boom would have been far more inflationary.)

The real casualty of the later 1980s was the MTFS itself. Targets (or consistent forecasts) for public borrowing were stabilizing against inflationary shocks (a focus on the nominal position meant that the real thrust of policy tightened if prices rose), but they were potentially destabilizing against most other shocks, including oil shocks and demand shocks—such as occurred with the credit boom (see Allsopp, 1985). Thankfully, the public-sector balance was allowed to go into surplus in the late 1980s. Imagine what would have happened if taxes had been cut further, for which some argued at the time.

To be sure, the situation in the late 1980s was difficult. It appeared that financial liberalization had let a genie out of the bottle, and control was difficult. Short of reversing the impact, the arguments above suggest that further substantial fiscal tightening was desirable to slow the rise in consumption, but tax increases would not have been easy politically. The housing boom was a seriously complicating factor, which ideally required more specific actions, including closer regulation. The choice, when it came, was for monetary action rather than for fiscal tightening or more direct measures to deal with the causes. With rather deft management of credibility effects, policy-makers managed to raise short-term interest rates to a peak of about 15 per cent in 1988/89, without an untenably large upward impact on the exchange rate. What this did was to send the credit boom into reverse. A policy which had been supposed to produce a stable financial environment had delivered just the opposite.

One way of looking at the problems with public borrowing and debt that arose during the recession suggests that the situation was not as serious as sometimes appears. If the credit boom was the origin of the fall in debt, and if that went into reverse, it can be argued that all that was being observed was a rise of national debt back to more normal or equilibrium levels. Official projections suggest a stabilization of national debt at around 50 or 60 per cent in the later 1990s—to just about the level of the early 1980s. Such a debt ratio would not involve an unsupportable burden: at a real interest rate of 3 per cent, the taxes needed to finance the interest would amount to 1.5 per cent of GDP, hardly enormous. If allowance were made for real growth as well, as it should be if we are talking about the maintenance of a stable debt ratio, the tax burden would be negligible.

The problem in 1993, as explained by Buiter and Currie (in this volume), was that the existing policy settings would not deliver such a relatively benign scenario. The implication was that *sometime* taxes and expenditures would have to change, to be consistent with future sustainability. A useful way of approaching the problem is to consider the medium term and then to consider possible ways of getting there. The

---

[9] The absurdity of views on fiscal policy is well illustrated by the fact that, in international circles, the UK position in the late 1980s was widely admired as exemplifying successful public-sector financial consolidation.

dynamics, in real economies, may be very hard to manage for the reasons discussed above.

The public sector in the medium term

Many analyses of the fiscal problem start by considering the position of the *public* sector in the medium term, aiming to meet some target for the primary deficit deemed to be sustainable. From the discussion above it is clear that this is inadequate. The concern here is with the principles involved rather than the detailed numbers.

A natural starting point would be to assume some level of GDP corresponding to a realistic target for unemployment and to work out the effects on revenues and expenditure and hence on the deficit. An implied medium-term tax or expenditure adjustment would then be needed to meet the deficit target. Thus, if GDP were expected to be able to rise by, say, 10 per cent over a span of years and if the percentage effect on the deficit were 70 per cent of that (using the rather high Treasury estimate reported in Buiter and Currie) the deficit would be reduced by 7 per cent of GDP, other things being equal. Other things would not be equal, of course, (the public expenditure trend is not zero, for example) and the effect of variations in the level of output on the deficit could be a good deal smaller. There could be much argument about the needed fiscal adjustment. That, however, is not the point that is being made here, which is that this method of looking at the fiscal issue could be dangerously misleading.

The basic problem is that there is nothing in the method which even asks whether the *private sector* would be in financial equilibrium: alternatively, there is a danger that the assumed future position would not 'add up' economically. Yet it is safe to conjecture that, in many political discussions, the public borrowing problem is seen in just these terms.

A prime example of the dangers is, of course, the late 1980s boom itself. A rigid focus on medium-term targets for the budget balance would have led to even greater instability. It would be extremely perverse if the lessons of that period were not learned: they need to be applied not just to the recessionary aftermath but also to the problem of managing the recovery. The prime need is to reframe medium-term policy for the public sector so that it is actually about policy for the private sector.

The private sector and the public deficit

It is easy to think of changes in private-sector behaviour which would soon eliminate the problem of public borrowing. The most obvious would be a sudden change in consumer sentiment leading to a debt-financed spending spree on the late 1980s' model. It is also easy to imagine situations where an improvement in the public finances would be well nigh impossible; if firms refuse to borrow and consumers go on saving at high rates then the accounts tell us that the public sector has to be in matching deficit unless the economy runs an improbably large external surplus (induced, perhaps, by major recession).

Neither scenario is attractive, and most policy-makers would start with some better medium-term scenario in mind. This would involve some view about the desirable level of the economy in relation to potential, as well as, within that, the scale of investment in relation to consumption. Real aspects of the public-sector position would also need to be worked in, such as the amount of publicly financed investment, benefit levels, and so on. Obviously, political views about the desirable shape of the economy in the medium term differ markedly—but that is not the point. Implicit in any such real-side scenario would be some pattern of financial flows (as well as wealth stocks and debt) which would have to be consistent with likely economic behaviour. (If they were not, then the scenario itself would not eventuate).

To continue the thought experiment, the next stage would be, so to speak, to juggle with available policy instruments, including fiscal variables such as taxes and expenditures, but also others such as interest rates, to make the desirable real pattern economically plausible, taking into account likely savings behaviour and so on. *Inter alia*, this would involve some assessment of the public-sector balance. Notionally, a third stage might take into account cumulative wealth effects from public borrowing, or return to stage two, considering alternative policies (e.g. if the implied public-sector position were judged unacceptable, perhaps because of expected financial market reactions), or even to stage one, altering targets in the light of feasibility. All that has been described is an iterative process for solving the implicit medium-term policy-optimization problem. (Perhaps optimization sounds too dignified, but policy-makers presumably do the best they can.) A model with assumed or empirically estimated relationships would, of course, allow the process to be made explicit: policy-makers know, however, that model uncertainty is an important part of the problem.

Models and simulation studies

The recommendations of Buiter and Currie in 1993 were, in part, based on simulations with an empirical economic model and involved the combination of tax increases and a cut in interest rates. In principle, if the model captures the system properties adequately this is the best way of analysing the policy problem, especially as other interactions, such as inflationary pressure, and other policies, such as interest rates and exchange rates, are given due weight. Effectively the problem is to find a set of policies that produces a reasonable dynamic track for the economy. Uncertainty remains a problem (see the discussion in Buiter and Currie on the desirability of erring on the side of public-sector caution).

One of the main conclusions is that it is not easy to 'square the circle' and meet standard policy objectives for the UK over the medium term with the instruments available—see, for example, Bray *et al.* (1993). Whilst a number of caveats are necessary (the results are sensitive to the models used and to the projections of the different model groups used as a base case) the general conclusion that there is limited macroeconomic room for manoeuvre is all too plausible. Alternatively, relationships that have operated in the past would have to change in the future quite a lot to justify very much optimism.

Alternative strategies

Not all strategies are as difficult or as potentially unstable as the fiscal expansion with lagging investment described in the theoretical section above. In fact, if private investment expenditure could be directly and reliably influenced, much of the difficulty would go away. There are possibilities here which have been tried, with at best mixed success, in the past. Temporary grants or tax allowances can alter the timing of investment. Public investment could concentrate on humped projects which tail off, and so on. In the past in the UK, but not since privatization, changes in cash limits could affect the investment of the utility industries. Treasury officials could no doubt think of more schemes if they were asked to.

In fact, however, the UK's strategy for recovery in the 1990s was based on low interest rates and a substantial devaluation. There is no doubt that this is an attractive strategy: export- and (hopefully) investment-led expansions have always seemed likely to have more favourable dynamic properties than consumer- or public-sector-led booms. The main disadvantage is also well known: devaluation tends to lead to inflation. In most econometric forecasting models the pass through to prices is reasonably quick and eventually complete. In practice, the reaction of wages—the most basic, and arguably the best, indicator of inflation trends—was notably small. If the wage reaction is absent, the room for manoeuvre for macroeconomic policy is greatly increased.

For the UK in the early 1990s, the attractions of the strategy were particularly strong, for two reasons. The first was the concern over the fiscal deficit. The second was the continuation of an external deficit even in recession, suggesting the need for increased competitiveness if it could be achieved. (The private sector's aggregate surplus is, of course, as a matter of accounting, equal to the public deficit *minus* the current external deficit). Despite the attractions of the policy, there are risks and difficulties which need to be taken into account.

One important issue concerns the role of interest-rate policy. It was suggested in the theoretical discussion that there were reasons for preferring a fiscal to an interest-rate response to swings in private-sector financial behaviour. Relying on interest rates to offset high savings and generate recovery could be to repeat the errors of the late 1980s in the opposite direction. In the medium term what is needed is a stable financial policy, and manipulation of interest rates and the exchange rate does not look like a way of delivering it. Of course interest rates needed to come down and the exchange-rate fall was helpful. But the policy dilemma can be sharpened by considering whether interest rates and the exchange rate should be further lowered and balanced by tax increases. Apart from the inflation risk, there is the problem highlighted by Artis and Lewis, in this volume, that, in the UK, it is consumption and asset-price inflation that are particularly sensitive to the short-term interest rate. If the taxes were raised or the tax base widened there are alternatives to interest-rate falls which would be more conducive to investment, such as certain types of public investment or support to private investment. There are also alternatives which might well be preferred on social grounds.

Issues of monetary and exchange-rate policy are likely to come to a head if interest rates fall further in Europe. The UK's policy change after September 1992 was effectively forced. But it still amounts to a sort of beggar-thy-neighbour approach and derives much of its impact from the relative devaluation compared with core European countries, which, in 1993, were also suffering from recession. It may well be the best policy available. But it can hardly be regarded as a model solution to the longer-term problem of finding a framework for UK policy within the emerging European economy.

3. Fiscal problems in Europe

One of the risks to the UK recovery is a continuation of recession in continental European countries. Most countries have been, however, reluctant to apply a fiscal

Table 1.2. Germany: Sectoral Balances, 1987–94

	1987	1988	1989	1990	1991	1992	1993	1994
General government financial balance	−1.9	−2.2	0.1	−2.1	−3.3	−2.8	−3.5	−2.6
Current-account surplus	4.2	4.2	4.8	3.3	−1.1	−1.1	−0.8	−1.0
Private-sector surplus[a]	6.1	6.4	4.7	5.4	2.2	1.7	2.7	1.5

Note: [a] Private-sector surplus derived as current-account surplus plus general government deficit.
Source: OECD.

stimulus—because of worries over debt or deficits or the Maastricht fiscal convergence criteria, or a combination of these. Thus (discretionary) fiscal policy has been tightening, despite rising unemployment and despite little room for manœuvre on interest rates.

With the effective break-up of the ERM, a number of countries, notably Italy and Spain, followed policies not dissimilar to that in the UK, hoping for a recovery based on lower exchange rates and lower interest rates. It is clear that, as in the UK, this was an attractive strategy for individual countries, worried simultaneously about rising unemployment and fiscal deficits. The exception has been France, which maintained the 'franc fort' policy after August 1993, preferring to wait for falling German interest rates. In France there has been a fiscal response with a rapid rise in the budget deficit, implying mounting problems for the later 1990s, particularly in meeting the Maastricht targets.

Many of the issues that arise are similar to those already outlined for the UK. Here, two rather more European aspects are briefly discussed. These are: the fiscal effects of GEMU and the longer-term issues raised by trend increases in debt levels.

GEMU and fiscal policy

The effects of GEMU have been extensively analysed. Here it is argued that Germany, too, properly analysed, may need larger, not smaller budget deficits as well as lower interest rates. This is contrary to the prevailing conventional wisdom which sees further fiscal restraint in Germany as a necessary condition for the interest-rate falls needed within Germany and in the rest of Europe.

There is nothing wrong with the conventional view that GEMU amounted to a large, essentially fiscal, stimulus. The federal government was suddenly committed to ensuring large expenditures in the eastern Länder. The total transfer from West to East in the early 1990s was about 5 per cent of 'West German' GDP, much of which fell directly on the budget.

Most analyses of the German policy problem look at the implications of such an impact, *ceteris paribus*.

But other things were not equal. The 1990s also saw the ending of the European investment boom and the development of recessionary forces elsewhere. As a result, largely for demand reasons, West German exports collapsed, which was only partially compensated for by increased demand from the eastern Länder.

The discussion of fiscal policy, above, emphasized the position of the private sector, rather than the public sector. The basic point is that, as far as the former West German private sector is concerned, the *net* effect of the external and internal impacts was strongly negative: i.e. taking the two shocks together, there was a negative impact, not a positive one in the early 1990s.

One way of looking at the overall picture is through the sectoral balances—the net financial flows between sectors. As a matter of accounting, the private sectors financial surplus is equal to the sum of the budget deficit and the external surplus—i.e. the current-account surplus (see Table 1.2). Over the period from 1989 to 1993, using OECD estimates, the German budget deficit rose by 3.6 per cent of GDP. The external surplus declined from +4.8 per cent of GDP to −0.8 per cent of GDP—a swing of nearly 5.6 per cent of GDP. Even on this crude basis, the fiscal impact on the private sector was strongly negative. Since much of the recent rise in the government deficit was, in any case, recession induced (2–2.5 per cent of GDP according to OECD figures) it is clear that the fiscal impact of the two changes taken together was strongly deflationary. (Alternatively, the private sector's surplus declined, despite recession, which would normally be accompanied by a rise.)

The above provides an answer to a puzzle which is why, if reunification amounted to a large fiscal *stimulus*, Germany went into deep recession in 1993 rather than continuing to boom. Bundesbank overkill does not seem to explain enough. The answer is that it does not need to, since, as far as the private sector was concerned there was, in effect, a strongly negative 'fiscal' impact from the trade sector as well. (This was masked for a time by the short-lived spending boom immediately after reunification.)

In fact, over the longer term, through most of the

1970s and 1980s, Germany has had a problem in achieving financial (or fiscal) balance. The reason is well known: Germany (like Japan) is a high-saving country, and, with relatively slow growth the savings of the private sector have exceeded domestic investment. As a matter of accounting, excess savings have to be balanced by either a balance-of-payments surplus or a public-sector deficit. In the early 1980s Germany ran budget deficits despite the traditional dislike of deficit finance. In the late 1980s, owing to world growth and, especially, the European investment boom, the internal deficit was replaced by an external surplus as Germany was dragged up by expanding export demand. (Automatic effects on government revenue and expenditure, together with some 'consolidation' of the fiscal position, formed the mechanism by which the budget deficit was reduced.)

From this point of view, the impact of reunification was highly fortuitous. (This may seem controversial given the problems it has caused.) Without it, the ending of the export boom would have sent Germany into really major recession—which, no doubt would have been only partly offset by lower interest rates and a lower exchange rate. (Budget deficits or a renewal of the export surplus would have automatically ensued.) And, in the longer term, it is probably helpful for other European countries that Germany has a domestic need for its domestic savings to finance investment in the eastern Länder. The only catch is that the transfer to the east will in all likelihood have to continue to be financed through the budget.

It was argued above that, unless there are good reasons to suppose that public borrowing is excessive in the sense that it is leading to inflationary pressure or crowding out of private capital accumulation, attempted solutions to a perceived problem of budget deficits should concentrate on the characteristics of the private sector. Applying this to Germany, it would imply that some change to either investment or savings behaviour would have to occur. After GEMU there is a clear need for investment in the eastern Länder. It is easy to conceive of a situation where private entrepreneurs and businesses tap the domestic (and international) capital markets and invest to the required extent. The need for deficits to match Germany's savings potential (or to go beyond it) would disappear. It is widely agreed, however, that this is unlikely to happen, at least in the near future. There is, it may be said, a pervasive 'market failure'. The diagnosis of some market failure is the normal case for intervention—to correct the failure.

But the nature of the failure needs to be diagnosed and alternative ways of alleviating it need to be weighed.

One problem as is well known, is the wage level which is currently about 60–70 per cent of that in the West. A wage subsidy—directed towards that 'market failure'—reducing over time, would be more favourable both to the continuation of production and to investment than present arrangements (Flemming, 1993). There would continue to be budgetary costs and, of course, much public investment in infrastructure, as well as social security costs, will inevitably fall on the budget. Nevertheless, privately financed investment would begin to take a greater share in the required transfers during transition.

The issues are not pursued here. But the main points are worth reiterating: Germany can afford the high costs of transition in the eastern Länder, as is shown by high savings and the extremely large balance-of-payments surplus before reunification. Ideally, the private sector might invest enough to effect the needed transfers, but, whilst this is a legitimate objective, it is unlikely to happen in the near future. To the extent that policy-makers want to check the deficit, attention needs to be directed to the processes involved in transition and the private sector's response. Ill-thought-out attempts to cut the budget deficit to 'finance' transition (e.g. by increasing taxes) risk continuing recession, lower investment, and further problems in the east, and almost certainly would not actually lower the deficit unless the recession was so marked that the balance payments went back into surplus—exactly the opposite of an appropriate response to a domestic need for heavy investment.[10] This view of the macroeconomics is, of course fully consistent with continued attempts to contain costs and improve efficiency at all levels of government. And, of course, circumstances could arise where aggregate claims on the German productive potential do become excessive and tax rises or public expenditure falls, to lower private or public demands on the economy, would become appropriate (For example when private investment and exports revive.)

The longer-term issue of rising debt levels

Public-debt trends in Europe are roughly correlated *negatively* with performance over time. To state the obvious, it would be quite wrong to infer on that basis that fiscal restraint to slow the trend increases seen in the 1970s and 1980s would improve performance.

As a matter of accounting identity, the upward debt trend is showing something about the balance

[10] The German upswing in 1994 depended on increased export demand and some improvement of net trade. Recessionary forces reappeared in 1995.

between savings and investment over the longer term. One story is that the public sector is, by some mechanism, crowding out investment by the private sector. Equally plausibly, poor growth leading to low investment combined with continued high savings in many countries (reflecting, *inter alia*, an ageing population) could be the cause of public borrowing. It is difficult to arbitrate between the two opposing views.

It should be clear, however, from the discussion so far, that this may not, for most policy purposes, be necessary. There should be agreement on the desirability of a strong focus on the underlying real resource-allocation issues, including the mounting problem of unemployment.

Beyond that, however, the debt trend does raise questions about the *policy mix*. Real interest rates were high during the 1980s and may have been one factor behind low investment and the need for deficits. (There are many other possible explanations, of course, including the monetary authorities' response to inflation, and so on.) One positive aspect of the mounting concern over public borrowing and debt is that it is highly likely, in a situation of low inflationary pressure, to lead to a *policy* response in terms of lower interest rates. This is discussed in the next section.

The real question is not about forecasts—whether the European recovery is strong or weak. It is about whether the policy framework could cope with a situation in which slow growth and rising unemployment continue. The indications are not good. Policy statements are mostly about the need for fiscal restraint, especially to meet the Maastricht criteria, and there is no sign of recognition that fiscal tightening could make deficits and the debt trend worse rather than better because of the dynamic difficulties outlined above.

III. Monetary and exchange-rate policy

1. Monetary policy in the UK

Artis and Lewis (in this volume) draw attention to the wild swings in the underlying basis of monetary policy that have occurred since the inception of the MTFS, noting that the present basis, after the UK's exit from the ERM, remains unclear. There is a target range for inflation, and attention is paid to a number of indicators. Beyond this, the Bank of England has an increased role at least in what it says about inflationary trends. But this hardly adds up to a strategy. The only thing that is clear is that short-term tactics, as the UK at-

tempts to maintain recovery, clearly involve a combination of low (compared with the early 1990s) interest rates and devaluation and that some trade off between fiscal restraint and other aspects of policy is recognized.

Monetary policy cannot be looked at without considering the wider framework. Monetary (interest rate) and exchange-rate policy amount effectively to one instrument rather than two—which, of course, does not mean that targeting the exchange rate is, in practice, the same as targeting domestic monetary conditions. Moreover, the role of fiscal policy depends crucially on the monetary regime and vice versa. Under EMU, for example, fiscal policy would be the only domestic (regional) macroeconomic instrument available for stabilization (and its use would be circumscribed). Thus, putting aside short-term concerns, there are real issues about the assignment of instruments which arise whether or not the UK goes back into some European exchange-rate system.

Many of the difficulties with UK monetary policy since the early 1980s have arisen because the monetary system does not work in the easy way expounded in economic text books. Thus, it turned out that the 'technical monetarist' views that underpinned early versions of the MTFS were unrealistic on almost all counts, and the framework was abandoned. It is now widely accepted that, in a highly developed financial system such as exists in the UK, the Bank's main monetary instrument is short-term interest rates and that its control over the financial system is indirect. Recognition of complexity (and of the less pivotal role of the banks) also means giving substantial weight to the regulatory regime and to the effects of changes in it—such as the credit boom and house-price rises in the late 1980s. The simple story of high-powered money leading to multiple expansion of deposits and of real balance effects from the money stock has little relevance to the UK: money is overwhelmingly 'inside money', created by banks and building societies, and the government's contribution to money has been very small (and negative during the period of overfunding in the 1980s).

A monetary-policy framework can be seen, in general terms, as the setting up of some reaction function, tying short-term interest-rate movements (as well as other aspects of monetary policy) to some objectives or intermediate targets or indicators. During the 1980s a large variety of anchors for monetary policy were tried (broad money, narrow money, the exchange rate, as well as others) separately and in combination—a process which culminated in the UK's entry into the ERM in 1990 and a formal commitment to remain within the parity bands. That framework was abandoned in 1992.

The focus of policy since has been is on targeting interest rates more directly to expected inflation, with support, however, from target ranges for the more indirect indicators of broad and narrow money. Other aspects of the 'reaction function' remain, however, unclear. In particular, how is monetary policy going to link with the exchange rate, with fiscal policy, and with the real economy?

Looking through the variations in the framework of monetary policy over the last several years, it is clear that, in practice, the monetary/exchange rate nexus *has* been used first to fight or offset the credit boom, and then to offset its aftermath. Interest rates and exchange rates were raised in the late 1980s; they were then lowered to aid recovery. Thus, peering through the fog, interest-rate policy has been assigned, basically, to managing nominal spending, or demand. Such an assignment is obviously inconsistent with formal exchange-rate systems, such as the ERM, unless policy-makers turn out to be very lucky—and they were not. The question that arises is whether it is a sensible assignment, and whether there are alternatives.

The obvious alternative to such an assignment is to use fiscal policy more actively to offset demand shocks, such as the credit boom. (In principle, other policies of a more regulatory kind, such as credit controls, could be used, though there are obvious difficulties in open financial markets.) That assignment was the one used in practice in the Bretton Woods era, though, because of exchange controls, some monetary autonomy remained. With a more active use of fiscal policy, tying monetary policy to the exchange rate becomes more feasible, and hence, more credible.

Though the issue appears most starkly in the context of exchange-rate regimes such as the ERM, the same problems arise with softer systems and with greater orientation towards the domestic economy. Assigning monetary policy to inflation is a relatively uncontentious option if the private sector is assumed to be basically self-stabilizing and something like the classical monetarist dichotomy between real and nominal influences applies. It becomes much more contentious if shocks have to be coped with, emanating from abroad or from the domestic economy (whether policy-induced or endogenous). If there are important elements of instability in the private economy, it becomes very contentious indeed, unless there are *other* ways of dealing with the problems (such as, for example, fiscal policy). And short-term problems are likely to have long-term effects owing to hysteresis.

It has been argued above that there is a case for giving fiscal policy a greater role in offsetting certain shocks— such as the credit boom—where the fiscal impact may seem closer to the problem that needs to be offset than alternatives such as interest-rate changes. A greater role for fiscal policy would take some of the pressure off monetary policy. In practice, however, most governments do in fact feel highly constrained by the need to cut deficits and debt.

2. The international game

This paper has pointed to the perceived advantages to individual countries, given their concern over fiscal problems, in pursuing strategies based on low interest rates and low exchange rates. So far, the process is partial in Europe, but, especially if recovery stops in Germany, interest rates will fall, perhaps by a lot. As the process generalizes, the exchange-rate advantage to an individual country—which is stimulating to demand (and in some cases may be needed for competitiveness reasons as well)—is eliminated and the system is left with low interest rates.

This is not just a European issue. The US is pursuing a policy of deficit reduction (and is also digesting the aftermath of high private borrowing). Interest rates are low and the dollar competitive. In Japan, worried about the recession that followed the ending of the 'bubble economy' and the extraordinary rise in the Yen, short-term interest rates were lowered to 1.75 per cent in 1993 and 1.0 per cent in 1995. If European interest rates fell far, there could be exchange-rate tensions between the US and Europe. Short of continuing rapid recovery and inflationary pressure (e.g. in the US) the logic points to continuing fiscal restraint and low interest rates—a scenario reflected in financial markets over the world.

For the UK, low interest rates elsewhere could pose further policy dilemmas. If interest rates fall far in Europe, then, with unchanged interest rates, much of the devaluation gain might be lost. But lower interest rates, in the UK, risk starting off a consumer boom and stop–go. Thus, the UK, which is relatively sensitive to short-term interest rate changes, could find itself out of step with the rest of Europe.

For Europe more generally, the risks seem rather different. The main danger is that interest-rate falls, especially if combined with attempts at fiscal consolidation, might not be enough to revive investment and stimulate recovery.

3. ERM, EMU, and the regional analogy

By common consent, the ERM would be hard to put together again. With credibility lost, the potential costs to individual countries of remaining in an exchange-rate system when speculative pressure starts are very great. There are possibilities of a return to some softer system, such as existed before 1987. It is also clear that some 'core' European countries may be attracted by the alternative of entry into EMU with a common currency, by fiat.

In a common currency area, monetary policy would be passed to some supra-national central institution. (Under the Maastricht treaty, national central banks are supposed to become independent on the way to giving up their powers—a strange bit of sequencing.) There are obvious constitutional and accountability questions which arise. Clearly, however, if the European Monetary Institute were to develop into a proper central bank, monetary policy would then be run with an objective of producing stability for Europe as a whole. For any one country, this would be unlikely to be ideal. (For example, German monetary policy since 1990 would have had to be different.) This raises again the crucial issue of what alternative policies there might be to deal with shocks, should they arise.

The fundamental issues come out most clearly using the regional analogy—European countries can be looked at as regions within the wider grouping, which might or might not have the power to use differential monetary policy. Under EMU, they would have no power.[11]

The regional analogy

One respect in which Europe is not like present federal states or regions within a single country is the lack of net cyclical flows from the central budget (arising automatically from the tax and benefit system) to regions (countries) experiencing difficulties. The EU regional funds do not perform this role. The possibility of substantial inter-country flows of this nature is not even on the agenda.

What this means is that individual countries facing shocks, or otherwise finding they were out of phase with the rest, could face serious 'regional' difficulties, difficulties which, because of the lack of automatic regional transfers from the centre, would be expected to be greater than those applying to existing federal systems or to regions within individual countries.

But, important as this point is, it does not mean that stabilization within a common currency area would be impossible. Starting from some equilibrium situation,

if a region fluctuated (relative to the rest) stabilization would occur through net trade and the current account of the balance of payments. The greater the integration (e.g. the smaller the region), the greater the stabilization. (In the example, the current account would fluctuate with positives and negatives with no net longer-term effect.)

But it is possible to go beyond this. Counter-cyclical regional fiscal policy could raise spending and stabilize the regional economy, even to the point where the balance-of-payments impact did not occur. The automatic stabilization through the balance of payments would then be being replaced by the discretionary rise in the budget. (Again, if the problem is one of fluctuation, there is no long-run effect on debt levels or private-sector asset stocks). Moreover, since there is, in principle, no difference between automatic stabilization and discretionary stabilization (the degree of automaticity is a design issue: or, perfect discretion could match any automatic effects if so desired), national or regional fiscal systems, could be expected to play as much of a stabilization role within a common currency area as they would outside. Fiscal policy does not need to be centralized, though it would need to be co-ordinated.

For some shocks, such as the demand (savings) shocks discussed for the UK, it has been argued that the fiscal offset looks appropriate. This kind of offset could be applied within a common currency area, and looks near ideal. It follows that, for a similar shock, it should also be the preferred strategy for a region which has monetary and fiscal freedom. The alternative would be first to raise then to lower interest rates and the exchange rate, disturbing intertemporal and international prices.

The argument generalizes, at least in theory. If each region carried out the appropriate fiscal strategy in response to cycles or demand shocks, not only would relative stabilization be achieved, but the collective as a whole would have the right fiscal policy. (So perhaps the doctrine of 'subsidiarity' applies to fiscal stabilization.) Moreover, the argument carries through substantially unchanged, in or out of a common currency area. This should not be surprising. If a fiscal offset is appropriate, the exchange-rate regime should not matter.

The fundamental issues

A tight exchange-rate regime, such as the ERM became after 1987, can be seen as having two objectives. One, of course, is exchange-rate stability. The other is that it ties down inflation, and anti-inflation policy, at least

[11] The regional perspective is also discussed in previous issues of the *Oxford Review of Economic Policy*—see Vol. 5 No. 3 (Exchange Rates) and Vol. 6 No. 3 (Balance of Payments). See also Vol. 11 No. 2 (Regional Policy).

in relative terms. This second aspect was a major feature of the ERM. For most countries (other than Germany) it defined the medium-term framework of monetary policy.

Despite Alan Walter's stricture that the system was 'half-baked', it was in fact a heavy discipline on inflationary trends: a country with relatively high inflation would become increasingly uncompetitive and end up with unemployment (as in an uncompetitive region). One major pressure, therefore, was upward pressure on costs in some countries. If the pressure became intolerable, credibility would be lost as the risk of a step devaluation mounted. Countries in the ERM, of course, knew this. They were highly reluctant to devalue, since devaluation in those circumstances had become widely regarded as dangerous: not only would it accommodate inflation, but it would make any future medium-term framework against inflation increasingly difficult to put in place. It may be conjectured that financial-market operators knew this too, which decreased the perceived risk of devaluation, and hence of speculation. The UK entered the system as a discipline, accepting the risk of uncompetitiveness.

One possible reason for the debacle is thus competitiveness pressures and the effect on individual countries. The system, it might be argued was simply too hard for inflation-prone countries. There is truth in this. It arguably applies particularly strongly to Italy. In the case of the UK it is not quite so convincing, since, despite a loss of competitiveness, wages were still relatively low (in manufacturing they were about 60 per cent of West German levels—they are now about 50 per cent). Since devaluation is just a way of lowering wages with inflationary risks it was quite clear that the authorities, at least, did not want this.

Another standard reason for the problems was, of course, German interest-rate policy after GEMU. Outside Germany, interest rates all over Europe became inappropriate as far as domestic objectives are concerned. But the reason that was so serious is that other countries were in recession and, given the prevailing policy framework, had no alternative policy to deal with the problems, other than hoping for a change in German interest rates. The Bundesbank, however, was prepared to see the whole system fall apart rather than react to foreign pressure and the French were not prepared to countenance an exchange-rate realignment.

The point of going over this familiar ground is to stress that the system failed because it had no alternative to monetary policy to deal with shocks and disturbances, and that is simply incompatible with the assignment of monetary policy to longer-term anti-inflation aims as well. The exchange-rate regime implied that relative monetary-policy stances could not change; domestic problems meant that they had to. It was not just a matter of Germany being out of line (though that clearly was very important). Under a common currency, such as EMU, Germany would not have been able to use monetary policy to offset the pressure that built up owing to the joint effect of the European boom and GEMU. It would have had to use alternatives. But problems elsewhere would still have been serious as recessionary forces developed. Italy and Spain arguably needed to devalue. More debatably, so did the UK. Beyond that, the UK was suffering the aftermath of the credit boom. And nearly all countries in Europe were suffering the effects of the ending of the investment upswing. The situation would not have been at all easy, even without GEMU.

Since 1992/3, countries such as the UK have abandoned medium-term strategy in favour of short-term demand management via interest rates and the exchange rate. We have suggested that combined with fiscal consolidation, this could lead to low interest rates throughout Europe. But it leaves these countries without a medium-term anchor for a non-accommodating policy against inflation. A longer-term strategy is, however, needed. But most long-term strategies—and not just those based on an exchange-rate regime or target—require monetary policy to be directed to the longer-term problem of establishing a credible strategy against inflation, and not to the offsetting of cyclical or other shocks. So the dilemma will return. A longer-term framework against inflation will require a significant shift in viewpoint towards the more active use of fiscal policy, difficult though that might be.

Much discussion has focused on ways of building in additional flexibility to a future exchange-rate regime, or on the pros and cons of big-bang-type strategies for entering EMU. What the above suggests is that some of this is beside the point: more attention needs to be given to the shocks and difficulties likely to be faced, for the EC as a whole, and within individual countries, with a view to finding ways of dealing with the problems which do not involve the diversion of monetary (and exchange-rate) policies away from their proper purpose.

IV. The supply side

With slow growth in Europe in recession, and with high unemployment, direct measures to alleviate the problems are again being considered. Some scepticism is appropriate, however: many measures (such as job-sharing) were tried in the 1970s without much success. Similarly, measures to increase 'flexibility', whilst important, are unlikely to have much effect, except in the long term.

Thus the prime need is to generate sustainable expansion. This, as suggested above, may need to involve co-ordinated action, including fiscal action. Such a policy, however, appears unlikely, given prevailing views about the need to lower public borrowing and debt. The danger is, in fact, not just a lack of needed action, but of further fiscal restriction. The argument for fiscal relaxation is not that fiscal deficits and debt do not matter. On the contrary, the argument is that fiscal expansion may be needed in order to bring down borrowing and debt in the longer term. A clear vision of the way in which fiscal policy fits into the overall framework of policy is badly needed, but seems lacking.

That said, there is no doubt that Europe has structural problems and that there is a limit to the amount that could be achieved by macroeconomic action. Macroeconomic actions need to be directed towards the supply side, favouring investment rather than consumption. Measures to improve labour-market flexibility and lower cost pressures in the medium term are also desirable. But many of the desirable policies, such as retraining, education, etc., involve increased public expenditure. Supply-side policies, too, are constrained by the prevailing views about fiscal policy.

One of the main dangers is inappropriate actions. An obvious risk, Sinclair (1993), is that of protectionism. The pressures, especially from high unemployment, are obviously building up. This is a further reason why a change in macroeconomic policy is needed.

Whilst the focus of policy in continental Europe is on the labour market, in the UK, where recovery has progressed further, the main concern is with productivity and competitiveness. One indicator of the underlying problems is that balance-of-payments deficits continued even in the midst of recession. But it is very hard to argue that Britain needs lower labour costs. As already noted, manufacturing wages in the UK are about half those in west Germany (which means that they are substantially lower than in the eastern Länder). Thus, the UK is already one of the lowest wage areas of the EC (Greece and Portugal are, of course, the lowest wage areas). In other words, the UK's problem is primarily one of low productivity.

The policy framework of the 1980s put the accent firmly on market solutions to the problem of low productivity, focusing on liberalization, especially of labour markets and financial markets, as well as on privatization. There are, however, few signs of the 'miracle' that was hoped for, and UK wages in relation to other countries in Northern Europe are probably lower now than they have ever been. Many people argue for more direct measures to improve productivity. There are a number of examples, in Europe and elsewhere, of successful industrial policies, aspects of which could be helpful in the UK.

It is more or less generally agreed that international comparisons suggest that the UK has fallen behind for many decades in the areas of industrial training and education. One thing, however, is clear. Any serious attempt to improve education, training, and skills would be expensive: it would require a substantial increase in budgetary resources. There could not be a clearer example of the way in which concerns over public debt and borrowing limit needed supply-side policies, despite unemployment and the continuing under-utilization of the nation's resources.

V. Conclusion

The UK is trying to maintain recovery, whilst containing public borrowing and debt. Since being forced out of the ERM macroeconomic policy has gone through a U-turn. The previous focus of policy on medium-term objectives for inflation, to be brought about by the maintenance of a stable financial environment, has been abandoned to be replaced by a concentration on the shorter-term problem of generating and sustaining recovery.

One of the main issues relates to the mix of policy between fiscal support for recovery and changes in the interest rate—which feed through to the exchange rate. Given high borrowing, there is immense pressure to rely, in the face of contingencies, on interest rates, accepting the inflationary risks. This emphasis is the more likely to gain support since many believe that further competitiveness gains are desirable.

There may, indeed, be no alternative to this policy: it is perhaps the best one in the circumstances. But as a strategy for the future (or for Europe) it is seriously flawed. The main difficulty is that, with monetary and exchange-rate policy assigned to short-term demand management, there is no longer-term framework for inflation control akin to the framework provided by the ERM. There is wide agreement that some such framework is needed and should be put in place.

But there is an underlying contradiction that needs to be resolved. This is that, with such a framework, short-term policy of the kind that is being followed at the moment would not be possible. Even without returning to some new form of exchange-rate mechanism, the pressures would re-emerge. Monetary policy would be being asked to do too many things.

This paper has stressed the potentially offsetting and stabilizing role of fiscal policy. For shocks such as the credit boom, it has been argued, a fiscal offset is not only possible, but appropriate. With a more active fiscal policy, some of the pressure would be taken off monetary and exchange-rate policy.

The lack of alternative instruments to control the economy and to offset shocks can be seen as an important reason for the break-up of the ERM. Even without problems such as GEMU, a tight version of the ERM, such as existed after 1987, was bound to accumulate strains. Competitiveness problems were perhaps a price worth paying for downward pressure on inflation, though in some countries they became too great in the period of 'excess stability' since 1987. But the lack of alternative ways of dealing with the recession that followed the ending of the European investment boom was arguably more serious and more symptomatic of underlying deficiencies. On top of that, GEMU meant that the only instrument that governments appeared to be prepared to use—monetary policy—moved in the wrong direction.

One lesson is that monetary policy needs to be co-ordinated. But a more subtle one is that an exchange-rate system that pins monetary policy is more or less bound to fail in the face of shocks if it has no other set of instruments for dealing with them. Thus a much more active use of fiscal policy as part of the recognized framework of macroeconomic policy would seem a prerequisite for future schemes to stabilize exchange rates and to use exchange rates (for countries other than Germany) as a longer-term anchor against inflation. The same is true for EMU should it eventuate: in the absence of cyclical inter-country fiscal transfers, *regional* fiscal policies would be needed for stabilization purposes. Even outside an exchange-rate mechanism, individual countries would find their longer-term aims for the exchange rate, for interest rates, and for inflation, easier to achieve if fiscal policy played its proper part.

The main difficulty is that though such an assignment of instruments would make other aspects of European integration easier to achieve (and would be helpful for individual countries), it is not on the agenda. Almost universally, governments are now concerned to lower deficits and slow the rise in debt ratios. There is a severe danger not just of inadequate responses to the present recessionary tendencies in Europe, but of actively perverse policies being introduced.

This paper has suggested that the problem of public debt and borrowing is wrongly framed. It is usually looked at in terms of the public sector's position. There are cases where the problem does lie with the public sector. Normally, however, and in Europe at the moment, the problem that leads to deficits and debt lies with the private sector, and offsetting policy by the public authorities is needed. Investment needs to rise. If investment does not revive rapidly with declining interest rates, fiscal policies will be needed. But this may well involve raising deficits in the short run in order to bring them down in the future. The longer expansion is delayed, then the greater will appear the problems, and the harder it will be to introduce the needed measures in the future.

Such a strategy for Europe, involving a significant change in the basis of macroeconomic policy could not be expected, on its own, to solve Europe's problems of high unemployment and structural difficulties. Supply-side policies are also needed. But these could not work on their own, and without a co-ordinated expansion, they too will tend to be blocked off by worries over public deficits and debt. There is thus the possibility of a dangerous spiral.

Such an outcome is not certain. It may be that falling interest rates and the natural forces of recovery will generate a sustained revival in Europe. But such a hope, or even such an expectation, is not an adequate basis for policy—which must be robust against contingencies. The real problem is that policies which may be desperately needed in the future if Europe again starts to suffer from slow growth, look impossible given prevailing attitudes.

Much of the problem, it has been argued, arises from the way in which fiscal policy is viewed. One major theme is that fiscal policy needs to play a larger role in the short-term stabilization of demand shocks. Not only is it the appropriate offset, but some of the pressure would be taken off monetary and exchange rate policy, which, in the absence of fiscal stabilization, is almost bound to be used instead, destabilizing interest rates and the exchange rate. Such a change would be more-or-less essential under EMU, but would be desirable anyway. But one of the reasons fiscal stabilization is out of style is the legitimate concern over the longer-term problem of rising levels of debt in most European countries.

In this conclusion, two aspects of the longer-term problem need to be stressed. The first is that direct attacks on the problem via increased taxes and lower public expenditure risk being counterproductive. Public borrowing is endogenous and, by inducing recession or slow growth, the policy risks making the problems worse. There are situations where pump-priming in the form of fiscal expansion involving larger deficits in the short term, help (with appropriate fiscal adjustments over time), to lower deficits in the longer term. But the dynamics are likely to be difficult, and this is an additional reason why alternative strategies, using interest rates and exchange rate depreciation, may appear very attractive. More fundamentally, however, the problem seems wrongly framed. Rising debt ratios are usually taken as a signal of the need for public sector 'consolidation'—that is cuts in public expenditure. But the counterpart identities imply that the problem can equally be seen as arising from inadequate private-sector investment (in relation to the savings potential). A switch of emphasis towards the stimulation of private investment as a response to public debt problems is needed. With a revival of investment, the medium-term fiscal policy adjustments that are required would become relatively straightforward.

References

Allsopp, C. J. (1985), 'The Assessment: Monetary and Fiscal Policy in the 1980s', *Oxford Review of Economic Policy*, 1(1), 1–20.

Barro, R. J. (1974), 'Are Government Bonds Net Wealth?', *Journal of Political Economy*, 82(6), 1095–117.

Bray, J., Kuleshov, A., Uysal, A. E., and Walker, P. (1993), 'Balance-Achieving Policies: A Comparative Policy-Optimization Study on Four UK Models', *Oxford Review of Economic Policy*, 9(3), 69–82.

Dornbusch, R. (1986), *Dollars, Debts and Deficits*, MIT/Leuven.

Flemming, J. (1993), 'The Policy Framework in the Medium Term', *Oxford Review of Economic Policy*, 9(3), 26–35.

Miller, M. (1985), 'Measuring the Stance of Fiscal Policy', *Oxford Review of Economic Policy*, 1(1), 44–57.

Sargent, T., and Wallace, N. (1981), 'Some Unpleasant Monetarist Arithmetic', *Quarterly Review*, Federal Reserve Bank of St Louis, Fall.

Sinclair, P. J. N. (1993), 'World Trade, Protectionist Follies, and Europe's Options for the Future', *Oxford Review of Economic Policy*, 9(3), 114–25.

Stiglitz, J., and Weiss, A. (1981), 'Credit Rationing in Markets with Imperfect Information', *American Economic Review*, 71, 393–410.

Tobin, J. (1980), *Asset Accumulation and Economic Activity*, Oxford, Blackwell.

2

Options for UK fiscal policy

WILLEM II. DUITER

Cambridge University

DAVID A. CURRIE

London Business School

I. Introduction

This article is written as the UK's Chancellor of the Exchequer, Kenneth Clarke, prepares his November Budget. It will be the first Unified Budget, with taxation and expenditure dealt with at the same time. In framing this Budget, the Chancellor must make a fine judgement between the need to restore stability to UK public finances and the need to maintain economic recovery. The choice is, in essence, the familiar one between sound finance and the use of fiscal policy to manage demand.

The background to this judgement may be briefly summarized as follows. The UK economy has been recovering from its low point of the first quarter of 1992, and has shown slow but positive growth for six successive quarters up to the latest 1993 third-quarter numbers. But this slow recovery comes after a long and deep recession (though less deep, despite popular perception, than the 1980/81 recession). Thus total output has still not recovered its previous peak of the second quarter of 1990, and will not make up the remaining 1.3 per cent shortfall until early next year, implying no growth for 4 years. Moreover, the recovery looks vulnerable. With Europe, particularly Germany, in recession, UK export growth may fade despite the competitive advantage gained from last September's sterling devaluation. With other components of demand sluggish, most forecasts of continued recovery are posited on growth in consumer spending coming from a continued

decline in the savings ratio and a revival of stockbuilding. There is clearly the risk that either or both will not materialize. Moreover, there is the concern that Kenneth Clarke inherits from his predecessor, Norman Lamont, considerable delayed tax increases from the March Budget: these will amount to some £6.5 billion in the financial year 1994/95, and may act to halt the revival of consumer spending.

These considerations suggest that Mr Clarke should avoid any additional tax increases or further spending cuts in this Budget, or should even cut taxes (or abort Mr Lamont's announced increases) to boost recovery. But there are also strong arguments to set against this. First, there are risks that inflation will rise over the next year or so to breach the government's 1–4 per cent target range. An expansionary budget, or a fiscally neutral Budget accompanied by interest-rate cuts, increases the risks of such a breach. Second, there is concern about the very high level of public borrowing, currently around 8 per cent of gross domestic product (GDP). Even with continued recovery, the public-sector borrowing requirement (PSBR) is projected to remain high at over 6 per cent of GDP in 1994/95 and only marginally less the following year. Continued public borrowing at this level raises serious questions about the sustainability of current fiscal policies, and suggests that Mr Clarke should make it a priority to curb borrowing in this Budget.

In the following, we examine these arguments in more detail. We then go on to examine the question of debt sustainability. We then examine alternative options for the Chancellor. We then conclude with our own recommendations for fiscal policy, and macroeconomic policy more broadly.

First published in *Oxford Review of Economic Policy*, vol. 9, no 3 (1993).

Table 2.1. Required Primary Surplus (as % of GDP) with 60 per cent Debt Ratio

		Real rate of interest, r				
		2.5	3.0	3.5	4.0	4.5
Real rate of growth, g	1.5	0.5	0.8	1.1	1.4	1.7
	2.0	0.2	0.5	0.8	1.1	1.4
	2.5	—	0.1	0.4	0.7	1.0
	3.0	—	—	0.1	0.4	0.7

Note: This required primary surplus as a percentage of GDP for a debt ratio d is given by $s = (r - g)d - m$, where m is the increase in base money as a percentage of GDP, given an equilibrium velocity of base money (assumed constant). In the above calculations, base money velocity is assumed to be around 30, and inflation to be 2.5 per cent.

II. Debt sustainability

A minimal requirement of any fiscal programme is that it should be sustainable, in the sense that it should not lead to an explosive growth in the public debt/GDP ratio, so that government solvency is ensured.[1] An explosive path for the debt ratio would certainly generate instability in the longer run; it is also likely to generate instability in asset prices in the shorter run if forward-looking financial markets anticipate future problems and adjust portfolios accordingly. If instability is present and the government does not wish to resort to monetary financing, then the government will be forced to change its fiscal plans, whether by cutting spending or raising taxes, to halt the rise in debt. If such action is required, it may be better taken early than later: delay will usually mean that the required adjustment is bigger, because of the need to service a larger debt stock through higher interest payments, and the adjustment may be forced at a time when it is unwelcome on other grounds. Early action may also be justified in terms of microeconomic efficiency, which suggests the need for a smooth profile of tax rates to avoid intertemporal substitution resulting from changes in tax rates through time.

A sustainable debt position requires a primary fiscal surplus in the medium to longer run (if real interest rates exceed real rates of growth as they have done over the past decade or more). This contrasts with the current primary deficit of around 5 per cent of GDP. The size of the required surplus varies with the differential between the real rate of interest and the real rate of growth, the debt ratio itself and the revenue from

base money creation.[2] Table 2.1 sets out the permanent required primary surplus as a percentage of GDP on different assumptions about the real rate of interest and the real rate of growth. The calculations assume a 60 per cent debt ratio and a 2.5 per cent inflation rate. The 60 per cent debt ratio is chosen somewhat arbitrarily for illustrative purposes: it is above current debt levels in the UK, but represents the level to which it is expected to rise over the next 5 years on consensus forecasts. The market real rate of interest, as measured by the real gross redemption yield on 2.5 per cent Treasury index-linked 2024, stands at the end of October 1993 at 3.2 per cent, so that our assumed variations are around the current market expectation.

The table reveals a fairly small variation in the permanent required primary surplus with different assumptions about the growth rate and real interest rate: with a growth rate in the 2.5–3 per cent range, the primary surplus need be no more than 1 per cent of GDP on the most pessimistic view of real interest rates, and may even be close to zero for low real interest rates. With growth in the 1.5–2 per cent range, the required primary surplus may be as much as 1.5–2 per cent of GDP, but could be substantially less with a low real interest rate. Of course, these variations become larger if the steady-state debt stock is higher, but still remain quite small.

However, the variations shown in Table 2.1 understate the extent to which sustainability is made more difficult by adverse outcomes on growth and interest rates. To see this, we must also consider the permanent actual primary surplus. This may be thought of as the long-run average primary surplus–GDP ratio that is

---

[1] Strictly speaking, the growth rate of the debt must be less than the interest rate in the long run, if non-distortionary and costless taxes are available. In practice, taxes are distortionary and costly to administer, so that the debt ratio must be bounded.

[2] The relationship is given by the formula $s = (r - g)d - m$, where s is the permanent required primary surplus as a percentage of GDP, r is the long-run real rate of interest, g is the long-run rate of growth, d is the ratio of debt to GDP, and m is the increase in base money as a percentage of GDP.

expected to emerge under unchanged policies.[3] The permanent primary gap, defined as the difference between the permanent required primary surplus–GDP ratio and the permanent actual primary surplus–GDP ratio, measures the scale of the fiscal adjustment required to ensure fiscal sustainability.

The first half of this calculation, that of the required primary surplus, is rather more precise than that of the second half, the permanent actual primary surplus. Predictions of movements in government borrowing are extremely uncertain, as the history of the past 10 years reveals: few forecasters predicted that the public finances would move into surplus in the late 1980s, nor the scale of the movement into deficit in the early 1990s. Moreover, these calculations are very sensitive to variations in the rate of growth. Slower growth means inevitably that tax revenues rise more slowly and transfer payments on unemployment and other benefits rise more quickly. This means that the permanent actual primary deficit will be larger and the required fiscal adjustment, measured by the permanent primary gap, will be correspondingly bigger.

The Treasury estimates that the cyclical effect of a 1 per cent shortfall in GDP on the public finances is about 0.7 per cent of GDP (see HM Treasury, 1991). This very large ratio results from the interaction of the tax and benefit system, which builds in a very high sensitivity of net taxes (taxes less benefits) to income levels. However, it is not appropriate to apply this ratio to a shortfall in GDP arising from a medium-term shortfall in GDP, since the consequences of this for unemployment, and consequently benefit payments, are likely to be rather different. To the extent that lower medium-term growth results from a weaker productivity performance, the consequence will be less unemployment and lower benefit payments.[4] As a result, there will be a rather smaller effect on public borrowing, possibly of the order of one-third to one-half of the Treasury estimate for the cyclical impact. None the less, assuming even this lower estimate, the prospects for public borrowing on the pessimistic growth projections look decidedly poor. On reasonable projections, the actual primary deficit will show rather little improvement

from the 6 per cent of GDP expected next year if the UK economy experiences slow growth in the 1.5–2 per cent range over the next few years.

This slow growth outcome is not the central view, at least as measured by the consensus of forecasters, or by the views represented on the Treasury's Panel of Independent Forecasters (HM Treasury, 1993). The halt on growth over the past 4 years should mean that the UK economy can enjoy above average growth, at least for a period, which should restore a more healthy position to the public finances. But with the uncertainties associated with the economic outlook, it remains only a possibility. There are also risks on the upside, in terms of higher than expected growth, but this possibility should weigh less heavily in determining policy, and certainly should not be used to dismiss worries about low growth. The possible outcomes may be uniformly distributed around the consensus view, but the risks associated with the low growth outcome and the consequent problems of sustainability are much greater than those associated with the more favourable outcome.

Because of this asymmetric distribution of the costs of alternative outcomes, there is a strong case for further action in the November Budget to curb public borrowing, to remove the risk of an unsustainable outcome. This action should be phased, but should cumulate over two or three years to around £6 billion or about 1 per cent of GDP. Even with this action, forecasts, including the recent London Business School October forecast (London Business School, 1993), suggest that the PSBR will remain at around £30 billion, or 5 per cent of GDP, in 1995/96; without it, the PSBR will fall more slowly, and may even rise in the event of slow economic growth, giving rise to an unsustainable debt path. The recent experiences of Belgium and Italy illustrate the dangers of allowing debt to rise excessively.

A high level of public borrowing also carries other dangers. With high borrowing, the interest paid on new borrowing cumulates rapidly, leaving a heavy burden of debt interest to be financed in the future. High borrowing leaves interest rates more vulnerable to shifts in sentiment in financial markets, and may make it more difficult for the authorities to avoid an interest rate rise, particularly if the inflation background deteriorates. Loose fiscal policy, in the form of a large PSBR, also places more reliance on monetary policy to restrain inflation, risking higher interest rates. Moreover, the combination of a loose fiscal policy and tight monetary policy makes it more difficult to maintain a competitive level of the pound without inflationary pressures emerging, impeding adjustment of the external current account deficit.

---

[3] More precisely, it is that constant primary surplus–GDP ratio which has the same present discounted value as the stream of expected future primary surplus–GDP ratios on unchanged policies, where the relevant discount rate is the real interest rate minus the real rate of growth.

[4] However, there is one gloomy scenario in which this is not the case. Global competition, particularly from the Asian Pacific Rim, may force UK industry to maintain a strong productivity performance to compete, involving aggressive labour-shedding, and the closure of inefficient firms. This scenario could combine slow growth with strong productivity performance, resulting in high unemployment and a very adverse outcome for the public finances.

Moreover, there is a need to bring down borrowing with recovery, in order to give fiscal room for manoeuvre in the future in the event of an unforeseen economic downturn. The recent experience of the US illustrates the point that a long succession of fiscal deficits may eliminate the room for fiscal relaxation in recession. If we cannot bring down borrowing in the growth phase that the UK economy is entering, then fiscal policy will be immobilized in the future. That would be an unfortunate legacy to leave to the future.

For these reasons, the Chancellor should take further measures to reduce the PSBR, over and above the measures put in place by his predecessor. This offers the best chances of maintaining low interest rates and a competitive pound without breaching the government's inflation target, and of correcting the imbalance of excessive consumption (private and public) and inadequate investment in the UK economy.

III. Budget options

In view of these considerations, we therefore recommend Budget measures to curb borrowing, amounting to some £6 billion over and above the measures announced by the previous Chancellor. These measures should be phased in over two or three years. This leaves open the question as to how these cuts in borrowing should be distributed between public spending cuts and tax or revenue increases.

A distinctive feature of the November Budget is that it is the first of the new Unified Budgets. The motivation behind its introduction was the aim of looking at public spending and revenues together, rather than considering taxes in the spring and spending in the autumn, as before. Although this might have disadvantages (including the possibility that spending ministries might seek to fight off proposed public spending cuts by recommendations for tax increases), it was felt to permit more rational planning and control of public finances.

It is, therefore, one of the curiosities of the present position that the Chancellor appears to have ruled out a truly unified consideration of public spending and taxation. The public expenditure Control Total was set in July. There is now perhaps a little scope to lower this Control Total further to reflect lower inflation and the continued policy of tight control on public-sector pay. This could cut public borrowing by £2 billion, requiring only modest additional action to raise revenues. This cut may, however, be hard to sustain beyond 1994/95 without radical reform on the spending side if in-

flation picks up by more than the government has allowed. It is, of course, possible that the Chancellor is keeping his powder dry on this question, and still has in mind a reduction in the Control Total of this kind. A preferable alternative might be to maintain the Control Total and redeploy resources within the total from public consumption to public investment. This would allow support of those areas of public spending that help strengthen the longer-run supply-side performance of the economy, including education, training, and R & D. In that case, he will need greater action on the revenue side.

On the revenue side, one option for the Chancellor is to restrict tax allowances further to the 20p rate of tax, accompanied perhaps by an extension of the 20p tax band: this has the merit of raising income tax revenues without an adverse effect on incentives. Other measures that have economic merit include the abolition or further reduction of mortgage interest tax relief and a further widening of the VAT net. Medium-term measures that merit serious consideration include the introduction of new user charges, notably road charges; and a move to a carbon tax on environmental grounds. All of these measures broaden the tax base without adversely affecting incentives and, indeed, are helpful in reallocating resources efficiently.

IV. The budget and recovery

If the Chancellor pursues the line of action that we have proposed, what of the other horn of the dilemma that he faces: the danger that cuts in public borrowing will stall the fragile recovery, which European recession is making more vulnerable? There are two answers to this, and the Chancellor would be wise to accept both. First, borrowing reductions should be phased, to reduce their impact on immediate recovery prospects. Second, the Chancellor should accompany his Budget measures on the fiscal side with a further cut in interest rates of 0.5–1 per cent, and stand ready to cut rates further if the recovery shows signs of faltering. Equally, he must stand ready to raise interest rates if inflation threatens a serious and sustained breach of the 4 per cent target ceiling.

Any budget measures to curb public spending should necessarily be tapered, simply because of the lags in effecting adjustments to spending programmes and the microeconomic inefficiencies that can arise from rapid change. More controversial is the case for the phasing of any tax increases. This case was argued by

David Currie and Gavyn Davies in the February Report of the Treasury Panel of Independent Forecasters, and was adopted by Mr Lamont in the March Budget. It rests on two propositions: that specifying and committing to specific tax measures will carry greater credibility with markets concerned about high levels of borrowing than general promises to curb borrowing in the future; and that delayed tax increases will have a smaller dampening effect on demand than immediate tax increases. Indeed, a good case can be made that the effect on current demand of credible announcements of future restrictive policy is expansionary. (See Blanchard, 1981; Turnovsky and Miller, 1984; Giavazzi and Pagano, 1990.)

Both propositions may be disputed, but in our view stand up. It may be argued that future commitments can be reversed, and indeed it is the case that one option for Mr Clarke is to reverse the measures announced for the future by Mr Lamont, notably the extension of VAT to household fuel. But while that option exists, it is not an easy one to follow: there would undoubtedly be damage in terms of the credibility and consistency of policy were Mr Clarke to go back on Mr Lamont's promises. Thus, while future commitments do not absolutely bind future government decisions, they limit the room for manoeuvre. In so doing, they add credibility to the promise to curb borrowing. Nor do they eliminate the room for fiscal discretion were events to show that the tax increases were not needed, either because the fiscal position improved more or the economy proved weaker than expected. Most Chancellors are not equally willing to raise and lower taxes. So while few doubt the willingness of a Chancellor to cut taxes if circumstances warranted (or even if they did not), many are sceptical about the willingness to raise taxes except in the face of compelling evidence that it is needed.

Some argue that forward-looking consumers and companies will anticipate the impact of future tax increases, so that their effect on demand will be felt immediately. Some anticipation should, of course, be expected, but it is unlikely to be large enough to be equivalent to the demand effect of an immediate tax increase. This could be so only in a world of so-called Ricardian equivalence. But in such a world, the tax increases needed to curb excessive budget deficits will already be anticipated and, therefore, reflected in private savings. In such an imaginary world, lump-sum tax increases, immediate or phased, will have no effect on demand anyway. In the real world, anticipation effects of this kind will be much less than complete, so that phasing of tax increases will reduce the impact on demand.

However, the main instrument for offsetting any stalling of recovery should be interest rates. If growth falters and if inflationary pressures remain subdued (which they may not), then interest rates should be cut to sustain growth. The greater use of monetary policy in this way will help to rebalance the UK economy between overall consumption and investment, and will also help to sustain a competitive exchange rate without inflation, to the benefit of the UK's international trading position and the balance of payments deficit.

V. The strategy for macroeconomic policy

The Budget also provides an opportunity for the Chancellor to restate the strategy for macroeconomic policy after the demise of the exchange-rate mechanism (ERM) in its hard, narrow-band form.

After the events of the last year, and the move to wide bands in the ERM, it seems most unlikely that any reconstituted ERM could provide the required framework for macroeconomic policy. This is not the same thing as saying that the UK should not consider re-entering a reconstituted ERM. Rather it is to suggest that, since any new form of ERM will need to have sufficient flexibility to ensure survival, it will not act as an effective constraint on UK domestic macroeconomic policy. Any future move to greater monetary integration or monetary union is likely to depend more on the co-ordination and convergence of coherent domestic-policy frameworks than on the return to a tight external constraint of the kind that the ERM had imposed in its hard form since 1987. For these reasons, there is a need to establish a sustainable domestic strategy for macroeconomic policy, and this is one of the challenges facing Mr Clarke. He and his predecessor have been right to reject a return to reliance solely on monetary targets, however specified, but there needs to be greater openness about how the range of relevant indicators is interpreted. The Bank of England inflation report is helpful in this. It sets up an arrangement whereby the Bank examines recent indicators and produces a best forecast as to inflation in the future. Interest rates (and other policy instruments) are then adjusted to influence future inflation. Because monetary policy influences inflation only with an appreciable lag of around 1.5–2 years, the inflation forecast must look this far ahead. If the forecast is that inflation will go outside the target range, this is a signal that the authorities should adjust policy, notably interest rates.

It is interesting to note that this framework places economic forecasts right at the centre of the determination of macroeconomic policy. For a long while in the 1980s it was fashionable to dismiss the role of forecasts on the grounds that there was no need to fine tune the economy; all that was needed was to establish the appropriate medium-term framework for policy. But this presupposed the existence of an effective and objective nominal anchor for policy. In practice, the two anchors that were tried, first monetary targets and then the exchange rate, slipped, and slipped badly. We have now returned full circle to a recognition of the key role of forecasts; what has changed is that it is inflation, not output growth, which is being placed centre stage in the policy debate.

The Bank of England framework for policy is a helpful one, and has similarities to the framework proposed earlier by London Business School (1992). But so far it omits one crucial element, namely greater powers, with accountability, for the Bank of England. The durability of the framework will remain in doubt so long as the setting of interest rates remains with the Treasury, because the present division of responsibilities leaves open the question as to how the Treasury will choose to interpret worries on inflation expressed by the Bank of England. Increasing the powers of the Bank does, of course, reduce the scope for co-ordinating monetary and fiscal actions. But, in our view, that advantage has been ill-used in the past, and any possible benefit is more than outweighed by greater policy consistency and credibility.

To illustrate this vulnerability, consider the possibility that in 1 or 2 years' time, ahead of a prospective election, the Bank becomes concerned over future inflation trends, but the Chancellor is concerned about the effect of a rise in interest rates on the standing of the government with the electorate. It is possible to imagine that the Treasury may use its weight to disagree with the analysis of the Bank, and to dispute the inflation fears. It is, of course, always possible for reasonable people to disagree over forecasts, as the new Treasury 'Wise Men' Panel illustrates. Such a disagreement would, of course, pose certain risks for the Chancellor, since a disagreement between the Bank and the Treasury might unsettle financial markets. But with present institutional arrangements, such a disagreement might be patched up in the form of an agreement to 'wait and see'. This might serve the political need, though at the risk of delaying a needed policy adjustment. Under present arrangements, there is, therefore, scope for dodging necessary but unpopular policy actions. By contrast, if the Bank had the power to set interest rates, this scope would be appreciably reduced.

The vulnerability of the present arrangement would be greatly reduced if the Bank of England were given the power to set interest rates but were also held accountable for its actions. The Prime Minister has indicated that the primary objection to a more independent Bank of England is the issue of parliamentary accountability, but this could be addressed by a form of the New Zealand contractual arrangement between the government and central bank. In this arrangement, the objectives of monetary policy are set out explicitly in a contract, and the central bank is held accountable and subject to parliamentary scrutiny. Without further institutional reform in the direction of a more independent Bank of England, the present framework for policy lacks credibility. With it, there is the opportunity to lock into sustained low inflation. The Chancellor has the opportunity in the November Budget to signal the intention to move in this direction.

References

Blanchard, O. J. (1981), 'Output, the Stock Market and Interest Rates', *American Economic Review*, 71, 132–43.

Giavazzi, F., and Pagano, M. (1990), 'Can Severe Fiscal Contractions be Expansionary? Tales of Two Small European Countries', NBER *Macroeconomic Annual*, MIT Press, 75–111.

HM Treasury (1991), 'Fiscal Developments and the Role of the Cycle', *Treasury Bulletin*, 2(1), Winter 1990–91.

—— (1993), Panel of Independent Forecasters, Reports, February, July, October.

London Business School (1992), *Economic Outlook*, 17(1), October.

—— (1993), *Economic Outlook*, 18(1), October.

Turnovsky, S. J., and Miller, M. M. (1984), 'The Effect of Government Expenditure on the Term Structure of Interest Rates', *Journal of Money, Credit and Banking*, 5, February, 16–33.

PART II

EXCHANGE RATES AND
MONETARY POLICY

3

Après le déluge: monetary and exchange-rate policy in Britain and Europe

MICHAEL ARTIS

European University Institute

MERVYN LEWIS

University of Nottingham

I. Introduction

Europe today stands at a cross-roads, in politics, in technology, in economics, in culture . . . But my thesis today is that Europe stands at a monetary cross-roads. (p. 143)

The money markets of Europe have not been as disturbed as they have over the past three years since the chaotic periods of the 1930s . . . The only way to establish a unified money market is to kill the sporadic and unsettling speculation over currency prices that ravaged the European markets. (p. 147)

The arguments for changing the exchange rate are short-run arguments based on a money illusion that is increasingly disappearing . . . I am speaking here of the currency fluctuations between areas as closely connected through trade and lending as the different currencies within Europe . . . In matters of finance it is better for Europe to move towards the US pattern of high capital and labour mobility than to break up . . . the degree of integration already achieved. (p. 149)

These extracts, written by Robert Mundell (1970) on the eve of the breakdown of the Bretton Woods system, serve to remind us that currency crises in Europe are not new, and often act as a stimulus for later initiatives—in that case the Werner plan for monetary union of March 1971 and the Snake agreement of April 1972 whereby European countries sought to limit fluctuations of bilateral exchange rates to a ± 2.25 per cent range, i.e. half the size implied by the width of the dollar fluctuation bands of the Smithsonian agreement of December 1971.

First published in *Oxford Review of Economic Policy*, vol. 9, no. 3 (1993).

In many ways the choices facing Europe—and Britain—in 1971 differed little from today's. Then, as now, Europe faced the choice of narrow or wide exchange-rate parity bands against the backdrop of plans for a single European currency and an embryonic European central bank (in the form of the European Monetary Co-operation Fund). In the event, the Werner plan was effectively scuttled when Germany and Holland floated their currencies in the face of strong opposition from other European countries. While the Snake lasted until March 1979, when the European Monetary System (EMS) was formed, parity realignments were frequent and countries entered and exited until, in the end, only a Deutschmark zone of Germany, Belgium/Luxembourg, the Netherlands, and Denmark were still members. Britain's dilemma then, as now, was how fully to participate in these moves. In fact, Britain lasted only six weeks in the Snake before striking out alone in currency matters for the next 18 years.

Over the intervening two decades conditions have altered. Those European countries which participated in the currency arrangements of the EMS have had greater exchange-rate stability than other major OECD countries, at least up until recently. As a consequence, trade between the European countries has not been exposed to the massive fluctuations in nominal and real exchange rates which have occurred, for example, between the United States and Japan. If such variations were to take place within Europe, trading patterns would have to change. In addition, Britain is now far more closely enmeshed in developments in the rest of Europe than was the case before. For example, British exports to the EC have increased from 30 per cent of exports in 1971 to 57 per cent in 1993. The next stage of European union is scheduled for 1 January 1994 when the European Monetary Institute (EMI)—the precursor of the European central bank—is to start work. The remit of the EMI requires it—in an advisory capacity—to intensify the co-ordination of national monetary policies and also to design the institutional framework for monetary union. Britain is committed along with the other EC countries to participate in this work.

This paper examines the options for Europe and Britain in this very different, and still very fluid, context.

II. Breakdown of the ERM

Sustained speculative pressure on the French franc and other currencies, exacerbating internal pressures within the EMS, led EC finance ministers on 2 August 1993 to allow member currencies of the exchange-rate mechanism (ERM) to float within bands ± 15 per cent of their central rates (although the mark–guilder trading range remained the old ± 2.25 per cent band). The crisis of July 1993 was a second edition of the speculative havoc of September 1992 which saw the lira devalued by 7 per cent (12 September 1992), then the pound sterling (16 September 1992) and the lira (17 September 1992) forced out of the mechanism, and a devaluation of the peseta by 5 per cent (17 September 1992). Earlier that month (8 September 1992), Finland cut the link of the markka to the ECU. In following months, the peseta and the escudo were devalued by 6 per cent (22 November 1992), the punt was devalued by 10 per cent (30 January 1993), and the peseta devalued by 8 per cent along with an escudo devaluation of 6.5 per cent (13 May 1993). Speculation also forced Sweden (19 November 1992) and Norway (11 December 1992) to sever their currency links to the ECU.

Parity realignments are a feature of the ERM and were common in the early years of the system; there were two realignments in 1979 and five between March 1981 and March 1983. The period of 'excess stability' did not begin until 1987; from February 1987 until September 1992 there were no realignments other than the technical adjustment of the lira on 5 January 1990. Floating is also a characteristic of the system, for the ERM combines aspects of fixed, but adjustable, parities

with that of a flexible rate regime.[1] Bilateral exchange rates are 'fixed' in the sense that the central bank of each member country undertakes to maintain them, by compulsory intervention and other policy measures including access to Community financing, within bands around a fixed central parity, and in conjunction with other central banks can choose the magnitude and timing of changes in these parities. (Central parities are established for each currency vis-à-vis the ECU, and the bilateral central rate for each pair of participating currencies is the ratio of the two ECU central rates.) Exchange rates are at the same time 'flexible' in that fluctuations within the bands are a result of market forces. Previously the ranges of fluctuation were either 5 or 12 per cent. The 30 per cent bands are unprecedented. It is these broad bands—as much as the fact that the pound and the lira have left the ERM— which define the extent of the currency upheaval. The Snake has taken on boa constrictor proportions!

Speculation forced the issue in both 1992 and 1993, but this fed upon a number of tensions in the system (BIS, 1993; Portes, 1993; Williamson, 1993a, 1993b). These are examined below under four headings: the phase of 'excess stability'; the German reunification shock; the trend of German interest rates; and the dependence of the exchange-rate structure upon expectations. The contribution of these factors seems clear enough now, but it must be said that it was far from clear at the time. One of the shrewdest observers of the problems of the ERM and its inherent contradictions is T. Padoa-Schioppa, Deputy Director-General of the Banca d'Italia. He observed in a survey of the ERM for the *New Palgrave Dictionary of Money and Finance*:

The system has also remained relatively stable in the face of major external disturbances, such as the large fluctuations of the dollar and the increase in oil prices in the second half of 1990.

The consequences of German economic and social unification have also been absorbed smoothly. In spite of its being claimed in some quarters that a revaluation of the DM was necessary, the German currency has remained relatively stable within the band, even after German short-term interest rates were raised by more than one percentage point in the second half of 1990.

The commitment to monetary union and economic convergence among member countries is a powerful stabilizer of exchange-rate expectations. Diverging from the common stance, or even modifying central rates, has a high cost in terms of the credibility of member countries' policies. The commitment to defend EMS central rates is bound to

---

[1] The origins and workings of the system are examined in van Ypersele and Koeune (1984).

represent the core of policy co-operation among European countries until the passage to full economic and monetary union. (Padoa-Schioppa, 1992, pp. 826–7)

1. The phase of 'excess stability'

An adjustable-peg system has to strike an appropriate balance between 'fixity' and 'adjustment'. The former is designed to foster free trade in goods and capital resources, and promote wider community goals. Realignments are required when deviations from purchasing power parity prevent a satisfactory mix of price stability and employment in individual economies. The balance between fixity and adjustment tilted strongly towards fixity in the ERM from 1987: capital controls were progressively eliminated by the larger countries and nominal exchange rates remained very stable.[2]

From the date of the decision to remove exchange controls it was clear to policy-makers that they would need to find a new solution to the problem of the 'inconsistent quartet': free trade; free capital movements; national policy autonomy; and fixed exchange rates. This policy dilemma, first clearly stated by Henry Wallich (1972), has been restated by Padoa-Schioppa in the context of European monetary arrangements more than once (e.g. Padoa-Schioppa, 1988). The point is well taken: it is not possible to combine all these four desirables in one system, and successful systems involve some trade-offs. Under the gold standard, monetary autonomy was sacrificed. The architects of the Bretton Woods system envisaged that official controls over capital flows would reconcile the inconsistency. As restrictions upon capital were progressively removed in the 1960s, a more complex solution resulted in which all four of the objectives were compromised to some extent and in different ways by member countries. When control over international capital flows effectively passed from national authorities to the markets, and national monetary autonomy was threatened, fixed parities gave way to floating rates.

In the first phase of the ERM, the compromise was on freedom of capital movements (with capital controls) and also on fixity of exchange rates (since there was a number of realignments as well as the ability to float within the bands of fluctuation). This allowed countries to retain a degree of policy autonomy. With the withering away of capital controls and an increased desire for stability in exchange rates, members of the system were in much the same position as countries in the latter stages of Bretton Woods. The interim solution arrived at in the Basle–Nyborg Agreements of 1987 specified in particular an extension of credit lines to intra-marginal intervention, an increase in the credit (repayment) period, and commitment to 'greater co-ordination' of interest-rate policy. In the Autumn of 1987 and in early 1988 the new arrangements were used successfully to squash two speculative raids—one on the French franc, the other on the Italian lira. Afterwards there ensued a period of unparalleled stability.

Central banks became less disposed towards realignments, encouraged initially by convergence of macroeconomic policies and performance, which reduced the need to realign, and by the conviction that continued counter-inflationary success required unswerving commitment to a fixed nominal exchange-rate peg against the Deutschmark (DM). These features meant that the other admonition which it was supposed that EMS participants had learnt from experience—the need to ensure that realignments are timely ones contained 'within the bands' and thus smaller but more frequent—was completely neglected.

The commitment to exchange-rate stability in the post-1987 period was, it seems, also greatly increased by the relaunching of the idea of an Economic and Monetary Union (EMU)—the ultimate radical solution to the 'inconsistent quartet' problem, in which national policy autonomy is replaced by a sharing of collective monetary sovereignty. This project acted as a powerful stabilizing force: for countries to be seriously discussing such a project implied the feasibility of *irrevocably* fixed exchange rates, after all. Finally, in foreign exchange markets, nothing succeeds like success. The stability of the markets led to more stability. All of this became a magnet for non-members with Spain and Britain joining the system with ± 6 per cent bands (in 1989 and 1990 respectively), while Norway (1990) and Sweden and Finland (1991) pegged their currencies to the ECU with narrow ± 2.25 per cent margins.

In other respects, the 'excess stability' was unfortunate. Even though inflation differentials against Germany were declining for the high inflation countries, they remained positive and resulted in a cumulative rise in relative national price levels: with fixed nominal exchange rates this implied declining competitiveness and a drift in the real exchange rate (see Figure 3.1). A second problem was that the excess stability of the system, together with the growing commitment of central banks and governments to the 'credibility' model

---

[2] Many writers spoke of the 'new' EMS, see de Grauwe and Papademos (1990).

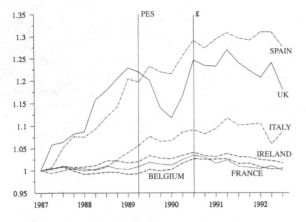

Figure 3.1. Real Exchange Rates Versus the DM

Note: Inflation-adjusted using consumer prices. Vertical line indicates dates when Spain (PES) and the UK (£) joined the ERM.
Source: IMF, International Financial Statistics (various).

of economic policy[3]—i.e. using a fixed exchange rate and the anti-inflationary reputation of a nominal anchor currency in a disinflationary programme—increased the premium, *for any one country*, on avoiding a downward realignment of the currency. Then there was the German reunification shock.

2. German reunification

A key argument for exchange-rate flexibility is that the exchange rate can act as a short-run cushion to absorb shocks that are asymmetric between countries. German reunification looks like a classic example. The shock itself initially involves a phase of excess demand in Germany. A DM appreciation would dampen the excess demand and divert Germany's export surplus to domestic consumption. This, apparently, was sought by the Bundesbank, along with more flexibility in the exchange-rate grid so as to reduce the need for central-bank interventions. But the credibility argument blocked adjustment as soon as France made it known that it could not accept a DM revaluation. It is of course no secret that the Bundesbank has always favoured floating exchange rates because of the freedom of

action they give it to pursue domestic price stability (Funabashi, 1989); in the end, that has been the result.

The most sophisticated argument for not revaluing the DM against the other currencies is perhaps the observation that once the economic reunification process is complete it is likely that a real depreciation of the DM will be needed (Adams *et al.*, 1993). The pay-off to the massive investment in and restructuring of the former GDR is to produce a position of excess supply. A real depreciation is needed to liquidate the excess supply and to secure a trade surplus to repay foreign borrowing. Hence if the adjustment period were short enough, it might be possible to 'drive through the middle' of these two contrary exchange-rate adjustments.

But, of course, the adjustment period for reunification has proved much longer than first anticipated and without the aid of a nominal exchange-rate adjustment the real appreciation of the DM had to be brought about by movements in prices and wages. German prices need to rise relative to other countries: this is a difficult process because of the German aversion to inflation. The implication is that much of the burden needs to be in the form of deflation in other countries. Some countries have proved more capable than others of accepting this.

3. Interest rates

As German inflation increased, tighter monetary policy bore the brunt of the adjustment with the result that domestic German considerations have required interest rates higher than has been desirable in other countries in Europe. Moreover, the inflationary pressures and the interest-rate consequences lasted longer than many forecasts of the time seemingly suggested (e.g. Central Planning Bureau, 1990).

In the absence of capital controls, uncovered interest-rate parity implies that the difference between interest rates on comparable instruments denominated in different currencies will equal the sum of the risk premium and the expected (average) exchange-rate change over the maturity concerned.[4] In a system in which exchange rates fluctuate in bands around a central parity rate, which itself can be adjusted by discrete realignments, expectations of exchange-rate changes consist of two elements: the expected change relative to central parity and the expected realignment.

---

[3] In view of the amount written about this model in the context of the EMS, the evidence for it in terms of market responses appears to be quite weak (see Egebo and Englander, 1992). Perhaps the main beneficiary of any credibility effect is the central bank. Spaventa (1989) argues that getting inflation down and keeping it down has become an electoral desideratum and politicians have responded to it by giving central banks a freer rein to pursue anti-inflationary policies. In the process, the external anchor has been strengthened as central bankers have competed for rankings in the anti-inflationary stakes.

[4] The risk premium is an elusive concept since it embraces those exchange-rate risk and political risk factors which lead to different risk characteristics among assets. Svensson (1992) argued that risk premia among EMS currencies are likely to be low because the volatility of EMS exchange rates is low.

For countries credibly pegged directly to the DM, exchange-rate variability relative to central parity is low. They had no alternative but to fall into line and raise rates proportional to the German interest-rate change. Since the nominal interest-rate increases are not tied immediately to increased inflation, this means a rise in real rates as well. The alternative was to allow the exchange rate to take as much of the adjustment as possible—a flexible and attractive option while expectations remain 'fixed' by the central parity and the credibility of the commitment is intact. But as the exchange rate depreciates towards the lower limit of the band some loss of credibility may occur; then domestic interest rates may have to be raised or the central bank may have to intervene, which if unsterilized will tend to raise domestic rates with a lag because of the reduction in the monetary base.

Countries that have committed themselves to an exchange-rate peg always run the risk that the interest rates required on domestic grounds may differ from those required by the maintenance of the peg. The British government, in its 1991 restatement of the Medium-Term Financial Strategy appeared well aware of this, stating, 'There may be occasions when tensions arise between domestic conditions and ERM obligations, with domestic conditions pointing to interest-rate levels either higher or lower than those indicated by ERM obligations.' Whether this 'tension' is bearable depends on the extent of the interest-rate rise, the stage of the economic cycle when it occurs, and the sensitivity of the economy to interest rates. Admittedly, few anticipated the extent of the German monetary and budget overshoots and the implications of these for interest rates. Yet the likely direction if not the magnitude of German interest-rate increases was clear enough. In the UK in particular, consistent predictions of an upturn in the economy (see Figure 3.2) may have encouraged the authorities to believe that rate increases would coincide with the upswing phase of the economy, and to that extent be more bearable. Once again the contribution to policy failure of forecasting uncertainties and the immense difficulty of knowing where the economy is likely to be headed may have been under-rated.

4. Speculation

This then was the background to the events that unfolded between the first ERM crisis of September 1992 through the turbulent ensuing months to the speculative crisis of July 1993. The adjustments that took place in this can be seen to be related to the correction of misalignments and differences in the policy and interest-rate cycle. In the first case, the medium-run 'fundamentals' are wrong and the exchange rate needs to be changed. This is not always so obvious for the second case. Some countries—France, for example—had reduced inflation below German levels thus accommodating the needed real appreciation of the DM; others had not done so. Doubts about the political commitment to EMU removed a key factor underpinning stability of the system, raising the spectre of realignments.[5] Once credibility was eroded and traders were presented with the prospect of a one-way bet, a speculative crisis reminiscent of the final years of Bretton Woods unfolded.

Speculative pressures took on a momentum of their own. As currencies left the system or devalued, an early convergence to EMU seemed less likely. As one currency after another fell, the competitiveness of those remaining on the old parities worsened. This raised doubts about the sustainability of high interest rates. Even in countries where 'fundamentals' looked about right—France, Belgium, Denmark—the determination of the governments proved to be insufficient to overcome these doubts.

In order to preserve an exchange-rate parity, it might seem that a country has only to consider whether it will put up with somewhat-higher-than-desired interest rates for a time, waiting for a fall in German interest rates to solve the problem, perhaps changing its own

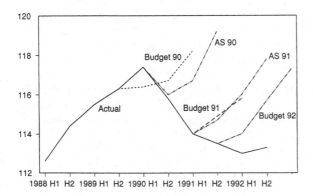

Figure 3.2. Treasury Forecasts and Actual GDP (1985 = 100)

Note: As refers to Forecasts made in Autumn Statements.
Source: Midland Global Markets.

---

[5] Unfortunately, the more convincing are the explanations offered for the collapse of the system in September 1992 and July 1993, the more difficult it becomes to explain why the crises did not occur much earlier. To pin the whole explanation for the earlier stability (from 1987 to 1992) on the markets' belief in the countries' commitment to EMU and convergence, a belief only shaken by the Danish referendum, does not seem entirely convincing—although it is an account which explains the lack of market activity in the preceding years and the pivotal position of the referenda. Eichengreen and Wyplosz (1993) commend it.

policy mix in favour of fiscal relaxation in the meanwhile. But speculation dramatically alters the costs. While countries declared in effect that they were prepared to reconcile the 'inconsistent quartet' by forgoing monetary independence, speculators could bet that when it came to the crunch matters might be different and, in so doing, turn an unpleasant, but acceptable, trade-off into an undesirable and possibly politically unacceptable one. Since there is a doubt about the exchange rate, interest rates now have to cover the expected rate of realignment, which depends on the probability of a realignment per unit of time and the expected size of a realignment if it occurs. Buying off speculation means that a country is not asked simply to accept the higher German interest rates but interest rates that are a multiple of Germany's. How painful the higher interest rates are depends in part on the structure of the financial system and the state of the economy. Economies which are highly indebted and where the debt is predominantly at floating rates, as in the British case, find the trade-off especially adverse. So also do countries in which unemployment is already too high.

III. The future of the EMS

An arrangement in which most currencies can fluctuate 15 per cent either side of their central rates might be seen as indistinguishable from a float, although much larger swings than that have occurred outside Europe. From February 1985 to January 1988 the dollar prices of the yen and mark both rose by 110 per cent. In the year to August 1993, the yen appreciated 24 per cent against the dollar and 43 per cent against the mark. Nevertheless, fluctuations on this scale seem unlikely just now amongst the European currencies. Krugman (1989) argues that such volatility is the result of an interactive process which has taken place over time whereby trade flows and production decisions become desensitized to exchange-rate movements, which then permits those rates to become still more volatile because they have so little effect on the real sector.

Retention of the system, if only in name, means—apart from allowing some official face-saving—that narrower bands can be re-introduced readily. But, lest any attempt prematurely to reconstruct tighter parities suffers the fate of the Smithsonian agreement and the Snake, there are some fundamental issues which need to be addressed as to where Europe is going in monetary affairs. It is also important that the Community rediscovers a 'shared vision' of where it is

going politically, if only because a strong political commitment to Europe has been a force cementing economic co-operation and institutions. Equivocation over Bosnia and squabbling over the ERM are still fresh enough in people's minds to give the 'all-is-well, integration-as-usual' line a hollow ring. In a very real sense the Maastricht approach of 'deepening' before 'broadening' has been run down by events. The enlargement process under way brings together the EC and seven European Free Trade Association (EFTA) countries to form the European Economic Area (EEA), requiring changes in the way European institutions are run. Any extension of trading relations to Central and Eastern Europe will accelerate that trend.

On monetary matters, the 'old' ERM sought to combine fixity of exchange rates with flexibility. One option now is to stay with flexibility; another is to strike out for complete fixity of rates (and so move on to EMU); the other is to seek, again, some middle path.

1. How desirable are stable rates?

To stay with floating exchange rates begs the question of why the ERM was established in the first place. Foremost among the reasons was to prevent currency instability. It is often forgotten that Continental European countries have an historic aversion to exchange-rate volatility which is nearly as strong as the German dislike of inflation. The aversion also dates back to the interwar years when, in the aftermath of the collapse of the gold standard, competitive devaluations produced a rise of protection and the erection of capital controls and barriers to trade (Nurkse, 1944). Exchange-rate stability has been one of the pillars of the unification process, paralleling the increase in trade integration and the single-market programme.[6]

Are floating exchange rates incompatible with the single market? Trade, financial, and monetary integration are discussed in one breath in the European context, but in principle are separable. Substantial trade can take place under freely floating rates, as now occurs between Canada and the USA, while virtually complete financial integration can exist alongside flexible rates, as shown by the Eurocurrency markets. Yet many would question whether a truly free economic market can exist between countries without a common currency, or fixed exchange rates, to facilitate

---

[6] Surprisingly, no explicit desired exchange-rate regime features in the Treaty of Rome. It must be recalled that the Treaty was drafted when Bretton Woods operated and fixed exchange rates were the norm. Both the Snake and the EMS sought to *restore* some roughly equivalent stability of exchange rates amongst European currencies.

cross-border trade and investment by non-financial enterprises and intensify financial links. As Alan Greenspan (1991) notes: 'since the United States is both a free trade zone and a single currency zone, I cannot dismiss the proposition that a single currency is an important ingredient in a successful free trade zone'. Admittedly, fixed exchange rates are not the same as a common international currency. Nevertheless, as we move from floating exchange rates to credibly fixed exchange rates, information costs, transactions costs, and exchange-rate uncertainty can be expected to fall.

However, these benefits of fixed exchange rates rest on currencies being credibly fixed. It is not so clear what is the balance of benefits and costs when the choice is between exchange rates which vary by (not so) small amounts on a daily basis as under floating rates and those which under an unstable ERM may be subject to intermittent, large jumps, often after intense speculative pressure. This leads us back to the issue of speculation.

2. The problem of speculation

Undoubtedly the excessive rigidity of the ERM arose as some countries used fixed parities as a means of enforcing convergence of inflation rates and economic performance via declining competitiveness and 'imported credibility'. The system was also used as a vehicle for achieving monetary and political union— under the usual interpretation (but see below), the Maastricht Treaty has been seen as requiring a period of 'narrow bands' exchange-rate stability as a precondition for entry to EMU. These conditions as we have seen, opened up the ERM to speculation.

However, there was another reason for the excessive stability: paradoxically, that was the fear of not being able to control speculation resulting from market realization that realignments were still likely. Whenever a realignment occurs, there must be confidence that the new, lower parity will be maintained indefinitely. This is to ensure that speculative positions will be unwound after a parity change and not become a basis for further speculation, as devaluation creates the expectation of further devaluation. An adjustment of parity may merely encourage speculators to expect that it will be altered again in response to future difficulties, so making matters worse 'next time round'. In these circumstances, future policy credibility and the avoidance of increasingly large and destabilizing speculation may seem to come from making realignments as seldom as possible, so making markets believe that they are truly exceptional events.

This was the conclusion reached by ERM members after the events leading to the January 1987 realignment. As in July 1993, French and German short-term interest rates diverged despite broad convergence of the economic 'fundamentals', and intervention could not stop the speculative run. Under the Basle–Nyborg agreement of September 1987, members strengthened intervention provisions, and 'agreed to reduce the scale and frequency of exchange rate realignments' (Collins, 1992). This was an echo of Bretton Woods:

Gradually tacit agreement had emerged among the central banks and finance ministers of the world that the devaluation clause of the Bretton Woods agreement should not be used because it was disruptive and would lead to speculation. Instead of the adjustable peg envisaged by the Bretton Woods agreement we should have 'fixed and immutable exchange rates', to use the words of Robert Roosa who was Undersecretary of the Treasury during the early 1960s. (Houthakker, 1977, pp. 10–11)

But immutability is not possible. So long as separate exchange rates exist they can be changed; thus there are no means by which the possibility of devaluation can be excluded.

Wider bands offer some reprieve. One advantage is by making speculation a 'two-way street'. When there is a one-way bet, speculators who guess wrongly simply liquidate their positions at the parity rate, losing only the transaction costs. When the rate can go either way, an expectation of profit from depreciation carries with it the risk that the currency might in fact appreciate. Another advantage of widened bands is to increase the possibility that when realignments do occur, they are 'within the bands' so that market rates need not change, and a discrete jump in the exchange rate is avoided.

Further, if the bands are perceived as highly credible, exchange rates will be nearly fixed. When the exchange rate is near the bottom of its band, the market will expect the rate to drift back towards central parity; and vice versa, when the rate is near its ceiling. Speculation will be 'helpful' and stabilize the rate. It is a measure of how rapidly things can change that this was the British experience up to February 1992:

the ERM has weathered other episodes of tension in the past few years, and will be able to do so again if the right policies are followed. That surely is the key to ERM stability. If the monetary authorities and governments of member states are genuinely committed to sustainable non-inflationary policies, and can convince the markets of their determination and ability to keep to them despite economic disturbances and periods of political pressure, market speculation will tend to reinforce exchange-rate stability, not upset it, This has been our experience in the sixteen months since we joined the ERM. (Bank of England, 1992)

The rub, of course, is that once credibility is questioned, speculation can quickly push the exchange rate over the edge of the band. Very wide bands reduce that possibility, but then may do little to stabilize the exchange rates. Worse, they may encourage the markets to believe that there must be some reason for the wide bands, namely that big variations are going to happen, and act accordingly.

Sometimes observers say that the volume of funds available to speculators is so huge that central banks are bound to lose. Perhaps they are right—as the unsuccessful defence of the French franc suggests. Yet no one has a bigger supply of DM than the Bundesbank. It follows that the Bundesbank can, if it wishes to, see off a speculative raid on any bilateral DM exchange rate.[7] It has only to keep its nerve in the face of the inevitable—but temporary—increase in DM-denominated bank deposits, much of it nominally, if not actually, in the hands of 'residents'. What this ignores is the extent to which the new conditions prevailing in European financial markets have subverted the operation of the conventional model of monetary policy formation in Europe, based around the Bundesbank as the anchor of the system.

3. The question of the anchor

A fixed exchange-rate system with fiat currencies needs an anchor to tie down nominal magnitudes, but does not necessarily need an anchor currency and the EMS was not designed with one, but rather as a collaborative and symmetric system—which in many respects it still is. Germany's anchor role has come about because of its size, the Deutschmark's reserve-currency status, and policy responses. The Bundesbank sets monetary policy for itself; it sterilizes the effect of foreign-exchange inflows and outflows when it is obliged to intervene, as at the margins of the fluctuation band, whilst leaving the bulk of intervention—in the normal way—to the so-called 'intramarginal' intervention operations of other central banks. By not realizing the consequences for control of bank liquidity, money base, and money supply which would appear to be inherent in the obligation to intervene at the margins, and thus operating more or less independently from the rest of the ERM,

the Bundesbank has forced its monetary policy on to other members. In this way the EMS is similar to earlier fixed exchange-rate systems which also revolved around leaders or hegemons—sterling under the gold standard and the US dollar for Bretton Woods. To some extent this may be intrinsic to the working of international monetary systems: co-ordination of monetary policies jointly to determine communal inflation and macroeconomic outcomes may render a symmetric system unmanageable, while competition for leadership may make for instability (witness the consequences of competing claims for leadership from the franc fort in 1993).[8]

Hegemonic systems are usually viewed as involving 'implicit contracts'. The 'hard' EMS which emerged from the 1987 Basle–Nyborg agreement was based on two assumptions:

(a) Germany would continue to have low inflation and low interest rates;
(b) the Bundesbank would defend exchange-rate parities 'unstintingly', in the words of Kenen (1992), provided that members of the system made the necessary interest-rate adjustments to protect their currencies.

The first assumption was a reasonable one at the time—few, if any, predicted that in two years' time the Berlin Wall would be pushed down. The second overlooked that at some point the Bundesbank could be faced with a situation in which its obligation to intervene would be incompatible with control over German monetary conditions— or would be felt to be so. The two crises of September 1992 and July 1993 may well exemplify this.

A straightforward reading of the original EMS agreements might lead one to suppose that the Bundesbank could be obliged to intervene *ad infinitum* when an exchange rate hits its floor against the DM (and the DM hits its ceiling). Intervention obligations are supposed to be symmetric. Indeed, since the Basle–Nyborg agreements extend the financing facilities to intramarginal operations, it might appear that the Bundesbank could be obliged to intervene in unlimited quantity within the bands. Clearly, to judge from events, these obligations do not exist in such unlimited form. Nor, on reflection, could they. Otherwise a strong-currency country could be obliged to support a weak currency

---

[7] Speculation can also involve non-EC currencies such as the dollar. This might lessen the gains to speculators as compared with a direct franc–DM switch, for example, if realignments were symmetric. But this has not typically been the case, and EMS realignments involving appreciations of the DM relative to other EMS currencies have not usually involved appreciations of the DM *vis-à-vis* these currencies. In short, realignments have centred on the DM—the anchor currency.

[8] The Bank of France cut its intervention rate nine times between April and June, stimulating discussion of the role of the franc as a new anchor for the EMS, replacing the DM. During June, the French Minister of Finance invited his opposite number in Germany to discuss co-ordinated reductions in interest rates at the forthcoming France–Germany economic summit. This resulted in cancellation of the scheduled meeting. This event seems particularly significant for subsequent developments.

Table 3.1. Monetary Policy in Germany During Periods of Speculation

	Foreign exchange purchases by Bundesbank[1] DM billion	Changes in money stock (%)[2] M2	M3[4]	Change 'due' to foreign exchange movements (%)[2,3]
1983 Feb–March	+13.3	+6.7	+10.8	+4.7
April–May	−19.3	−3.1	+2.8	−9.1
1985 Dec–April	−0.7	+7.3	+6.8	+14.9
1986 April–May	−9.5	−3.6	+2.1	−4.0
1987 Jan–Feb	+15.6	+10.2	+9.1	+18.3
March	−5.9	−1.0	+2.4	+3.1
1992 September	+92.2	+43.2	+27.0	+42.7
October	−43.3	−3.5	−2.4	−25.2

Notes: [1] Includes interventions by foreign central banks financed through the Bundesbank and DM repayments within the EMS. Excludes swaps with domestic banks.
[2] At annual rates.
[3] That is, change in the net external assets of the banking system (including the Bundesbank).
[4] Seasonally adjusted.
Source: Deutsche Bundesbank, *Monthly Bulletin*.

ad infinitum, however inappropriate the particular exchange-rate value in question.

So long as the DM is the other side of a speculative run, the functioning of the system depends on the discretion of the Bundesbank. Whether or not at some point in either or both of the recent crises the Bundesbank called a halt to the support process forcing the other countries to adjust, it is clear that the scale of the problem for Germany was far beyond what had been previously experienced within the EMS. Table 3.1 illustrates the difference in scale between the September 1992 crisis and some earlier ones. Given interruptions of this scale to the course of monetary growth in Germany the presumption must be that the Bundesbank will 'take the risks on the cautious side', i.e. faced with an expansion of M3 above target (which may be a currency substitution effect with no inflationary consequences) the Bundesbank is liable to err on the overdeflationary, rather than the opposite, direction. Everyone knows what that means. Whenever the Bundesbank has been confronted with the choice between an appreciation of the DM and large-scale intervention which poses a threat to monetary targets and domestic inflation, it has jettisoned fixed exchange rates and chosen appreciation. While the method has varied—withdrawal from foreign exchange markets, letting its views on realignments be known, or interest-rate changes that do not happen—the result has been the same, and that will continue to be the case (correctly) while the Bundesbank's charter requires it to focus on domestic goals.

Stated differently, the real asymmetry in the system is in terms of the distribution of benefits and costs. Germany's partners have gained in terms of 'imported

credibility' and inflation convergence when they affixed themselves to the DM anchor. But it is not clear how Germany benefits. Some prestige and seigniorage come from the use of the DM as a reserve currency, but these must be small in comparison with the tiresomeness of frequent exchange-rate intervention. There are presumably some political advantages to Germany because the EMS is part of the movement towards European integration which the German government supports—although this goal may not be one that is shared by the Bundesbank.

Any attempted reconstruction of the system must address this imbalance and provide for greater assurance of automaticity and symmetry. It may be that the bilateral character of the exchange-rate commitments in the EMS need to be changed so that the burden of intervention is more widely shared.

4. A reconstructed ERM

Any thought of a renarrowing of the bands immediately provokes two questions. The first is how to get there—how soon, and at what parities; the second, how to stay there. It is the second on which we concentrate, although the two are obviously inter-related. If it is necessary to endow a reconstructed ERM with features not found in the old model then it would be especially perverse, not to say dangerous, if the narrowing were to take place at misaligned rates. Exchange-rate-band narrowing is a prospect that should be resisted until the recovery is well under way and fiscal positions are improving.

Because the system has been shown to be vulnerable

to speculation, the incentives for future speculation have been raised. With the market now sensitized to policy inconsistencies, the system has to work in a manner which does not invite a further crisis. We can distinguish at least three distinct types of reform. The first consists of disabling the market's capacity for destabilizing speculative responses by re-introducing capital controls of some kind. The second softens the target for speculation by introducing elements of greater flexibility into the system. The third seeks to attenuate the likelihood of policy discrepancies arising.

5. Capital controls

It is standard in analyses of the history of the EMS to give capital controls a critical role in the successful management of the system during its early period. They are credited with allowing other countries to 'opt out' of the discipline of following German monetary policy (though only for a time) and enabling the authorities of those countries to manage realignments in an orderly way.

In the new circumstances the reimposition of capital controls is again being spoken of. Eichengreen and Wyplosz (1993) have proposed the imposition of a 'Tobin tax' on foreign-exchange-market transactions, designed to hobble speculation. Alternative arrangements, such as two-tier exchange rates, might be designed which would do less damage to foreign trade and the world allocation of capital. The administratively most feasible option, however, appears to be the imposition of incremental reserve requirements on financial institutions taking open positions in the foreign exchange market, an alternative which Eichengreen and Wyplosz also favour.

The Maastricht Treaty is hostile to the imposition of controls over capital flows (but the 'Tobin tax' and the reserve requirement schemes would escape the relevant article (73) since they are not an administrative measure of the type envisaged in the article); though there is always the 'safeguard clause' in the 1988 Directive allowing countries to use short-term protective measures.[9] We do not pursue these issues further, on the grounds that the political (and economic) objections to the reimposition of controls are considerable.

[9] Article III states that 'where short-term capital movements of exceptional magnitude impose severe strains on foreign exchange markets and lead to serious disturbances in the conduct of a member state's monetary and exchange rate policies', a member state can take 'protective measures' in respect of certain short-term capital movements.

6. Increased flexibility

One way of increasing the flexibility of the system would be to provide for a regular review of central parities, and for a presumption of automatic indexing on relative inflation. Another would be to have instead a 'softer' commitment to 'try' to maintain the exchange rate within the margins announced. These adjustments would relieve the system of the danger of fostering cumulative misalignments and of offering 'one-way bets'—but at the cost of converting it from a counter-inflationary nominal bands system into a Williamson-style real exchange-rate system. (Williamson, 1985)

Such a system gives the market some information about the authorities' intentions, which probably strengthens their hand in intervention operations and helps to free the system from 'irrational' runs and fads. And it would help preserve the achievements of the Single European Market against the threat of disruptive—even competitive—devaluations. But the central parity-adjustment procedure combined with softness in the bands could just lead to an accommodation of inflationary pressures.

7. More symmetry

In Stage Three of Maastricht, monetary policy will be made by the European Central Bank. By contrast, in Stage Two, monetary policy will continue to be set by the national central banks. With narrow bands the setting of policy would remain in the hands of the Bundesbank, other central banks falling into line to the extent required by their commitment to remain with the bands. Whilst the European Monetary Institute will be preparing the way for Stage Three and will be encouraging the co-ordination of monetary policies between the individual central banks, it will not have the power to execute policy.

It follows that if a 'more symmetric' system is to develop, the Bundesbank must be persuaded to set a more European policy. An immediate difficulty is that the Bundesbank is bound by statute to follow policies that privilege price stability in Germany—not elsewhere. This suggests that there are two—possibly complementary—routes. One argues that, because of trade and other links, the Bundesbank will make better policy (for Germany, and also for Europe) by targeting European conditions to some degree; the other is a more frankly political approach which would seek to secure from the German government an assurance that where the Bundesbank takes actions which are guided by 'European' considerations it will not be deemed to have violated its statutory duties.[10] Being 'guided by 'European' considerations' might in this context be identified with accepting a recommendation of the EMI. Obviously, this advice could not be binding upon the Bundesbank; but without more symmetry, a restoration of the German leadership model merely invites a re-run of July 1993.

8. Monetary union

All of these problems—exchange-rate instability between member countries, the asymmetries of the system, speculation, and intervention obligations—can be removed at a stroke by a quick move to EMU. Currencies which are permanently locked or which cease to exist cannot fluctuate or be realigned, while a European central bank would substitute a European monetary policy for a German one, targeted at price stability in Europe and not in Germany alone. Moreover, this route to Stage 3 is not ruled out by the Maastricht Treaty. Although that Treaty specifies an ERM 'convergence criterion', this requires that the member countries should have participated in the 'normal' bands of the ERM without 'severe tension' and without provoking a devaluation for a period of two years. While the phrase 'normal' was intended to be read as ± 2.25 per cent, it would be open to the present participants, should they wish, to redefine the so-far explicitly 'temporary' ± 15 per cent band as the new 'normal' band. The other convergence criteria of the Treaty would still need to be met.

Avoiding the perils of a narrow bands apprenticeship for ERM would remove one set of problems, but there is the danger of creating some new ones. While recent events make out a case for monetary union as the only sure way to peg exchange rates, they illustrate the difficulties of getting there by an engineered convergence process over time, polarizing the options into those of a 'big leap' toward unification on the one hand and a more natural evolution on the other, much like that amongst the Benelux countries.

The costs of a rapid push to EMU are tied up with whether or not Europe can be said to constitute an 'optimal currency area'—a question which has been given a thorough airing recently (Eichengreen, 1992a; Smaghi and Vori, 1993). In terms of wage and price flexibility, labour mobility, or even cross-border investment, there may be many grounds for arguing that Europe is not an optimal currency area. Then, again, perhaps the United States and Canada and some other federations are not optimal either, or at least *were* not until unification got under way. On this comparison we record some of the omitted lines from the article of Mundell quoted earlier—if only because it was he who invented the concept of optimal currency area:

The size of the European currency domains have become too small relative to other assets to make continual fluctuation among the exchange rates a desirable system. Nor do I believe it would be desirable to break up the currency system of the United States in order to allow fluctuations in exchange rates between different districts. Both the United States and Europe possess enough of the characteristics of an optimum currency area to move towards narrower, rather than wider fluctuations in exchange rates. (Mundell, 1973, p. 149)

Politically, the attraction of accelerating the movement towards EMU may turn out to be a perception that there is no better alternative. Even so, the necessary institutional framework cannot be created overnight. It is sobering to recall that the organization of decimal currency conversion in Britain took a good two years—in fact, over seven years elapsed between the Halsbury Committee's report and the introduction of decimal coinage.[11] Opposition of the Bundesbank and the German people to giving up the Deutschmark for an 'Esperanto' currency would be considerable. In fact, the easiest and quickest way to achieve monetary union would be to adopt the DM as the single European currency, as Sir Alan Walters (1993) has mischievously suggested, using a currency board system to effect the conversion. At the very least, the firm rejection by Holland, Belgium, Luxembourg, and Denmark of the French request on 1 August 1993 for the DM to leave the ERM temporarily, and the two-tiered ERM which then resulted, suggests that the formation of a Bundesbank-dominated inner core group of countries may not be as unlikely as it once might have seemed.

9. Alternatives to EMU

EMU requires a single currency or immutably fixed exchange rates. An alternative is to have a parallel or common currency, the most notable being the British proposal for a hard ECU. Although (briefly) revived again, it is not apparent that the hard ECU has features which address the problems behind the recent crisis. The plan also suffers from being seen as an all too

---

[10] The suggestion that the German government might in effect indemnify the Bundesbank against violating a legal requirement has a parallel (if not such an explicit one) in present practice. For it appears that the German government in effect indemnifies the Bundesbank against violating its obligation to intervene to support a weak currency in the ERM.

[11] And it was almost 300 years after Sir William Petty's original proposal!

transparent attempt to stave off the loss of monetary sovereignty under EMU—this has always been a British sticking point. Some of the concerns about EMU on the grounds of national sovereignty might be allayed by recalling the example of the gold standard. Under gold, there were immutably fixed exchange rates based on what was in effect a uniform currency—gold—yet nationalism (and cultural diversity) was strong.

But this example prompts a question: if it is the case that monetary union is now the only way to ensure a credible exchange-rate commitment, how was it possible to stabilize exchange rates within narrow bands under the gold standard? Perhaps the sheer volume of internationally transferable funds these days creates a new problem. Yet capital flows were extensive under the gold standard and they were also highly responsive—Bagehot once remarked that '8 per cent will bring gold from the moon'. Why were these movements 'helpful' to the authorities when maintaining parity? One suggestion (Eichengreen, 1992b) comes from an institutional characteristic of the time—private central banks. Governments' commitments to gold parity were supported by the prestige and independent status of the central banks. This is a feature which it might be possible to reconstruct. A prerequisite for any restoration of the ERM might have to be the independent status of central banks—and it is the case that the Maastricht Treaty requires that central banks become independent during the course of Stage Two. Central bank independence is discussed further below.

The gold standard, however, rested on more than the independence of central banks: there was the 'mystique' of gold. Countries on gold were expected to be so for ever, and moreover at the same parity. Such was the attachment to the historic parities that when countries did suspend convertibility temporarily (usually because of wars), resumption invariably took place at the old levels. The pound effectively had the same official metallic value from 1717 to 1931, the dollar from 1834 to 1934. People regarded this as unchangeable, as evidenced by the remark of Sidney Webb (Lord Passfield) following Britain's suspension of convertibility in 1931: 'Nobody told us we could do this'.[12] That mystique has gone.

IV. Britain and the EMS

Despite sterling's current 'safe-haven' status, Britain is formally a member of the System and will shape its

future. It will not wish to be seen to be turning its back on Europe for eighteen years, as before, and in any case the range of alternative outcomes for the EMS is so wide as to impinge on British policy-making.

1. Available options

As a hybrid of fixed and floating exchange rates, the ERM—like Bretton Woods—held out the promise of providing the best of both, and for a number of years it in fact did so. Ironically, the very asymmetries which made the system a source of attraction to a number of countries, including Britain, also proved to be its undoing. Now that it has been unlocked by speculation, the available options polarize towards permanently fixed or floating, and the viability of a middle path between these two poles is sharply reduced. Capital controls aside, stable parities seem likely to succeed only if member countries maintain—and are seen by markets to maintain—a uniform monetary policy. In effect, a federation of European central banks would need to be formed. But this would be the result under EMU anyway! EMU is certainly feasible, but cannot occur overnight even if there was general agreement to an early union, which there is not; the UK in particular has special problems with the idea.

Floating exchange rates are a safe choice in two senses: they can work and we know their foibles, whereas EMU would be a leap in the dark and a reconstructed ERM just now would remain at considerable risk.

2. A floating Europe?

A period of floating exchange rates could be seen as a way of putting the EMS back together with more certainty. It would give Germany time to sort out reunification and overcome its domestic problems in order to resume delivering low inflation and low interest rates. Other countries would have the opportunity to loosen the tight monetary policies which are pushing up unemployment and aggravating the fiscal situation. In fact it is probably only in the context of a low-interest-rate recovery scenario that a sufficiently large number of countries would be able to satisfy the fiscal convergence criteria laid down in the Maastricht Treaty, which would presumably still apply to any move to Stage 3 (EMU) even if the ERM convergence criterion is regarded as being satisfied by a new interpretation of 'normal bands'. The fiscal criteria call for budget deficits of less than 3 per cent in ratio to GDP and for gross debt to GDP ratios of less than 60 per cent.

---

[12] Quoted in Taylor (1965).

On the latest data only three countries (Germany, Denmark, and Luxembourg) satisfied these criteria in 1992 and the situation will clearly have deteriorated in 1993.

After such a period of fiscal consolidation it might become possible again to resume progress towards EMU on the basis of narrow bands—though if so, the co-ordination advice of the EMI would need to be seen to be accepted and policy would need to appear more symmetric than hitherto.

Alternatively, floating exchange rates could signal a change of direction for Europe, by in effect converting it into a free-trade zone—literally the EEA. The EEA came into effect with the European single market in 1993 and is an agreement between the EC and EFTA which guarantees the 'four freedoms', i.e. free movement of goods, capital, services, and labour, but does not carry any monetary or political obligations. Most Britons would probably be happy to settle with that, and the less-than-fullsome endorsement of Maastricht by European voters suggests the same may be true elsewhere. By accepting looser ties, the EC could focus on an objective very much in the spirit of the original European movement—that of broadening the market area to include the 'old' Europe of Centre and East.

There is the worry that floating rates might undo some of the integration which has already been achieved; the conditions which led to excessive swings in funds may now lead to excessive swings in exchange rates. The example of the NAFTA countries shows that flexibility of exchange rates need not be incompatible with increasing integration; nevertheless, Canada gives considerable weight to US developments when setting its monetary policy, while Mexico maintains a crawling zone link between the peso and the dollar. A more co-operative arrangement could be sought in Europe to stabilize the behaviour of floating rates, via the co-ordination of monetary policies. This, in fact, is what the EMI is designed to do. The history of tripolar policy co-ordination is not encouraging but if the EC cannot be more cohesive than that, then it should not be contemplating EMU.

3. British aversion to ERM

Britain's reluctance to commit itself even to the ± 15 per cent bands may be for the practical reason that it does not want to get locked into narrower bands should an attempted revival of the ERM begin soon. But it also shows how far public opinion in Britain has shifted. At first, the ERM was seen as the saviour. But as the recessionary forces became evident, opinion changed

—even though the economy was well into the downswing phase at the time when Britain joined.

Admittedly, much of Britain's 'bad experience' was of its own making. When Britain joined the ERM, it signalled a switch in policy from *internal* price stabilization to *external* price stabilization. It is in this respect that the exchange rate is not just 'a price like any other price' because it signals the underlying form of price stabilization. In the context of the ERM this can be overstated. Countries such as France did not achieve low domestic inflation during the 1980s merely by pegging to the DM; rather, the exchange-rate link flanked the use of domestic policies to produce low inflation, with both gaining credibility in the process. From the viewpoint of British government policy a signal of sorts was badly needed: economic performance deteriorated during 1990; an election was looming; and with the inflation rate headed towards double digits the anti-inflationary credibility which had been so hard won in the first half of the 1980s was slipping away rapidly.

4. The hard currency phase

Further, the entry rate for sterling (during the Gulf oil crisis) and existing inflation-rate differentials *vis-à-vis* the 'mark bloc' were important: stabilization of the nominal exchange rate implied the adoption of a 'hard currency' option by the British authorities. Some observers suggested that sterling may have been out of line by as much as 20 per cent! Moreover, the ERM that Britain entered in 1990 was much different from the system that it might have joined in the mid-1980s, in that parity realignments were thoroughly discouraged. To that extent, it looked as if Britain was almost trying to back-date entry. It is not altogether clear that sterling was over-valued to quite the extent thought at the time (see Wren-Lewis *et al.*, 1991, and Owen and Parkes, 1990–1 for conflicting views). Nevertheless, from the inflation trends shown in Figure 3.3, it was apparent that competitiveness *vis-à-vis* EMS countries would soon decline markedly unless wages and prices in Britain fell rapidly. (Figure 3.3 compares the integral of inflation differentials against Germany for Britain and some other EMS countries.)

Inflation control rested heavily on the direct effects of price arbitrage in tradeable goods markets, the consequences of diminishing cost competitiveness upon profit margins and wages, and the impact of the declaration of a new monetary standard in the form of the ERM commitment upon wage and price-setting mechanisms. The latter initiated a debate which bore

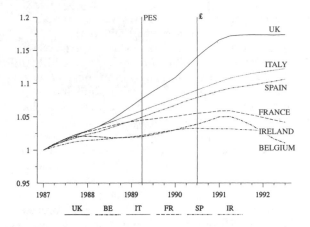

Figure 3.3. Cumulative Inflation Differentials

Notes: Inflation measured using consumer prices, based on January 1987 = 100. Vertical line indicates dates when Spain (PES) and the UK (£) joined the ERM.
Source: IMF, *International Financial Statistics* (various).

an uncanny similarity (but with roles reversed!) to those of 1979 when monetarists argued that inflation could be brought down without undue costs in terms of output and employment by having the government declare its unswerving commitment to money-supply targets. Then it was more the rediscovery of the Phillips Curve and not the new monetary rule book, which squeezed inflation from the system. Those who looked to a different result in 1990 pointed to the greater credibility of the ERM commitment over money-supply targets, and to econometric evidence suggesting that 'a firm exchange rate policy can help to reduce inflation without undue costs in terms of output and the balance of payments' (Bank of England, 1990). In fact, a very rapid convergence of UK inflation to the ERM average did occur while Britain was in the mechanism, but the added 'ERM effect' cannot be readily disentangled from the recessionary forces already in train.

5. Problems of interest rates

The 'crunch point' for Britain came from the divergent needs of German and domestic interest-rate policies. The two economies were clearly out of synch. But quite apart from this, the different structures of the financial systems make it difficult for British and German interest rates to move in tandem, especially in the face of speculative pressures.

Two features mark out the British financial system. First, bank reserves ('cash') are supplied by open-market operations in bills or in ways such as through the discount market which are transmitted quickly to money-market rates across the spectrum. Since the retreat from the 1981 'market-based' procedures (which many consider were a sham anyway), the authorities have developed an explicit signalling system which has allowed a rather aggressive use to be made of interest rates (see Artis and Lewis, 1991). Second, money-market interest rates are rapidly reflected in commercial bank base rates which in turn govern the cost of borrowing both on new and existing loans from a wide range of lenders. By contrast, many Continental European central banks rely less on market operations to supply reserves (Wilson, 1993). By means of repurchase agreements, privileged discount window facilities, tendering arrangements, or simply low interest-rate loans, those central banks have at times been able to push up money-market and offshore rates while limiting the impact on bank interest rates.

In addition, on the Continent a larger proportion of business loans and home-mortgage loan rates are long-term and either fixed-rate or indexed in terms of a long-term interest rate than is the case in the UK (see Table 3.2). German and French households also have mortgage debt to income ratios which, at 21 per cent of GDP, are less than half that in the UK, at 58 per cent of GDP in 1989. A further difference is that Germany has been largely immune from the cycle of deregulation, innovation, debt accumulation, asset price inflation and deflation which has taken place in the Anglo-Saxon countries.

Some of these differences are temporary; others are undergoing change or are likely to do so under the single financial market—already, much more mortgage lending in Britain is fixed-rate, although variable-rate loans are still the norm (see Table 3.2). So long as the UK remains the country in Europe where short-term interest rates matter most, membership of the ERM—with its implications for interest rates—will pose special problems for the UK.

V. Britain outside the ERM

Sterling's departure from the ERM is merely the latest turn in what can only be described as a series of somersaults in monetary policy since 1979. The Labour government was unwilling to participate in the ERM because the UK, with its propensity to inflate, might lose competitiveness *vis-à-vis* deflationary Germany. Three months later, the Thatcher government would not join because it wanted to deflate on its own by rigid adherence to money supply targets and reasoned

Table 3.2. Fixed and Variable Rate Lending in European Countries

Belgium	Fixed-rate loans are the norm. Some mortgages and long-term roll-over credits are at variable-rate.
Denmark	Mortgages and long-term loans are mostly fixed-rate. Shorter-term bank lending is usually variable-rate, linked to the discount rate and money-market rates.
France	Over 90 per cent of mortgages are fixed-rate. Over 60 per cent of corporate debt is estimated to be fixed-rate. Lending to the private sector by members of the French Banking Association in the second quarter of 1992 was distributed as follows: fixed-rate (51 per cent); variable-rate, linked to base rate (14 per cent); variable-rate, linked to money-market rates (34 per cent); other (1 per cent).
Germany	About 80 per cent of company debt is either fixed-rate or based on long-term interest rates. Less than 30 per cent of home mortgages are variable-rate, and only 20 per cent of household debt carries short-term interest rates. Much bank lending is medium- or long-term at fixed rates.
Greece	Long-term loan contracts provide for adjustment of interest rates at specified intervals.
Ireland	Bank lending to the private sector is usually variable-rate, linked directly to market rates. Mortgage rates can be adjusted with one month's notice.
Italy	55 per cent of housing finance consists of fixed-rate loans, and household debt is low. Nearly half of bank lending is on overdraft, with rates adjusted in line with the discount rate and other market rates. Of the remaining bank lending, less than 10 per cent is variable-rate.
Luxembourg	Most bank lending is at variable rate.
The Netherlands	Fixed-rate lending is the norm. Home mortgages are invariably fixed-rate.
Portugal	Except for short-term bank loans (less than 6 months), the rest is at variable rate with interest rates usually adjusted 6-monthly.
Spain	Most bank loans outstanding are at fixed-rate. Variable-rate loans currently represent about one-fifth of new bank loans. Half of all household debt and about 30 per cent of home mortgages are at fixed interest rates.
United Kingdom	Over 90 per cent of home mortgages outstanding are at variable-rate. In the second quarter 1992, 79 per cent of new mortgages were variable-rate, 15 per cent fixed-rate (over 1 year) and 6 per cent fixed-rate (up to 1 year). Some 27 per cent of banks' loans to UK business are on overdraft, all at variable-rate. The remaining loans and advances to UK business are also mainly variable-rate, often linked to LIBOR (London inter-bank offered rate).

Source: Based on Council of Mortgage Lenders, EC Commission data.

(correctly) that these were incompatible with exchange-rate targets. This then gave way, following the debacle of the Medium-Term Financial Strategy, to a period in which the exchange rate was in fact used as a conditioning factor in the implementation of monetary targets. Growing interest in the ERM option saw the policy harden into one of 'shadowing the Deutschmark'. When, after approximately one year, the pound burst through the (unannounced but informally made known) target ceiling of DM3 to the pound, Britain returned decisively to eclectic domestic monetarism, in the process seemingly raising additional hurdles to ERM entry. Then, a little over two years later, the UK entered the ERM without having cleared the principal hurdle of inflation convergence. Membership lasted slightly less than two years.

A first requirement for policy has to be to bring this series of about-turns to an end, and provide consistency to the prosecution of monetary policy. Too often these lurches in policy—each one of which is declared to be permanent and inviolable—have been prompted by the belief that there 'must' be a better way. But there is no magic wand, as recent debates about monetary policy in Germany show (has the M3 relationship changed, are monetary targets too low?). Indeed, the performance of the Bundesbank is not that remarkable:

it has missed its monetary objectives in eight out of the last 17 years; it has pursued a variety of objectives; and 3 per cent p.a. inflation is not the long-run price stability of the gold standard, or indeed even of Bretton Woods. Its record is good only in comparison with others, and that has come about because it has followed a relatively consistent 'simple monetarist' policy of low inflation and monetary targets (whatever their flaws).

A second thing that the UK must resolve is the exchange-rate regime. There is a considerable difference between freely floating, managed floating, conditional target, informal target, and membership of the ERM—all of which have been tried. Admittedly, to some extent, the UK does fall between two stools. It is generally accepted that an economy's size and openness to trade determine the appropriate exchange-rate regime. Thus Holland, the exports of which are equivalent to 47 per cent of output (30 per cent of which are to Germany), is in a different position to the US or Japan—large economies which have a small trade sector (they export about 10 per cent of GDP). Britain, which exports 18 per cent of GDP, does not fit neatly into one or the other category; but as integration with Europe proceeds, and production and trade become more specialized, the balance tilts towards the advantages of having exchange-rate stability with its continental trading partners.

For the moment, Britain must make the best of floating rates. If experience is a guide, then the long-term prognosis is not good: policies of internal price stabilization have had limited success on these shores. Britain built its reputation for monetary stability under gold. Inflation was held reasonably in check under Bretton Woods (average 4 per cent per annum from 1950 to 1970). While some countries have been able to control inflation under floating exchange rates the UK has not done so, and inflation averaged 14 per cent during the 1970s and 7 per cent during the 1980s.

A variety of technical considerations may account for different monetary experiences; these include differences in monetary tactics, stability of money demand, and differences in financial innovation and liberalization. But the institutional and political environment within which monetary policy operates must be considered. Gordon (1975) depicted inflation as the outcome of demand and supply. The 'demand for inflation' reflects the need for government revenue in the face of resistance to increases in tax rates and the desire of various groups to use inflation to obtain a larger share of real income. The 'supply of inflation' depends on election cycles, the structure of labour markets, and the insulation of the central bank from political pressures. Can the 'old' story of labour-market inflexibility and attitudes of trade unions in the UK still be blamed? Has financial deregulation encouraged asset price inflation? Homeowners benefited from inflation during the 1970s and 1980s and they may constitute a major political force for inflation (or constraining actions against inflation). Certainly, interest rates have become politicized. Interest-rate cuts timed to precede party conferences and by-elections or in the wake of bad unemployment figures and other adverse political news are not uncommon. A third requirement has to be to distance monetary policy from such pressures.

Another requirement comes under the heading of monetary indicators, for the removal of the external exchange-rate anchor for policy raises a question mark in another respect. When the UK was a member of the ERM, policy had a clear and unambiguous target at which to aim, namely maintaining sterling within its ERM bands. Indeed, one of the arguments for the discipline of the ERM is the visibility of an exchange-rate target. Under the present regime, policy is necessarily much more judgemental and ambiguous. All of the old issues of policy-making in the 1970s and 1980s have re-emerged.

1. The problem of objectives

The aims of monetary policy have remained unchanged since sterling's suspension from the ERM: 'policy is directed to the control of inflation as an essential requirement for a return to sustainable growth in output and employment' (HM Treasury, 1992).

This declaration of itself is not especially encouraging: objectives were also unchanged from 1979. Additional measures to support policy include:

(a) a quantified inflation objective of 1–4 per cent with the aim of 2 per cent or less in the longer run;[13]
(b) current annual growth rate targets of 0–4 per cent for M0 and 4–8 per cent for M4;
(c) publication of monthly monetary reports by the Chancellor and the Governor of the Bank of England;
(d) detailed accounts of reasons for interest-rate changes;
(e) regular *Inflation Report* by the Bank.

The notion that monetary policy should focus primarily, perhaps solely, on controlling inflation implies a large degree of acceptance of monetarist doctrine, the core of which is the 'importance' of money, the natural-rate hypothesis and thus the long-run neutrality of money. With this has come an acceptance that unemployment is best tackled by microeconomic policies, aimed at changing the flexibility of labour markets, rather than macroeconomic policies.

The phenomenon of 'hysteresis' in the natural rate of unemployment is an unpleasant reminder that this division of policies into 'micro' and 'macro' may not be quite so straightforward as it seems at first sight, raising the premium on policy flexibility. Such flexibility is best exercised against a background of consistency: otherwise it leads simply to short-term politicization of monetary policy.

In principle, there is no reason why the use of monetary policy to smooth fluctuations in output and employment need be incompatible with stabilizing the trend of prices. Consider three objectives of monetary policy: price stability, the stabilization of output fluctuations, and the prevention of liquidity crises. With one instrument and three targets there would seem to be a classic prescription for policy failure. However, as noted by Niehans (1978), these objectives have different time dimensions: one is secular, the second is

---

[13] Inflation, here, is measured with respect to a definition of the national price index which omits mortgage interest costs (the so-called RPIX). The measure brings the UK more into line with other countries and removes the 'perversity' that a rise in interest rates leads directly to a rise in prices.

business-cycle length, and the third is very short-run. This opens up the possibility of conducting monetary-policy operations of different durations for the different objectives. Variations in monetary policy, as reflected in (say) base money, could be imagined as being decomposed into three waves of different frequencies, each assigned in effect to a policy objective with a matching frequency.

In practice, however, this is a difficult game to pull off, the more so the closer the durations of the different cycles. Meeting a liquidity disturbance lasting weeks or months may be unlikely to compromise the long-run trend of prices should policy-makers get it wrong, e.g. supplying liquidity to meet a fluctuation in the demand for money which fails to transpire (although the aftermath of the stock-market crash in 1987 provides a warning). In the case of output fluctuations of several years, the combination of the difficulties of forecasting the economy, data inaccuracies and revisions, and the lag before the effects of policy take hold makes it likely that policy actions will be 'too little, too late', perpetuating the stop–go cycle.

If the case for focusing policy primarily on an inflation objective seems good, should the balance of payments also be admitted as a further target? (In current conditions, this presumably would mean a further *tightening* of monetary policy.) A current-account deficit according to one interpretation reflects a lack of competitiveness. Or it may be a symptom of macroeconomic demand imbalances, with domestic absorption running ahead of production. But the current-account deficit is also equal to the excess of private-sector investment over private-sector savings and the government budget deficit. On this interpretation the deficit is a reflection of savings–investment imbalances, but these need to be viewed in a global context. For example, if Japan has a high propensity to save and a large current-account surplus it follows that other countries must in aggregate have a current-account deficit. That is, the deficit is the consequence of the natural workings of the international capital market and a myriad of individual decisions about capital formation and savings. There is no more case for interfering with the process internationally than there is domestically.

Again, in practice things are not so simple. Do markets perceive it the same way? There are enough mercantilists around who see current-account surpluses as good and deficits as always bad to pose some concern to policy-makers. If, in consequence, a country running a large current-account deficit must be constrained by the worry that a downward shock to the exchange rate might lead to portfolio capital inflows drying up, the balance of payments cannot be ignored

and the exchange rate comes back into prominence. A final query involves the 'twin deficits' issue, for a current-account deficit can also indicate a fiscal-policy problem with the government sector taking too many resources.

2. The question of indicators

A novelty of the new arrangements is the absence of an intermediate target to perform the role that monetary targets and then the exchange rate carried before. The M0 target is described as a 'guide' and the M4 target as a 'monitoring range' to be assessed in conjunction with a broad range of other indicators including the exchange rate. Hence the policy is essentially one in which control of short-term interest rates is targeted directly to expected inflation.

There are some problems with this approach. First, there is the precision of the link from monetary policy to prices. Money is 'important' but what does that mean? Friedman once defined it as being that money is the single most important factor affecting nominal incomes and/or prices, 'important' meaning that at least 50 per cent of variations are explained (Friedman, 1970). This allows for considerable slippage in the relationship. Second, as a consequence there is considerable room for factors other than money to influence prices. For some types of inflation, particularly asset price inflation, the process can generate almost a life of its own. Third, policy lags are long and variable, so that the effectiveness of policy actions to control inflation can be discerned only after a long delay. There is also an inherent publication lag with prices data.

These problems are all well known and they are why intermediate targets came into use. The basic idea is for the central bank to choose a variable—such as the money supply—which theory and evidence suggests is 'midway' between its instruments and the target of inflation, so that this becomes the focus for day-to-day operations. This variable must be reliably related to the inflation target with some lead in timing and it must be responsive to policy actions. Disillusionment with monetary targets grew when it seemed that they met *neither* criterion.

The case for monetary targets is eroded further by changing views about the role of banks in the transmission mechanism. Control of bank output is no longer the centre-piece of monetary policy and alterations to the balance sheets of the banks emerge as a by-product of policy rather than a necessary precondition. Banks were the focus of policy when legal restrictions and regulations made them 'special'. With the blurring of the lines between banks and non-banks and between market transactions and intermediated ones, banks have lost much of their 'specialness'. As a result, the monetary system is evolving towards a new structure in which banks can no longer be considered part of the 'control core', but must instead be grouped with the other endogenously determined parts of the economy. The area of policy control shrinks to the central bank's own balance sheet.

In this conception, monetary aggregates remain as 'information variables' and, indeed, have tracked both the upswing in the economy from 1986 and the subsequent downswing. However, in this indicator role they have been joined—and increasingly supplanted—by market price data and asset yields as a richer view of the transmission mechanism has unfolded. Thus the old sequence of market operations, interest rate, money and credit, output, prices, has been augmented by the exchange rate, yield curve, interest-rate expectations, commodity prices, equity prices, housing activity and prices, property prices, and the income and capital gearing of households and business enterprises. These, along with new Divisia measures of money (Fisher *et al.*, 1993), all serve as indicators of policy in the new framework.

This more complex view of the transmission mechanism has difficulties, too. The often conflicting signals of the indicators have to be interpreted and this requires a model, either formalized econometrically or in policy-makers' heads. Econometric models are inevitably vulnerable to rapid changes in structural relationships: the forecasting errors shown in Figure 3.2, for example, are frequently attributed to the failure of the models to pick up the effects upon spending of the highly-geared asset acquisitions by private-sector entities. One way around this is to focus on methods of forecasting inflation directly from various leading indicators (see Henry and Pesaran, 1993), but this approach is still in its infancy.

'Looking at everything', while technically efficient in the sense that it uses all available information, may result in policy confusion and create a bias towards inaction (not necessarily a bad thing). It may also reduce accountability since there is no way to judge the appropriateness of policy for some years—and then it is too late. The new Inflation Report exercise is designed to overcome this, but some argue for a further step.

3. An independent bank of England?

If the UK rejoins the ERM with a view to making it to EMU, a more independent Bank is called for. Stage 3

of Maastricht requires both the European Central Bank and the national central banks to be free from all outside interference. Under Stage 2, the governors of the central banks must be independent when fulfilling their obligations as members of the EMI, but the central banks themselves need not be independent. Other countries such as France and Spain are none the less preparing the ground by new legislation for granting independence to their central banks.

Even if Britain opts out of moves towards EMU, the issue of an independent Bank re-surfaces as one way to supply more credibility. This is sorely needed. Until the recent 'inflation control' exercise, the responsibility for monetary policy was very clearly with the Treasury and the Chancellor and, while this line of command was consistent with the Bank of England Act 1946, viz. '4(I) The Treasury may from time to time give such directions to the Bank as, after consultation with the Governor of the Bank, they think necessary in the public interest', it was more the *manner* in which control was exercised in the 1980s (Lawson: 'I make the decisions and the Bank carries them out') which created the problem. The reputation of the Bank was undermined in the process, and some change is needed to restore the Bank's authority.

Independence can imply a number of different things: independence of thought and opinion; independent advice; independence over day-to-day operations, independence to formulate policy; or independence to set objectives. Obviously a certain degree of independence is necessary if a central bank is to provide an 'outside' input on government policy questions. Some separation from normal political processes is desirable for the execution of policies which affect market participants. But independence is rarely meant literally in the sense of a central bank having the right to conduct policy free of constitutional safeguards and political and institutional constraints. At the same time, laws and statutes governing the relationship between the central bank and the government of the day take us only so far. Informal influence may be more effective than formal independence. For example, the Bank of Japan was never intended to be independent of government and one of the legislated responsibilities of the Ministry of Finance is to supervise the Bank of Japan; nevertheless it is widely perceived as being highly independent (Cargill, 1989). Much then depends on the personalities of the officials involved and the unwritten conventions which evolve.

The Bank of England has the opportunity to form its own judgements, advise government and operate free of day-to-day interference, but has no statutory tasks of its own. Since the Bank of England Act brought a working entity into public ownership and did not create an institution *de novo*, the Bank's functions were not defined and accordingly accountability provisions were not written into the Act. The extent of independence is untested: does 'time to time' in section 4(I) mean 'continually' or 'only in exceptional or unusual cases'?

There are two arguments *against* a separation of powers in the case of monetary policy. One is the need for economic policies to be co-ordinated. Many forces impact upon inflation and economic performance, and an independent monetary policy increases the difficulty of achieving co-ordination between monetary and other government policies. Some of the most difficult dilemmas for monetary policy arise from a failure to co-ordinate fiscal and monetary policies; Reaganomics and German reunification are obvious examples. The other argument is Friedman's (1963) view that money is too important for it to be left to central bankers and unelected officials. The elected government must ultimately be responsible and the institutions engaged in the formulation and execution of monetary policy should not be operating on the basis of a fundamentally different principle than that applied to other government bodies.

Nevertheless, as with statutory authorities, governmental control over monetary policy need not preclude taking the *operational responsibility* for achieving the objectives of policy out of the hands of elected politicians and ministers, while leaving the setting of the objectives themselves to the government (or Parliament). The current arrangements in Britain make the Bank seemingly accountable for progress towards the government's inflation objective, while apparently giving it precious little freedom formally to accomplish that objective. The obvious contrast is with New Zealand, where a legislated 'sunshine clause' gives the Reserve Bank full responsibility over monetary policy for a specified period to bring about the 0–2 per cent inflation goal. This can be overridden only in exceptional circumstances, and then only by Act of Parliament.[14]

Those governments unwilling to hand over any real power should remember that there is always a higher court—the general public. Persistent actions by a central bank to frustrate government policies—like those by the Bundesbank—would likely see the removal of the central bank's independent power to act unless its actions coincided with public opinion. On this view, the essential factor underpinning the independence of the Deutsche Bundesbank—and low inflation in Germany—is the strongly held fear of inflation in Germany. The conventional wisdom has been that this antipathy is not obviously paralleled in Britain where 'a modicum of inflation is still seen by many as quite reasonable' (Blunden, 1990). But perhaps this is overdue for revision and the public is ready for a new monetary constitution which will permit the Bank to focus on the achievement of price stability separate from other shorter-term concerns. While the reasoning may be somewhat different, there is nothing very new about such proposals, as is betrayed by the following extract from Hugh Gaitskell's evidence in 1959 to the Radcliffe Committee:

There was a time when it was argued that it was desirable to have a central bank independent of the government because it would prevent the government from pursuing inflationary policies. I totally disagree with that attitude. I think that the government must have complete power and complete responsibility. I think that the democratic processes today are just as likely to produce in the government a fear of inflation as a fear of deflation. (Gaitskell, 1960)

Gaitskell's response is of interest too. He maintains that the electorate is capable of producing the right result unaided.

VI. Conclusion

Not so long ago it was possible to say: 'It is now, I think, widely accepted that the EMS has been a success' (Governor, Bank of England, November 1991). As a hybrid of fixity and flexibility of exchange rates, the system seemed to provide a satisfactory mix of both. It was held together by three things: the stability of the German anchor; the political drive to monetary and political union; and the commitment of extensive joint intervention resources. When the German anchor shifted, as it were, off the bottom, the momentum towards union faltered, speculators began to test the willingness of countries to maintain parities in the face of diverging domestic needs, and the intervention mechanisms were exposed as a bluff.

For the moment, at least, there seems to be no viable middle way for most countries between completely fixed and freely floating rates—the present wide bands are neither fish nor fowl and serve little real purpose. These polarized choices mirror those for the European

---

[14] A formal Reserve Bank of New Zealand Policy Targets Agreement, signed jointly by the Minister of Finance and the Governor in December 1990, covers a Price Stability Target, Measurement of Price Stability, Deviation from Targets, Renegotiation of Targets and Implementation. The Bank is required to 'formulate and implement monetary policy with the intention of achieving a stable [0–2 per cent] general level of prices by the year ending December 1993 and maintaining price stability beyond that date'.

Community in a wider sense, for the Maastricht Treaty itself increasingly looks like a relic of the Cold War. A looser but broader confederation embracing EFTA and the old communist countries, and stripped of many monetary and political ambitions, provides an alternative vision to Maastricht for the way ahead.

Be that as it may, a quick drive to monetary union under Maastricht is a way out of the present difficulties. Nevertheless, it has always appeared that France has most to gain from EMU and Germany the least (or the most to lose), and it is not fanciful to see in the present confusion an image of this deadlock. In the longer run, the process of European economic integration makes stronger the arguments for EMU. That might involve an attempted reconstruction of the ERM, perhaps under the umbrella of a federation of independent central banks. For Britain, however, September 1992 will remain for some time a trauma which it will be scared of reliving.

Floating rates are the best option just now but the prospect of unruly exchange rates in Europe and unpredictable conditions of competition would be most unwelcome and against the grain of European currency history. Co-ordinated floating under the supervision of the EMI offers a safer yet at the same time more demanding alternative, for it puts monetary co-operation in Europe to the market test. If successful, it is not inconceivable that this framework could provide a route to EMU.

The next few years are critical ones for British monetary policy. The ERM anchor seemed a way for Britain to break loose from the stop–go cycle, giving policy the continuity which it so obviously lacked and seemed incapable itself of providing. That support has gone, yet the low inflation remains. This provides a foundation for a stable monetary environment which has not been present since the mid-1980s. Past policies have been marked by an over-emphasis upon means rather than ends. Now the aims of policy are clear enough, but the policy framework for achieving them has to be questioned.

References

Adams, G., Alexander, L., and Gagnon, J. (1993), 'German Unification and the European Monetary System: A Quantitative Analysis', *Journal of Policy Modelling*, 15 August.

Artis, M. J., and Lewis, M. K. (1991), *Money in Britain*, Hemel Hempstead, Philip Allan.

Bank of England (1990), 'The Effects of the Exchange Rate on Inflation, Output and the Current Balance', *Bank of England Quarterly Bulletin*, **30**(3), August, 316.

—— (1991), 'European Monetary Arrangements', *Bank of England Quarterly Bulletin*, **31**(4), November, 516–20.

—— (1992), 'Monetary Aspects of European Integration', *Bank of England Quarterly Bulletin*, **32**(2), May, 199–204.

BIS (Bank for International Settlements) (1993), 'Anatomy of the European Exchange Market Crisis', *63rd Annual Report*, Basle, 191–200.

Blunden, Sir G. (1990), 'The Role of the Central Bank', Julian Hodge Bank Annual Lecture, London.

Cargill, T. F. (1989), 'Central Bank Independence and Regulatory Responsibilities: The Bank of Japan and the Federal Reserve', Monograph Series in Finance and Economics 1989–2, New York, Salomon Brothers Center.

Central Planning Bureau (1990), 'Consequences of German Economic Unification', Working Paper No. 34A, The Hague, Central Planning Bureau.

Collins, S. M. (1992), 'Exchange Rate Mechanism of the European Monetary System', *The New Palgrave Dictionary of Money and Finance*, Vol. 1, London, The Macmillan Press Limited.

De Grauwe, P., and Papademos, L. (1990), *The European Monetary System in the 1990s*, New York, Longman Inc.

Egebo, T., and Englander, A. S. (1992), 'Institutional Commitments and Policy Credibility: A Critical Survey and Empirical Evidence from the ERM', *OECD Economic Studies*, **18**, Spring, 45–84.

Eichengreen, B. (1992a), 'Is Europe an Optimum Currency Area?' CEPR Discussion Paper No. 478 in *The European Community After 1992: The View from Outside*, London, Macmillan.

—— (1992b), 'Should the Maastricht Treaty be Saved?', *Princeton Studies in International Finance*, No. 74, December, Princeton.

—— Wyplosz, C. (1993), 'The Unstable EMS', *Brookings Papers on Economic Activity*, **1**, 51–143.

Fisher, P., Hudson, S., and Pradhan, M. (1993), 'Divisia Measures of Money', *Bank of England Quarterly Bulletin*, **33**(2), May, 240–55.

Friedman, M. (1963), 'Should there be an Independent Monetary Authority?', in L. Yeager (ed.), *In Search of a Monetary Constitution*, Cambridge, Harvard University Press.

—— (1970), 'Comment on "Money and Income: Post Hoc Ergo Propter Hoc?" by James Tobin', *Quarterly Journal of Economics*, **LXXXIV**(2), May, 318–27.

Funabashi, Y. (1989), *Managing the Dollar: From the Plaza to the Louvre*, Washington, Institute for International Economics, 2nd edition.

Gaitskell, H. (1960), Evidence, *Committee on the Working of the Monetary System, Principal Memoranda of Evidence*, London, HMSO.

Gordon, R. J. (1975), 'The Demand and Supply of Inflation', *Journal of Law and Economics*, **18**, December, 807–36.

Greenspan, A. (1991), Opening remarks, *Policy Implications of Trade and Currency Zones*, Kansas City, Federal Reserve Bank of Kansas City.

Henry, S. G. B., and Pesaran, B. (1993), 'VAR Models of Inflation', *Bank of England Quarterly Bulletin*, 33(2), 231–9.

HM Treasury (1992), 'Recent Developments in UK Economic Policy', *Treasury Bulletin*, 3(3), 1–10, London, HMSO.

Houthakker, H. S. (1977), 'The Breakdown of Bretton Woods', Discussion Paper No. 543, Harvard Institute of Economic Research.

Kenen, P. B. (1992), 'Is There an EMU Doctor in the House?', *The International Economy*, 6(6), Nov/Dec, 57–9.

Krugman, P. R. (1909), *Exchange-Rate Instability*, Cambridge, Mass., The MIT Press.

Mundell, R. A. (1970), 'A Plan for a European Currency', Conference on Optimal Currency Areas, Madrid, reprinted in H. G. Johnson and A. K. Swoboda (eds.) (1973), *The Economics of Common Currencies*, London, George Allen and Unwin Ltd.

Niehans, J. (1978), *The Theory of Money*, Baltimore, The Johns Hopkins University Press.

Nurkse, R. (1944), *International Currency Experience*, United Nations, League of Nations.

Owen, D., and Parkes, I. (1990–1), 'Measures of Real Exchange Rates and Competitiveness', *Treasury Bulletin*, Winter, 2(1), 25–32.

Padoa-Schioppa, T. (1988), 'The European Monetary System: A Long-term View', in F. Giavazzi, S. Micossi, and M. Miller (eds.), *The European Monetary System*, Cambridge, Cambridge University Press.

—— (1992), 'Exchange Rate Mechanism of the European Monetary System, Problems of the', *The New Palgrave Dictionary of Money and Finance*, Vol. 1, London, The Macmillan Press Limited.

Portes, R. (1993), 'EMS and EMU After the Fall', *World Economy*, 16, 1–15.

Smaghi, L. B., and Vori, S. (1993), 'Rating the EC as an Optimal Currency Area', Temi di discussione, Banca D'Italia, Numero 187, Gennaio.

Spaventa, L. (1989), 'The New EMS: Symmetry without Co-ordination?', in *The EMS in Transition*, London, Centre for Economic Policy Research.

Svensson, L. E. O. (1992), 'The Foreign Exchange Risk Premium in a Target Zone Model with Devaluation Risk', *Journal of International Economics*, 33, 21–40.

Taylor, A. J. P. (1965), *English History 1914–1945*, Oxford, Clarendon Press.

van Ypersele, J., and Koeune, J. C. (1984), *The European Monetary System*, Luxembourg, Office for Official Publications of the European Communities.

Wallich, H. C. (1972), 'The Monetary Crisis of 1971—The Lessons to be Learned', The Per Jacobsson Lectures, Bank for International Settlements.

Walters, A. (1993), 'If you want Euromoney, the Mighty Mark will do', *The European*, 5–8 August, 8.

Williamson, J. (1985), *The Exchange Rate System*, Washington, Institute for International Economics.

—— (1993a), 'EMS and EMU After the Fall: A Comment', *World Economy*, 16, 377–9.

—— (1993b), 'Exchange Rate Management', *The Economic Journal*, 103, March, 188–97.

Wilson, J. S. G. (1993), *Money Markets, The International Perspective*, London, Routledge.

Wren-Lewis, S., Westaway, P., Soteri, S., and Barrell, R. (1991), 'Evaluating the UK's Choice of Entry Rate into the ERM', *Manchester School* LIX Supplement, 1–22.

4

Floating exchange rates in theory and practice

DAVID BEGG

Birkbeck College, University of London

I. Introduction

After nearly two decades since the end of Bretton Woods, what lessons should inform exchange rate policy in the 1990s? On the world stage the 1980s saw an important change of emphasis. The decade began as the high noon of freely floating exchange rates. Individual countries pursued domestic macroeconomic objectives and allowed foreign exchange markets to assess the consequences for the right price of different national currencies. By the mid 1980s the strains of wild swings in the international value of the dollar, the yen, and sterling had led to tentative steps towards the co-ordination of international economic policy. But how should a more stable exchange rate environment be achieved? Is concerted central bank intervention in foreign exchange markets sufficient, or must there also be broad balance in the fundamentals of national macroeconomic policies?

Similarly, within Europe, exchange rate issues are at the top of the policy agenda. How will the EMS be affected by the abolition of controls on capital movements? Is Monetary Union necessary to realize the goals of 1992?

To answer any of these questions it is necessary first to understand how foreign exchange markets work. Only then can we assess the likely outcomes under alternative policies such as free floating or various forms of exchange rate management.

My purpose is to set out what we know about the theory of exchange rate determination, to review the empirical evidence, and to draw out some of the lessons for economic policy.

First published in *Oxford Review of Economic Policy*, vol. 5, no. 3 (1989).

II. The theory of exchange rate determination

It is convenient to distinguish what must happen in the long run and the additional factors which influence short-run behaviour. I begin with the long run.

1. Internal and external balance in the long run

In long-run equilibrium, when all adjustment has been completed, an economy must be at both internal and external balance. Internal balance means that the demand for goods and services is equal to full capacity supply; external balance means that on the balance of payments the current account is exactly in balance, so the country is neither adding to nor subtracting from its net international wealth. Thus

INTERNAL BALANCE:
capacity output = domestic demand + net exports
EXTERNAL BALANCE:
net exports = interest on net foreign debt

Initially, I ignore interest on net foreign debt, which will be a reasonable simplification if either net debt is small or if (real) interest rates are close to zero.

In this special case, external balance requires that net exports are zero. Net export demand depends on income levels at home and abroad, and on competitiveness. In the long run, income will be at full capacity output, which I treat as exogenous. Competitiveness depends on the real exchange rate. If P is the price of domestic goods, P* the price of foreign goods, and S the spot exchange rate (units of foreign currency per unit of domestic currency), then the real exchange rate R is given by PS/P*. It measures the relative price of domestic and foreign goods when measured in a common currency. A rise in the real exchange rate makes the domestic country less competitive and reduces its net exports.

In long-run equilibrium, external balance requires that net exports are zero. Given domestic and foreign levels of real capacity output, there is only one level of the real exchange rate which secures the appropriate level of competitiveness to make net exports equal to zero. Given zero net exports, internal balance then requires that domestic demand (consumption, investment, and government purchases of goods and services) equals domestic capacity output.

Several observations follow from this simple description of long-run equilibrium. First, since the real exchange rate, like other real variables, is constant in long-run equilibrium, purchasing power parity holds. Any excess of domestic inflation over foreign inflation must be continuously offset by depreciation of the nominal exchange rate precisely to keep the real exchange rate constant, provided all other real things remain equal.

Viewing it in this way helps us recognize when purchasing power parity (PPP) will be valid and when it will not. PPP is no more than the assertion that behaviour is determined by real variables not nominal variables. PPP holds when nominal variables are changing but real variables are not; it cannot hold when real variables are changing. Thus, for example, if a country discovers major reserves of a natural resource, its supply capacity is altered and it requires a *change* in its real exchange rate to maintain net exports equal to zero. For example, it is fallacious to argue that North Sea Oil was accompanied by a decline in Britain's manufacturing trade balance and that without oil Britain's trade would really have been in trouble. A real appreciation of sterling was the market mechanism by which a vast trade surplus was prevented from emerging after the discovery of oil. Without that discovery, the real exchange rate would have been lower and the balance on manufacturing trade correspondingly higher.

Our simple framework also allows us to address the long-run consequences of other changes. Other things equal, an increase in the nominal money supply will eventually lead to an equivalent increase in domestic wages and prices, precisely to leave the real money stock and real wages unaltered when long-run equilibrium is restored. And it will require an equivalent depreciation in the nominal exchange rate to restore the original real exchange rate. In this carefully defined sense, the level of the money supply eventually determines the level of the nominal exchange rate. Again the caveat is necessary. The money stock is not the only determinant of the long-run exchange rate. Real changes, such as an oil discovery or a productivity shock, will alter the equilibrium real exchange rate requried for external balance, and the nominal exchange rate will change independently of the nominal money stock. Many things are relevant and all one liners are simplifications which are true only under special circumstances.

Finally consider a change in fiscal policy, for example a permanent increase in government spending on goods and services. Initially, this adds to the demand for domestic goods, which now exceeds full capacity output. One possible response is an exchange rate appreciation which induces a trade deficit, thereby allowing the additional demand to be met from foreign not domestic output. Although the goods market is back at internal balance, there is no longer external balance. Eventually domestic demand must return to its original level, for that is the only way internal balance and external balance can simultaneously be restored. Sooner or later domestic consumption and investment must be crowded out by the higher government spending. For example, the initial trade deficit, induced by a domestic boom coupled with a temporary exchange rate appreciation, means the domestic country's international wealth is falling. With less wealth, eventually domestic firms and households will be driven to reduce consumption and investment even at given levels of output and competitiveness.

In short, the internal and external balance framework allows us to work through the long-run implications of any shocks we may wish to address. It is a good place to begin our analysis not merely because it enables us to think about the determinants of exchange rates in the long run, but also because it will play an important role in the minds of short-term speculators who must bear in mind the direction in which the exchange rate is eventually heading.

Before proceeding to that short-run analysis, it is necessary to refine the long-run story in one important respect: recognition of the additional complications introduced by the dynamics of debt accumulation.

2. Debt accumulation and long-run equilibrium

Thus far I have assumed that there is a unique long-run real equilibrium to which the economy eventually returns. That equilibrium depends on the underlying supply capacity of the domestic and foreign economies but is not itself affected by the temporary path the economy follows whilst it is out of equilibrium.

Increasingly, economists are interested in the phenomenon of hysteresis, which means that where an economy begins, and the path it follows in the short run, may affect the long-run equilibrium to which it eventually returns. Let me give three examples.

In the labour market, a recession may permanently reduce the level of long-run employment. On the one hand, when workers become unemployed they may lose touch with the labour market. Their preferences may change or they may become branded as less suitable workers; in either case they may find it permanently harder to get a job and the effective labour force contracts. On the other hand, if insider workers in jobs have a large influence in wage bargaining, a recession reduces the number of insiders with jobs. Any subsequent expansion of demand may allow this smaller number of insiders to exploit their greater scarcity by pressing for real wage increases rather than employment increases. Bargaining power is permanently changed by the initial lay-offs.

Hysteresis may also be present in the capital market. Once installed, capital has only to cover its operating costs. Sunk costs are sunk. But if a recession leads firms to scrap capital or go out of business, any subsequent recovery means firms must think about initial costs of investment as well as subsequent operating costs before concluding that the capital stock should be restored to its former level. Again, a short-run recession may permanently reduce the supply capacity of the economy.

Here I wish to focus chiefly on a third channel of hysteresis, namely through the accumulation of net foreign debt. I now abandon the earlier simplification that the current account of the balance of payments is given simply by net exports. Properly it is net exports plus net interest receipts on foreign assets.

Imagine an economy which begins with zero net international assets. Any ownership of foreign assets is matched by foreign ownership of domestic assets. Initially, external balance does indeed mean net exports equal zero.Consider any change whose impact effect is to lead to a trade deficit. Under floating exchange rates this current account deficit is matched by an equivalent capital account surplus. Just like an individual or a government, the country as a whole has to pay for overspending its income on current account by selling off its assets or adding to its liabilities. That is what the capital account surplus records.

Suppose after some period of adjustment, the real exchange rate, net exports, and all other real variables return to their initial position. External balance has not been restored since the country has permanently to service the interest cost of the debt it has built up during the transition. In the new long-run equilibrium it has to run a permanent trade surplus just equal to the interest burden of its international net debt; only then is the current account in balance forever more. Hence a transient phase of trade deficits requires a permanent real exchange rate depreciation to ensure that the eventual long-run equilibrium has an adequate trade surplus for external balance. The extent of the eventual real depreciation to increase competitiveness will depend on the extent of the debt obligations accumulated in the meantime, and on world real interest rates which determine the burden of repayments any given debt level will entail.

Such complications make considerably more difficult the calculation of 'fundamental equilibrium exchange rate paths', the paths for nominal exchange rates which secure the levels of real exchange rates necessary for external balance in the long run. Notice, importantly, that for many decades world real interest rates were close to zero. In such circumstances, debt dynamics could be substantially ignored since the associated real burdens for the current account were tiny. One of the striking features of the 1980s has been the size of real interest rates throughout the world economy. For many countries, the implications of debt dynamics are simply too important to ignore. Correspondingly, crude simplifications such as PPP become increasingly inaccurate in such circumstances, even as a characterization of the exchange rate in the long run.

III. Exchange rate dynamics in the short run

I turn next to theories of exchange rate determination in the short run. Under a free float, the current account and capital account sum to zero. Whilst both are relevant, it is unsurprising that international economists have been tempted by simplifications which emphasize either the current or the capital account. Current account theories stress the market for goods and services; capital account theories stress the market for assets.

1. Current account theories

The most prevalent such theory of short-run exchange rate determination is purchasing power parity, which holds that even in the short run the best way to forecast changes in the nominal exchange rate is to consider how it must change to offset inflation differentials to keep real exchange rates and competitiveness constant.

This view is easiest to maintain if one believes in flexible prices for goods and labour, and hence in continuous market clearing. Apart from very temporary surprises which are quickly offset, economies are then at their full capacity levels of all real variables. Nominal changes quickly induce equivalent changes in all other nominal variables to leave all real variables unaltered. Even within its own terms, this argument requires that nominal shocks are large relative to real shocks. Earlier, I have explained that real supply-side changes necessarily require changes in the equilibrium real exchange rate and therefore violate the implicit assumption of constant competitiveness on which PPP is based.

It is moreover hard to accept the paradigm of continuous market clearing. There are powerful theoretical arguments why domestic wages and prices will respond only slowly to shocks, and the consensus of empirical research is that most major economies take several years fully to adjust to any nominal shock. In the short run, shocks alter real variables and we should expect this to show up in variables such as competitiveness and net exports. In short, even if the economy has a tendency to revert to a particular real exchange rate in the longer run, PPP is a very weak reed on which to lean in explaining short-run exchange rate changes. One need only think about the massive and protracted swings in the real value of currencies such as sterling and the dollar in the 1980s to see that PPP cannot be the most important part of the short-run story of exchange rate determination.

Before moving on, I must deal with a popular confusion between PPP and the Law of One Price. The latter has on occasion been used as a defence of the former, but this is entirely misplaced. Imagine a world with no tariffs and no transport costs, a kind of idealized version of Europe after 1992. Arbitrage in goods will equate prices of domestic and foreign goods competing in the same market. The Law of One Price holds for such goods. From this it is sometimes argued that domestic prices multiplied by the exchange rate must therefore equal foreign prices, whence the real exchange rate is constant (and indeed equal to unity).

This is not a proof that PPP holds. Competitiveness is an indicator of profit margins in the traded goods industry making exports or competing with imports. It is essentially the price of traded goods relative to domestic wages and hence relative to the price of domestic non-traded goods. Consider what happened in the UK in the early 1980s when the real value of sterling was so much above its long-run equilibrium value. If

British exporters had applied conventional profit margins to domestic costs, they would not have been able to sell in foreign markets; they were uncompetitive. One strategy would have been to withdraw from such markets and wait for the exchange rate to come down. But there are big costs of breaking into markets. Many exporters decided to match foreign prices in order to keep selling (obeying the Law of One Price), even though in order to do so they had to sell at sterling prices which were below sterling costs of production. In short, looking at output prices we conclude that the Law of One Price held (approximately) but it gives the wrong signal about competitiveness when profit margins are varying. Goods market arbitrage and the Law of One Price do not enforce constant competitiveness and cannot be used to justify the assumption of PPP in the short run.

2. Asset market theories: perfect capital mobility

Information technology and the progressive removal of controls on capital movements have led increasingly to a global capital market. Vast sums of footloose funds stand ready to move across the foreign exchanges in search of the highest expected return. An extreme, but very powerful, simplification is to assume perfect international capital mobility. This has two aspects. Complete absence of any legal or regulatory barriers to capital mobility, and risk neutrality on the part of national investors. Whereas risk-averse investors will recognize that assets in different currencies have different risks, and will make complicated trade-offs between risk and expected return, risk-neutral investors care only about the expected return: they will want to put all their funds in the asset bearing the highest expected return. Now the Law of One Price applies to asset returns since all assets are considered perfect substitutes by investors.

In such a world consider the foreign exchange market at 10.03 on a Monday morning. The market is clearing minute by minute. In this particular minute, there are a few transactions arising from current account transactions: ICI has sold 10 tons of fertilizer and is repatriating its dollar earnings. But in such a short time span the net currency flow from the current account is tiny. Specifically, it is tiny relative to the flow which might arise if all the footloose funds decide to move.

Suppose some bad news comes out about the United States economy. Everyone wants out of the dollar and into sterling. The foreign exchange dealers' trading screens are suddenly ablaze with potential customers. Dealers make money safely by quoting a small spread between buying and selling prices, and taking a small percentage out of balanced flows in both directions; that way they avoid taking a position themselves. If at 10.03 the exchange rate is 2 dollars per pound, dealers cannot afford to accommodate 20 billion dollars that wants to move into sterling; they might have to buy it back tomorrow at 2.10 and be completely wiped out. Essentially dealers have to move the exchange rate immediately to the new level which chokes off most of the desired flow, restoring foreign exchange business to small and balanced flows in both directions, with which dealers can cope. The central message is that the freely floating exchange rate is continuously moving around to keep trading volumes relatively small. It is not the actual movement of funds which changes asset prices, it is the threat that massive movements would otherwise take place which moves exchange rates.

To understand the implication of perfect capital mobility, we must look not at the magnitude of capital account movements but at the rates of return on internationally traded assets. The central prediction is that of Uncovered Interest Parity (UIP). Assume investors hold interest bearing assets in sterling and dollars for a year. UK interest rates are 10 per cent and US rates are 0 per cent. Initially there are 2 dollars to the pound. Beginning with 100 pounds, you can end the year with 110 pounds by investing in the UK. Perfect capital mobility means you will be indifferent about where you hold your funds only if you expect a dollar investment strategy also to leave you with 110 pounds at the end of the year. Initially your 100 pounds converts to 200 dollars, and that is its value at the end of the year since US interest rates are zero. Only if you expect the exchange rate to have fallen 10 per cent during the year, so that your 200 dollars convert back to 110 pounds, will you be indifferent about where you hold your funds. Since all other investors reason similarly, this is the only short-run equilibrium for the foreign exchange market. For if investors all expect to do better in one currency than another, they will all be trying to move massive sums through the foreign exchange market, and we have seen that the market will be unable to cope; it will already have moved the exchange rate to the level which restores investor indifference.

Note, crucially, that a currency with above average interest rates must be expected to depreciate by an equivalent amount. Total return is interest plus capital gains or losses, and it is the expected capital loss which offsets the above average interest payment to leave investors indifferent. If in addition market participants have rational expectations—they do not make systematic forecasting errors knowable in advance—then *ex post* change in exchange rates will differ from the interest differential only because of random errors reflecting new information which has meanwhile become available. When this relation obtains we say that the exchange rate obeys Uncovered Interest Parity. Speculators have judged the market correctly.

UIP says that high interest rate countries are expected to have falling currencies. But when the Chancellor raises interest rates, sterling rises. How do we square these two statements? In fact the second follows from the first. At any point in time the exchange rate embodies the market's best guess about the entire future course of interest rates. As in most smart decision-making, speculators work back from the far end. Eventually the exchange rate has to get to its long-run equilibrium, whose determination I have already discussed. One period earlier, the interest rate differential then ruling will determine where the exchange rate has to start from to provide the appropriate expected capital gain or loss to offset the interest differential. Knowing this, the market can judge where it has to be one period earlier still to offset the interest differential then expected to rule, and so on all the way back to the present. The initial exchange rate thus reflects the anticipated interest rate differentials over the future and the end point to which the exchange rate eventually converges. Interest differentials determine the speed of adjustment to long-run equilibrium; the long-run current account and external balance determine that equilibrium position itself.

Suppose that the market has been expecting a particular path for interest rates and the exchange rate. If interest rates evolve as expected, so will the exchange rate. But no change is ever predicted with certainty. If tomorrow the Chancellor raises interest rates, it is in part a surprise; the market had been discounting other possibilities too. UK interest rates are thus higher than previously expected. If the exchange rate does not adjust, sterling is now too attractive to international speculators and foreign exchange dealers will be washed away as the tide flows in. Instead they react immediately to the new information by raising sterling sufficiently that it is now expected to make a subsequent capital loss which is adequate to offset the attraction of higher interest rates.

Thus, to sum up, high interest rates should signify an expected fall in the currency. Surprise increases in interest rates require a surprise increase in the value of the currency precisely so that it is now expected to fall faster than before.

This story places the short-run emphasis squarely on asset markets and the capital account of the balance of payments. Why then do foreign exchange markets pay so much attention to monthly trade figures which reflect current account developments? That too can be explained within this paradigm. Nobody is quite sure of the long-run equilibrium exchange rate. Its required real value will depend on supply-side considerations at home and abroad, and on the accumulation of international assets or debt along the transition, which determines the long run level of interest flows on current account. The evolution of the trade figures contains two pieces of relevant information. First, it helps reveal how the supply side is doing and allows long-run guesses to be updated. Second, it conveys information about the accumulation of international assets or debts which will require permanent financing. Both allow the market to update its estimate of the eventual exchange rate. For a given path of interest rates, any increase in the estimate of the long-run exchange rate should be accompanied by an immediate and equivalent adjustment to the initial exchange rate in order to preserve the relation between interest differentials and expected exchange rate changes. Alternatively, if government policy has a clear view about the level of long-run net international debt it wishes, new information about the trade position may be expected to induce changes in policy, including interest rates, to enforce the government's objective. Either way, the market should pay attention to current account developments.

3. Asset market theories: imperfect asset substitutability

Perfect capital mobility is indeed a powerful simplification. It is one I am happy to use at dinner for back of the napkin calculations. But it is too strong for two reasons. First, not all currencies are free from controls on capital movements though increasingly we are moving to such a world. Second, investors may sometimes be brave but they are unlikely to be completely risk neutral. The whole of modern portfolio theory is about how unwanted risks are efficiently but incompletely diversified in markets for bonds and equities. It is not plausible that risk aversion should be prevalent in domestic financial decisions but not in international financial decisions.

When assets are perfect substitutes, asset demands are perfectly elastic at the going rate of return. The quantity of the asset supplied has no effect on its equilibrium expected rate of return. Risk makes different assets imperfect substitutes, and makes demand curves slope downwards. Increasing the supply of a risky asset may be expected to increase its expected rate of return in order to induce more people to hold it.

In such a world, the current account is more significant than under perfect capital mobility. If the current account is in deficit, the capital account must be in surplus and the country is reducing the quantity of foreign assets in its portfolio. This should now be expected to alter the balance between the expected return on domestic assets and foreign assets. For given paths of domestic and foreign interest rates, this implies that the expected exchange rate change must adjust to secure the required adjustment in expected rates of return. It provides a direct link from current accounts to exchange rates in the short run.

How significant is this complication? Note, first, that even under perfect capital mobility the foreign exchange market must pay attention to the current account for reasons discussed above; thus the difference is one of degree. Second, what we require is an empirical evaluation of the quantitative significance of risk aversion. If international investors are nearly risk neutral, we can go a long way towards understanding exchange rate determination with a small napkin and the simple framework of perfect capital mobility. For this reason, much of the empirical literature, which I discuss in section IV, is concerned with quantifying deviations from Uncovered Interest Parity which might be evidence of imperfect asset substitutability.

4. Other explanations for deviations from interest parity

Interest parity says that over any time period t to t + 1, the *ex post* observed percentage change in the exchange rate deviates only randomly from the interest differential on one period interest rates at t. Perfect capital mobility means expected returns are equated, and rational expectations mean that the actual exchange rate at t + 1 deviates only randomly from that which investors at t believed would hold at t + 1.

This suggests two immediate reasons why interest parity might not be empirically observed. First, investors may not have rational expectations. Second, assets may be imperfect substitutes so that their equilibrium expected returns need not be equal. Direct examination of the data implicitly tests both hypotheses jointly, and it is hard to know which hypothesis fails if the data rejects the joint hypotheses. Extra information may be gleaned by looking directly at survey data on expectations, as I shall shortly discuss.

First, I want to discuss an alternative interpretation of deviations from interest parity. This was introduced to the economics literature as the Peso Problem. Consider a country (say Mexico) which attempts to peg its exchange rate to another (say the dollar). Most of the time the exchange rate does not change at all, but occasionally there is a large devaluation and the exchange rate changes a lot. During any particular short time interval, the market knows there is a small probability that a devaluation will happen and a much larger probability that it will not. Thus the expected exchange rate change is small but not zero: the market must allow for the small probability of a large change. Suppose interest differentials do offset this expected exchange rate change. In most periods, when no devaluation occurs, *ex post* there will be zero exchange rate change but a small interest differential; occasionally there will be a large exchange rate change but a small interest differential.

If devaluations are relatively infrequent (say every 5 years), then it will take a very large data sample to reach the correct conclusion that on average exchange rate changes do indeed offset interest differentials. But a data sample of 14 years (with 2 devaluations) will yield a very different statistical conclusion from one with 16 years (and 3 devaluations). We think 14 years of monthly data is a big sample (168 observations) but it is not statistically adequate given the rare occurrence of a large event. It would be easy for a chimpanzee with a big computer to reach the conclusion that statistically exchange rate changes do not offset interest differentials on average.

The Peso Problem may be much more prevalent than one first thinks. Let me give two examples. First, the devaluation to which the problem initially referred is really just one example of a general question, the possibility of a major shift in policy. It could be a major change in monetary policy, fiscal policy, or anything that affects either the anticipated course of interest differentials or the long-run exchange rate to which markets will eventually converge. Particular governments have been known to make major U-turns in policy, quite apart from the possibility of a complete change of government at the next election. The market must always be making allowances for low probability events which will have considerable impact if they do indeed occur. Most of the time these events will not occur and never show up in the data. With rational expectations, the market is appropriately discounting all these possibilities. But in any particular sample, actual expectations may appear significantly biased because the rationally feared events did not in fact transpire.

Nor do such problems apply only to changes in government policy. The recent literature on speculative bubbles suggests that such 'changes of regime' can be manufactured within the psychology of the market itself. Let me take a simple example. Suppose the 'fundamentals' of economic policy suggest that the right price for the currency, both now and in the future, is 1.8 dollars per pound. One outcome is that the market sets this fundamentals price and sticks to it. Suppose, however, as in Keynes' beauty contest, every market participant believes every other participant has taken leave of their senses and embarked on a speculative bubble unjustified by any fundamentals.

Let us say that today the exchange rate is 2 dollars per pound and everyone believes there is a 50 per cent chance people will continue on the bubble and a 50 per cent chance the market will go back to the fundamentals level of 1.80. Suppose interest rates are the same at home and abroad so it is only capital gains that are of interest. At the current level of 2, there is a 50 per cent chance of a 10 per cent capital loss next period when the price falls to 1.80, so expected profits are zero if people expect a 10 per cent rise to 2.20 in the event that the bubble continues. Next period at 2.20, people are worrying about a fall back to 1.80 and need to believe they will get 2.60 if the bubble continues. At 2.60 they need the lure of 3.20 in the following period to keep them on the bubble. The bubble, if it persists, must accelerate because the price is getting ever further from the fundamentals price and the crash will be ever more painful when it comes. If there is a 50 per cent chance of the bubble bursting each period, sooner or later it will burst and the fundamentals price will be restored. But there is nothing irrational about being on the bubble while it lasts.

This is a version of the Peso Problem because there is a chance that an empirical investigator simply has an unlucky data sample which includes an unburst bubble. If so, it will appear that there are statistically significant deviations from interest parity, whereas in my example speculators were assumed to be risk neutral and have rational expectations.

These examples warn us that trying to establish the empirical magnitude of deviations from interest parity and the perfect capital mobility assumption it supposes will be a difficult task. It may be right to conclude that the data cannot be very informative. Nevertheless, since the implications for exchange rate policy depend on the extent of asset substitutability, it is important to see how much we can learn from empirical evidence.

IV. Empirical evidence on exchange rate determination

There are many excellent recent surveys of the empirical literature, including Hodrick (1987) and Goodhart (1988). Here I shall try to pick out the main themes of relevance.

1. Purchasing power parity

One need only plot the real exchange rate for a major currency such as sterling or the dollar to see that it exhibits huge short-run variations; no formal statistical tests are required to conclude that PPP is nowhere close to holding in the short run.

Of much greater interest is whether it holds in the long run. I have argued that it should if real things are remaining roughly constant, but should not if real variables are changing differently in different countries, or if hysteresis is important on the supply side or through debt accumulation.

My reading of the empirical literature is that there is no evidence that PPP holds even in the long run. Studies such as Adler and Lehmann (1983) and Taylor (1986) find little tendency of the real exchange rate to revert to any constant long-run value. Similarly, Taylor (1987a) finds that the relevant domestic and foreign nominal variables are not co-integrated, as would be required if, in the long run, the real exchange rate was constant.

Given my earlier discussion, I do not find these results surprising. They do not disprove the assertion that, other things equal, changes in nominal variables will eventually be offset by equivalent changes in the nominal exchange rate. Rather I prefer to interpret this evidence as indicating that changes in real variables have been too significant to ignore. Current estimates of the appropriate long-run real exchange rate are likely to follow a random walk. Current information should be included fully in current estimates of that eventual position. But new information is significant and will lead to major revisions in these estimates over time. One need only reflect on the debate about the appropriate long-run level for the dollar to appreciate the extent of uncertainty which inevitably prevails.

2. Forward markets and covered interest parity

Thus far I have focused on the spot market for foreign exchange. At this juncture it is convenient to introduce the parallel forward market for foreign exchange. A forward contract is an agreement today to exchange currency in the future at a price agreed today. Compare two riskless strategies. First, invest in a domestic one period asset paying a known return r. Second, convert cash into foreign exchange at today's spot rate S, buy a foreign one period asset yielding the known return r*, and also take out a forward contract to convert the proceeds back into domestic currency at the forward rate F. Arbitrage implies the two strategies should yield identical returns. Letting f and s denote the logarithms of F and S we thus have the Covered Interest Parity condition

$$f = s + (r^* - r)$$

as a characterization of market efficiency.

Does Covered Interest Parity hold? Several early studies suggested that it might not, but it is now widely believed that such results reflected data imperfections, for example comparison of interest rate data at one time of day with exchange rate data at another time of day. Taylor (1987b) reports that in a study of several thousand observations on interest rates and exchange rates at the same instant, Covered Parity invariably held. Indeed, as I remarked in Begg (1983), forward dealers frequently base quotations on the Covered Parity formula itself.

3. Uncovered parity and perfect capital mobility

Letting $E(s(t+1))$ denote the expectation at time t of (the logarithm of) the spot rate at $t + 1$, Uncovered Interest Parity implies

$$E(s(t+1)) - s(t) = r^*(t) - r(t)$$

and if expectations are rational the *ex post* spot rate $s(t + 1)$ differs from its rational expectation only by a random term $u(t+1)$ whose deviation from zero cannot be predicted using information at t

$$s(t+1) = E(s(t+1)) + u(t+1)$$

whence empirical tests may be based on

$$s(t+1) = s(t) + r^*(t) - r(t) + u(t+1) \qquad (1)$$

Notice that if covered interest parity holds, as I am happy to assume, we can also write

$$s(t+1) = f(t) + u(t+1) \qquad (2)$$

or

$$s(t+1) - s(t) = f(t) - s(t) + u(t+1) \qquad (3)$$

so tests of interest parity are equivalent to tests that the forward rate f(t) is an unbiased predictor of future spot rates s(t + 1). The literature has considered both forms of the test.

Early tests such as Levich (1978) found that equations such as (2) fitted the data rather well. Since then the evidence against Uncovered Interest Parity has steadily mounted. Hansen and Hodrick (1980), Cumby and Obstfeld (1981), Baillie, Lippens and McMahon (1983), Fama (1984), and Goodhart (1988) are merely a selection of those reporting statistically significant departures from Uncovered Interest Parity. Many of these studies confirm the finding of Meese and Rogoff (1983) that spot exchange rates are close to a random walk. These studies also imply that forward rates are biased predictors of future spot rates.

So how should we interpret these findings? One possibility is that they are explained by the Peso Problem which I discussed in section III.4. Our data samples may simply be too small to yield reliable statistical inferences. Even if individuals have rational expectations and are close to risk neutral, forward rates may deviate significantly from *ex post* future spot rates in our sample because market expectations were correctly discounting important but low probability events which did not in fact transpire. One way to check out this interpretation is to look directly at survey evidence on expectations, as I shall do shortly.

Alternatively, we can conclude either that markets are inefficient in the sense of making knowable mistakes, or that risk cannot be ignored. Unsurprisingly, the literature has focused on the latter possibility.

4. Time varying risk premia

Once we depart from risk neutrality there is no presumption that different assets should yield the same expected return. Risk premia refer to differences in expected returns when asset markets are in equilibrium. Risk premia may account for systematic and varying deviations from interest parity, but for this insight to have practical significance we must be able to construct a successful model of what determines variations over time in risk premia. Otherwise, an appeal to risk premia is simply a tautological account of our failure to explain departures from interest parity.

Suppose investors care about both the mean and the variance of returns on their asset portfolio. Different assets will be imperfect substitutes and modern portfolio theory tells us how optimal portfolio composition will be determined by investor preferences and the

perceived statistical properties of returns on different assets. One implication is that equilibrium expected returns will depend on relative asset supplies: since demand curves for individual assets are downward sloping, an increase in the supply of a risky asset will reduce its price and raise its expected return.

In a series of papers, Jeff Frankel has tried to establish whether variations in asset supply can explain variations in the departures from interest parity as the theory predicts. (See Frankel, 1979; Frankel, 1980; Frankel, 1982; Frankel and Engel, 1984.) His results are uniformly discouraging.

Other studies have sought not to relate risk premia to quantities of assets supplied but to variations in other rates of return. Fama (1984) argues that the data imply risk premia are correlated (negatively) with the expected rate of depreciation of the exchange rate, a finding for which Hodrick and Srivastava (1984) provide the following rationalization. Domestic investors care about real returns after deflating by the domestic inflation rate. If exchange rates respond in part to inflation, domestic and foreign assets offer different insurance against domestic inflation, and will therefore differ in equilibrium expected returns.

Hodrick (1987) contains a much fuller discussion of the vast literature on risk aversion and its implications for exchange rate determination. Some general conclusions emerge. First, it is not difficult to construct plausible models with time varying risk premia. Second, in such models both the expected return and the conditional variance of asset returns will generally be changing over time. The latter means that simple statistical techniques of data analysis may be inadequate and lead to misleading inference. Third, to date we have no robust and successful model which fits the facts. Fourth, in consequence, as yet we have no reliable way of quantifying the importance of deviations from perfect capital mobility, even though such a quantification is essential for policy assessment.

5. Evidence from surveys of expectations

Thus far, all the evidence I have discussed has been based on market data. Specifically, *ex post* spot exchange rates at time t + l have been assumed to measure with error the expectations at time t of spot rates at t + l. This amounts to imposing the behavioural assumption of rational expectations and the statistical assumption that data samples are sufficiently large that rational expectations will be unbiased predictors of future spot rates within the data sample.

Both assumptions may be invalid. If market participants are still learning how economies work or what policies are in fact being pursued, systematic errors may be made while the market is learning and converging to rational expectations. And the Peso Problem may imply that rational expectations are not unbiased predictors of future spot rates within a particular small data sample.

Frankel and Froot (1986a, 1986b, 1987) and Froot and Frankel (1986) bring survey evidence to bear on these issues. They examine several surveys asking market participants to forecast exchange rates and investigate whether expectations are unbiased. Their research seriously questions whether market expectations can adequately be represented as unbiased predictors within short sample periods.

Whilst in practice a random walk is a reasonable statistical description of the behaviour of spot exchange rates—current information gives no guide to systematic future exchange rate changes—all three of the surveys they study reject the hypothesis that market participants expect the exchange rate to remain constant. Market participants generally do have a view about exchange rate changes, and the difference between the forward rate and the spot rate does reflect the direction of expected exchange rate changes (though it underestimates the magnitude of expected changes).

These findings have two broad implications. First, they challenge the large literature which assumes that expectations are unbiased predictors and then seeks to model deviations from interest parity in terms of time varying risk premia. Risk premia may indeed be important but it is necessary to have a more accurate estimate of expected returns before discrepancies in these returns across different assets can be understood. Second, Frankel and Froot suggest that the assumption of unanimous expectations may be unduly restrictive, a theme also taken up in Goodhart (1988). Casual empiricism suggests that at least three types of belief have weight in market assessments: chartists, who rely on various forms of extrapolation; long-run fundamentalists, who emphasize the eventual exchange rate to which the market must converge; and short-run fundamentalists, who seek to integrate short-run speculative dynamics with eventual convergence on long-run equilibrium. In principle, we can write down models in which each group makes the smartest decisions it can, given its own beliefs about the truth and a recognition that market prices will also be driven by other groups with different views. Such models are now in the course of being worked out, and may prove rich enough to capture phenomena which our simpler accounts at present miss.

One final remark is in order. There is a good deal of extra information contained in market data on financial futures and options. Such data allows richer inferences about the implicit distribution of market assessments about the future of both interest rates and exchange rates. The development of such markets and the data they throw up is likely to be an important source for future research.

V. Some implications for economic policy

In this final section, I seek to draw out some of the policy implications of the preceding discussion, against the background of the actual policy debate in the late 1980s.

1. Exchange rate volatility

Whether capital mobility is close to perfect or merely substantial, we should expect substantial exchange rate volatility if different countries do not closely coordinate their macroeconomic policies. Some volatility is inevitable. As new supply-side information becomes available, markets will be revising their view of the appropriate long-run real exchange rate. In addition, periods of current account deficit or surplus will cause revisions to the estimate of the long-run flow of debt interest or asset earnings which individual economies will have to service. And while individual countries remain free, both now and in the future, to adopt very different monetary policies, the market guess about inflation and hence nominal exchange rates in the long run may be subject to substantial revision as time elapses.

These are all reasons why today's estimate of the eventual nominal exchange rate will continuously be revised. But exchange rates may be more volatile still, to the extent that changes in anticipated interest differentials along the path to long-run equilibrium cause the market to adjust today's exchange rate to preserve speculative balance along this path. Like other asset prices, we should expect considerable short-run volatility in spot exchange rates.

Economists remain divided as to whether volatility constitutes a major policy problem or not. To some extent firms can hedge exchange rate risk in the forward market. And empirical research has not established any clear adverse effect of exchange rate volatility on trade flows or levels of investment. Even so, industrialists and politicians have become increasingly convinced that measures to reduce exchange rate volatility are on balance desirable.

If capital mobility is close to perfect, it is straightforward to describe the policies required to achieve more stable exchange rates. First, whether through domestic reputation or by participation in an international arrangement which ties its hands, each country must convince markets that it will pursue a predictable monetary policy in the longer run, thereby allowing markets to assess the appropriate long-term path for nominal exchange rates with some confidence. If such a path is to involve not merely predictable but relatively constant nominal exchange rates, this requires in addition the co-ordination of national monetary policies and national inflation rates.

Second, to the extent that speculators are influenced by interest rate differentials, eliminating volatility of spot exchange rates requires not merely the harmonization of exchange rates in the longer term, but also co-ordination of interest rates to eliminate short-run speculative pressures. Together, these two aspects of policy co-ordination would provide the appropriate fundamentals for more stable exchange rates.

The tentative steps towards more explicit policy co-ordination during the 1980s have made clear the reluctance of individual governments to surrender as much domestic policy autonomy as the above would require. Is there any way out?

2. How much can we rely on intervention?

If foreign exchange market intervention by central banks is a powerful lever, it may be possible for governments directly to stabilize exchange rates without having to co-ordinate domestic macroeconomic policy. When is intervention likely to work?

Consider first the extreme case of perfect capital mobility. In such a world the private sector's footloose funds are vastly greater than the volume of intervention funds which have typically been at the disposal of central banks. More formally, since international assets are perfect substitutes the demand for each asset is perfectly elastic at the going expected rate of return. Changing the market supply of the asset, the purpose of intervention, will have no effect on the rate of return. In such a world sterilized intervention is effectively powerless.

It is important for the argument that the intervention be sterilized. In other words, when the Bank of England exchanges sterling for dollars in the foreign exchange market, it simultaneously exchanges sterling bonds for sterling in the domestic market to leave the domestic money supply unaltered. That is the appropriate way in which to assess the effect of pure intervention.

It may be of course that the intervention is not sterilized; then its effect is to change the domestic money supply, and this will affect domestic interest rates in exactly the same way as any other change in the domestic money supply. Unsterilized intervention can move exchange rates by influencing the interest rate incentives for speculators, but it is not an independent policy. It represents just as much a loss of domestic interest rate autonomy as if interest rates had been altered for exchange rate objectives through any other means.

In marked contrast, when market participants are risk averse and treat assets in different currencies as imperfect substitutes, the equilibrium price of each asset will depend on the quantity supplied and variations in relative supply, induced through intervention, will have an independent effect on the price and return of each asset. Intervention is then genuinely an additional policy implement.

In this context, two remarks are relevant. First, within the European Monetary System, the presence of pervasive foreign exchange controls, necessarily making international assets imperfect substitutes, has allowed central banks an extra degree of freedom and allowed considerable exchange rate stability without the complete sacrifice of domestic policy autonomy. Abolition of foreign exchange controls, now well advanced in the run up to 1992, will thus profoundly change the nature of the system. Much has been made of the increased intervention facilities to which EMS central banks will now have access. What has been stressed less is that, with substantially greater capital mobility, a given quantity of intervention will be much less effective than before. On balance I expect intervention to become progressively less effective, and EMS members may find that explicit macropolicy co-ordination, recognized in principle at the 1987 Basle and Nyborg Agreements as being the required counterpart of freeing capital movements, will have to bite rather hard and rather soon in practice.

Second, empirical research for currencies between which free capital movements have been allowed for some time broadly substantiates the claim that sterilized intervention is powerless (see, e.g. Rogoff, 1984), implicitly providing an indirect test of the proposition that, for policy purposes, capital mobility is not grossly at variance with the extreme simplification of perfect capital mobility.

3. Volatility transference and speculative bubbles

Suppose in a world of high capital mobility governments decide to stabilize exchange rates primarily by the use of interest rates. The world is a nasty place and unforeseen shocks are always occurring. It might be thought that, instead of the shocks showing up in exchange rates, their effect would simply be transferred to interest rates as governments worked ever harder to keep exchange rates steady. It is possible, but not obvious, that volatile interest rates are preferable to volatile exchange rates.

One interesting test bed in which to investigate this issue is the EMS where we know exchange rates were stabilized during the 1980s. Was there an accompanying increase in interest rate volatility? Artis and Taylor (1988) show that the answer is clearly no. There are two competing explanations of why not. The first is that foreign exchange controls allowed intervention to be quite effective for the reason discussed above. If so, we should expect a significant increase in interest rate volatility within the EMS, now controls on capital movements are being abandoned.

The alternative explanation is more intriguing. Earlier, I discussed the likelihood of speculative bubbles in freely floating exchange rate markets. There I remarked on the intrinsic logic of a bubble: as the price departs ever more from the fundamentals, it must be expected to accelerate ever faster if the bubble continues, for it requires an ever greater prospect of a capital gain to offset the ever more painful capital loss which will be incurred when the market reverts to the fundamentals. In this context, consider the impact of a credible exchange rate band as actually exists under the EMS and might exist if the G7 countries moved towards target zones for exchange rates.

A band provides a ceiling or floor on the exchange rate. Once the exchange rate gets close to the edge of the band, the market cannot expect a major continuation of the previous bubble path. Hence the bubble cannot be sustained close to the edge of the band. But that means that, one day earlier, the market cannot rationally expect an exchange rate the following day which is even close to the edge of the band. And so on all the way back. In other words, if the band will

credibly be enforced in the short run, this may be sufficient to prick bubbles before they get started. And this may explain why exchange rates within the EMS have been stabilized without a transfer of volatility to interest rates. If so, bands or target zones really do buy something for nothing. This line of research has recently been investigated in Krugman (1987) and Miller and Weller (1988).

Credible bands or target zones require a measure of international policy co-ordination. The contingent promise to defend the band is what pricks bubbles manufactured by market expectations themselves. But there may be other benefits of policy co-ordination. I have stressed that uncertainty about the policy of individual governments may be an important source of market uncertainty and peso-type problems. If policy co-ordination involves the adoption of more systematic rules for national policy, it may also serve to reduce considerably the worries about major regime changes which seem to underlie many of the empirical findings I discussed in the previous section. If so, it would be a recipe for greater exchange rate stability.

4. Misalignments of competitiveness

The view I have been elaborating places short-run exchange rate determination firmly within asset markets and the capital account of the balance of payments. Given any important degree of capital mobility, short-run exchange rates can be bid to levels which may be very different from their underlying longer-run levels. The empirical evidence is overwhelmingly that the tendency to revert to long-run levels is at best very weak. In such a world, real exchange rates and competitiveness can be, and have been, at inappropriate levels for several years on end unless domestic macroeconomic policy is specifically geared to ensuring that this does not occur.

Misalignment matters for two reasons. First, it subjects the traded goods industries to a disproportionate share of the economy's burden of adjustment in the short run. Second, and related, there may be important hysteresis effects so that initial changes in the traded goods sector—in the number of firms, employment, and capital stock—prove extremely difficult to reverse even if competitiveness is restored at a later date. In such a world, the gain to international policy co-ordination may be rather larger than when hysteresis is absent. Essentially, misalignments distort and reduce the long-run supply capacity of the world economy, and there is a gain to be shared out if misalignments can be minimized.

5. The future of the EMS and Britain's position

In conclusion, let me take up the implications of the foregoing analysis for the future of exchange rate policy in Europe.

I cannot forecast the extent to which European governments will pursue a path similar to that recommended by the Delors Committee. But let us suppose that Europe continues to follow the step by step approach which has characterized the evolution of the EMS to date. By 1990 most of the EC will have free capital movements and a somewhat tighter version of the EMS, with an eventual aim of moving towards Monetary Union. Even substantial concerted foreign exchange intervention will prove only partially successful, and EMS members will increasingly have to coordinate monetary and interest rate policy. There will be a common market for financial services, notably banking, supported by relatively light regulation agreed in principle in Brussels but implemented by individual member states, with mutual recognition of each other's enforcement procedure.

The British government has said repeatedly that it will enter the Exchange Rate Mechanism of the EMS when the time is ripe. The benefits of membership are clear. First, a reduction in exchange rate volatility, in part because bubbles will be less common given a commitment to defend the band and a wider commitment to a stable policy stance in general. Second, a reduction in exchange rate misalignment, both because of the short-term stability with the exchange rate band and because when realignments do occur they substantially offset any change in competitiveness since the previous realignment. Third, a measure of financial discipline. To be sure, the Thatcher government earned a substantial reputation for responsibility without the aid of the EMS, but, as Lord Cockfield is fond of saying, the time to sign up for discipline is the time at which you are not in need of it. These three benefits of membership are all important. They are essentially the same benefits which the British government has accepted in playing a leading role for greater policy co-operation at the level of G7 countries.

So what of the arguments for staying out? Many of the old ones are simply no longer relevant. Sterling is no longer an important petro-currency. There will no longer be any Bipolarity Problem, which referred to the possibility of free capital movements between deregulated London and deregulated Frankfurt when all other EMS countries were subject to foreign exchange controls; now all EMS countries are to have free capital movements. So any remaining objection must be based on the surrender of monetary sovereignty in order to stabilize exchange rates through concerted interest rate policy.

Other countries have taken the view that sovereignty pooled is not sovereignty diminished. The only case I can construct as to why the UK should take a different view goes as follows.

Economies, like horses, can be ridden with a tight rein. They will know where they stand, but progress will be sedate. Or the rein can be looser, which has the major advantage that the horse can run faster. But eventual discipline requires that the horse understands that bolting, or overheating, will be severely punished; the rider will rein in sharply. Once the horse understands this, the rider can revert to the loose rein, maintaining discipline but at a decent canter.

In the early 1980s, the UK economy was on a tight rein. Then, as the government built up a solid reputation for inflation aversion, the rein was loosened and policy eased; and the UK economy responded with faster growth. But the deal always was, and always had to be, that any abuse of this freedom, and specifically any renewed outbreak of inflation, would have to be promptly and severely dealt with. When in 1988 inflation turned up, policy was indeed tightened sharply. In due course, it may be possible to loosen the reins again once the lesson has been re-learned.

If this interpretation is correct, what is the stick which the government holds in reserve for administering beatings when they become necessary? For a government committed to further tax cuts and disinclined to manipulate budget policy for short-term demand management, that stick cannot be fiscal policy. It has to be monetary policy, which, moreover, has a fringe benefit: tightening policy through higher interest rates also drives up the exchange rate thereby directly contributing to inflation reduction through lower import prices. Thus a credible stick involves the possibility of sharp increases in interest rates, and sharp exchange rate appreciation during periods of punishment.

This the EMS would rule out. Without a credible monetary stick, the government would have two choices. The first would be to return to the straightjacket; safe but boring. The second would be to resurrect fiscal policy as a short term threat. Tax increases would not be politically attractive and would undermine a decade of asserting that fiscal policy is inappropriate as a short-run tool of demand management. For these reasons, the government has so far preferred to stay out of the EMS. And if this reasoning is correct, it suggests that the government would have preferred to remain outside if British entry had not become inextricably enmeshed with other negotiations about what Europe will be like after 1992.

References

Adler, M., and Lehmann, B. (1983), 'Deviations from PPP in the Long Run', *Journal of Finance*, 38, 1471–87.

Artis, M. J., and Taylor, M. (1988), 'Exchange Rates, Interest Rates, Capital Controls, and the EMS', in F. Giavazzi, S. Micossi, and M. Miller (eds), *The European Monetary System*, Cambridge, Cambridge University Press.

Baillie, R. T., Lippens, R. E., and McMahon, P. C. (1983), 'Testing Rational Expectations and Efficiency in the Foreign Exchange Market', *Econometrica*, 51, 553–64.

Begg, D. K. H. (1983), 'The Economics of Floating Exchange Rates', in *International Monetary Arrangements*, House of Commons Treasury and Civil Service Select Committee, HMSO.

Cumby, R. E., and Obstfeld, M. (1981), 'A Note on Exchange Rate Expectations and Nominal Interest Rate Differentials', *Journal of Finance*, 36, 697–704.

Fama, E. (1984), 'Forward and Spot Exchange Rates', *Journal of Monetary Economics*, 14, 319–38.

Frankel, J. (1979), 'The Diversifiability of Exchange Rate Risk', *Journal of International Economics*, 9, 379–93.

—— (1980), 'Tests of Rational Expectations in the Forward Exchange Market', *Southern Economic Journal*, 46, 1083–101.

—— (1982), 'In Search of the Exchange Rate Premium', *Journal of International Money and Finance*, 1, 255–74.

—— (1986), 'The Implications of Mean Variance Optimisation for Four Questions in International Macroeconomics', *Journal of International Money and Finance*, 5, S53–78.

—— and Engle, C. (1984), 'Do Asset Demands Optimise over the Mean and Variance of Real Returns?', *Journal of International Economics*, 17, 309–23.

—— and Froot, K. (1986a), 'The Dollar as an Irrational Speculative Bubble', *The Marcus Wallenberg Papers on International Finance*, 1, 27–55.

—— and —— (1986b), 'Understanding the US Dollar in the Eighties', *Economic Record*, special issue, 62, 24–38.

—— and —— (1987), 'Using Survey Data to Test Standard Propositions Regarding Exchange Rate Expectations', *American Economic Review*, 77, 133–53.

Froot, K., and Frankel, J. (1986), 'Interpreting Tests of Forward Discount Unbiasedness using Survey Data on Exchange Rate Expectations', NBER Working Paper 1963.

Goodhart, C. A. E. (1988), 'The Foreign Exchange Market: a Random Walk with a Dragging Anchor', *Economica*, 55, 437–60.

Hansen, L. P., and Hodrick, R. J. (1980), 'Forward Exchange Rates as Optimal Predictors of Future Spot Rates', *Journal of Political Economy*, 88, 829–53.

Hodrick, R. J. (1987), *The Empirical Evidence on the Efficiency of Forward and Futures Foreign Exchange Markets*, Harwood Academic Publishers.

—— and Srivastava, S. (1984), 'The Covariation of Risk Premiums and Expected Future Spot Exchange Rates', *Journal of International Money and Finance*, 5, S5–22.

Krugman, P. (1987), 'The Bias in the Band', NBER Working Paper.

Levich, R. M. (1978), 'Further Results on the Efficiency of Markets for Foreign Exchange', *Managed Exchange Rate Flexibility*, Federal Reserve Bank of Boston.

Meese, R. A., and Rogoff, K. (1983), 'Empirical Exchange Rate Models of the Seventies: do they Fit Out of Sample?', *Journal of International Economics*, 14, 3–24.

Miller, M., and Weller, P. (1988), 'Exchange Rate Bands and Realignments in a Stationary Stochastic Setting', mimeo, University of Warwick.

Rogoff, K. (1984), 'On the Effects of Sterilised Intervention', *Journal of Monetary Economics*, 14, 133–50.

Taylor, M. P. (1986), 'On Unit Roots and Real Exchange Rates', mimeo, Bank of England.

—— (1987a), 'An Empirical Examination of Long Run Purchasing Power Parity Using Cointegration Techniques', mimeo, Bank of England.

—— (1987b), 'Covered Interest Parity: a High Frequency, High Quality Data Study', *Economica*, 54, 429–38.

5

The case for stabilizing exchange rates

PAUL KRUGMAN

Massachusetts Institute of Technology

This paper presents the case for a return to an exchange rate system resembling that which prevailed in the Bretton Woods period, or which prevails within Europe at the present time. That is, it argues the case for a system of mostly fixed rates that can sometimes be adjusted.

The argument is not an enthusiastic one. I do not claim that an 'adjustable peg' system will solve all of the problems of the current system, or that it will be without problems of its own. However, the experience of floating rates has made the adjustable peg look much better in comparison, and greatly reinforced the case against letting exchange rates be determined just like any other asset price.

Fashions in exchange rate regimes go in obvious cycles. In spite of the vast additions both to our store of knowledge and to our theoretical sophistication since the 1940s, practical arguments against flexible rates are not very different from those advanced by the architects of the Bretton Woods system. The main point of the analysis in this paper is in fact to show why modern critiques of traditional logic here are in fact unjustified.

The paper is in five parts. I begin with some broad, even philosophical questions about the reasons for even considering a return to fixed rates. The second part documents what I believe to be the key thing that we have learned in the past fifteen years: that foreign exchange markets behave much more like the unstable and irrational asset markets described by Keynes than the efficient markets described by modern finance theory. The third part then asks what harm this instability does, and argues that conventional efforts to test for serious costs—which generally fail to show much

harm from exchange rate instability—are likely to miss the main action. The fourth part examines the advantages of exchange rate flexibility, and asks whether occasional parity changes are a good enough substitute. Finally, the fifth part tries to tie the whole argument together, offering a case for an adjustable peg system.

I. General considerations

1. Exchange rates and other asset markets

The post-1973 era of floating exchange rates, unlike the interwar experience, has taken place at a time when the intellectual pendulum has been swinging strongly to the right, and in which both faith in the marketplace and deregulation of markets have been growing. In this atmosphere, people are much more likely to make excuses for the behaviour of exchange rates than they might have in the past—after all, this is a time when belief that markets are always right is strong, and the foreign exchange rate is by no means the most exciting and volatile market around. Anyone who proposes exchange rate stabilization is likely to face the question, 'Why try to stabilize the exchange rate, when exchange rates are considerably less volatile than, say, stock prices? After all, you wouldn't try to fix the Dow Jones, would you?'.[1]

One answer to this question might be to say, 'Well, yes, maybe I would.' The great faith in financial

---

[1] For an example of a discussion that downplays complaints about the instability of exchange rates by pointing out that the exchange rate is less volatile than many other asset prices, see Frenkel and Goldstein (1989).

First published in *Oxford Review of Economic Policy*, vol. 5, no. 3 (1989).

markets that has become dogma in the 1980s is borne out neither by up-to-date theory nor by the evidence, as described briefly below; the idea that government policy should try to insulate the real economy from the animal spirits of speculators is by no means obsolete.

Even if we do not challenge the spirit of the times this sharply, however, we may still note that there is a fundamental difference between the case for untrammelled financial markets in general and the case for freely floating exchange rates. An exchange rate is the relative price of two fiat moneys—that is, of two artificial assets. The supplies and prices of these assets are not natural variables, they are set by policy. A world in which two countries share a common currency is at least as much a free-market economy as one in which the two countries have fiat currencies whose values fluctuate against one another; and a world in which one or both countries vary their money supplies so as to keep the exchange rate constant is equally market-oriented. Thus there is a fundamental difference in interpretation when we compare the volatility of exchange rates with that of stock prices. Any proposal to limit stock price variability is in effect a proposal to restrict and/or regulate markets, while a fixed rate system is just as much a free-market one as a floating rate system.

The point is that the argument over exchange rate regimes is not one of free vs. controlled markets; a floating rate regime has no better claim to the allegiance of the market faithful than a fixed rate regime.

2. Why worry?

Why would one want to tie the hands of the monetary authorities by fixing the exchange rate? Leaving on one side arguments that fixed rates enhance the usefulness of national moneys—a view eloquently expressed by Kindleberger (1981), in particular, but probably a better argument for currency unions than for fixed rates—there are two main classes of argument for fixity.

The first general class of argument has to do with the interlinked problems of co-ordination and credibility; the claim is that a fixed exchange rate can either help resolve potential Prisoners' Dilemmas among nations, or help governments convince private agents such as unions of their anti-inflationary resolve. The most commonly made arguments are the following. First, it is argued that under flexible rates, nations have an incentive to pursue tight monetary policies, because such policies if pursued unilaterally lead to currency appreciation that quickly reduces inflation. Unfortunately, if everyone tries to do this nobody gets the appreciation, and the result is an excessively tight

monetary policy around the world.[2] Second, it is argued that many countries have a problem convincing price-setters that they really mean to hold the line on inflation. Fearing inflation, firms and workers raise prices in anticipation, and the government then finds that it must ratify these wage increases with monetary expansion if there is not to be a recession.[3] In each case, it is argued that a fixed exchange rate system can resolve the problem. In the first example, a fixed exchange rate regime can rule out competitive appreciations. In the second example, fixing the exchange rate can purchase credibility by tying a country's monetary policy to that of other countries with low inflation.

Models of this kind are currently very fashionable, reflecting both the popularity of game theory as an analytical tool and the widespread view that the EMS is largely a way for France and Italy to borrow German credibility.

The other category is the traditional argument, most famously expressed by Ragnar Nurkse (1937), that foreign exchange markets are subject to destabilizing speculation that hurts the real economy. This view became very unfashionable among academics after the 1950s, both because of the general rise of efficient markets theory and the direct assertion by Milton Friedman (1953) that destabilizing foreign exchange speculation could not flourish, because it would be unprofitable. It nonetheless remains a popular argument among practical men.

My own view is that while co-ordination/credibility arguments are intellectually beautiful and have some relevance, the key argument against floating rates is still the traditional one, that such markets are subject to destabilizing speculation. In part II of this paper I will present the 'anti-Friedman' case, arguing that faith in the efficiency and stability of foreign exchange markets must come from strong a priori beliefs held in the teeth of the evidence. For now, let me simply define the problem as one of market-generated instability in exchange rates, and turn to the question of the costs of such instability.

3. What harm does destabilizing speculation do?

Suppose we hypothesize that foreign exchange rates do indeed fluctuate too much, because of destabilizing speculation. What harm does this do? Equally important, what would we look for as evidence of the damage?

---

[2] See, for example, Canzoneri and Gray (1985).
[3] This argument has been made by Giavazzi and Giovannini (1989), among others, in their analysis of the European monetary system.

The usual answer has been that exchange rate fluctuations make international transactions more risky, and should therefore have an adverse effect on international trade and investment. I will argue below, however, that this kind of cost is likely to be fairly small, and in any case the evidence does not support the view that it is there.

The more likely costs, I would argue, arise from two seemingly contradictory effects of uncertainty about future exchange rates. First, uncertainty about exchange rates makes it rational for firms to be very cautious about responding to shifts in relative costs of production in different countries. The uncertainty therefore reduces the flexibility and efficiency of the world economy. On the other side, the awareness that exchange rates may fluctuate wildly in the future provides an incentive for costly duplication of capacity, to allow firms to respond quickly to short-term shifts.

Both in the traditional arguments against floating rates by such authors as Nurkse and in more recent warnings, such as those of Stephen Marris (1985), heavy weight has also been placed on the *macroeconomic* risks of exchange rate fluctuations, with the danger being that sudden exchange rate changes can set off inflationary spirals that in turn lead to severe recessions. At least thus far, however, the experience of the 1980s has not borne out these more cataclysmic visions. Despite the huge exchange rate movements we have seen, the macroeconomic performance of the major industrial countries since the 1982 recession has been remarkably smooth. This could always change, but at the moment the macroeconomic instability argument against floating rates does not look like a powerful one.

Again, this discussion of potential costs is based on the starting presumption that exchange markets are indeed subject to instability arising from destabilizing speculation. This is by no means a consensus view. Indeed, not long ago it would have been a view so heretical that few would dare suggest it. So our next step must be to turn to the evidence that suggests that it makes sense to worry about the behaviour of foreign exchange markets.

II. Exchange market instability: the evidence

1. Characteristics of exchange rate fluctuations

The basic argument that I want to advance for a return to an adjustable peg system is that under floating rates,

destabilizing speculation leads to fluctuation that is both pointless and costly. The main thrust of this part will be to make the argument that the speculation that takes place in the exchange market is indeed destabilizing. However, as a preliminary step it is important to confront a view about exchange rates that has become widely influential among academics—the view that real exchange rate fluctuations largely reflect real shocks, and that the causation runs from real to nominal exchange rates, not vice versa. This view is important, because if it is right the instability of real exchange rates has nothing to do with the exchange regime.

The view that real exchange rates are moved primarily by exogenous real shocks has been most forcefully expressed by Stockman (1987a), who argues his case on both theoretical and empirical grounds. The theoretical basis is essentially that of new classical macroeconomics: nominal shocks, such as changes in money supplies, should not have real effects, and speculative shocks, such as sudden changes in investor opinion unwarranted by news about fundamentals, should not happen at all. The empirical basis is the observation that real exchange rates appear to follow something close to a random walk, whereas in just about any model in which nominal exchange rate changes drive real changes the real exchange rate ought to exhibit a tendency to return over time to some long-run equilibrium level. (The standard example is the Dornbusch (1976) overshooting model.)

This is an important argument, so it is essential to understand why it is wrong.

The first point is that the argument overlooks a second piece of evidence: the change in the behaviour of real exchange rates that occurred when the nominal exchange rate regime changed. By any measure, the volatility of real exchange rates since 1973 has been much higher than in the previous two decades, by a factor of perhaps five (see Mussa, 1987, for extensive documentation). It is difficult to reconcile this with the view that the causation runs from the real exchange rate to the nominal (although Stockman (1987b) has offered an ingenious, albeit implausible rationalization).

The second point is that the rates of convergence to long-run equilibrium implied by standard sticky-price models are sufficiently slow that tests of persistence will fail to reject the hypothesis that the real exchange rate is a random walk unless given very long time series.[4] Suppose, for example, that we imagine that the Federal Reserve were to increase the US money supply by one percentage point. How much would this increase

---

[4] This point was first made by Frankel (1988), whose work forms the basis for the argument that follows.

the US inflation rate over the next year? The answer from Cambridge Massachusetts is typically not more than 0.2 per cent—which implies that less than a fifth of the initial deviation from equilibrium will be eliminated per year. We can now ask, how many years would it take before an econometrician would reject the hypothesis of a random walk for a series that in fact eliminates only one-fifth of deviation from equilibrium per year? The answer turns out to be that more than fifty years of observations would be needed.

When very long time series are examined, there is pretty clear evidence of a tendency to converge to a long-run equilibrium value. Frankel (1988) has used 119 years of data on the dollar–pound real exchange rate and found that there is indeed a statistically significant tendency toward convergence back to a long-run level, with about 16 per cent of the divergence eliminated per year.

The point is that the behaviour of real exchange rates really does look as if they are continually being pushed away from a relatively stable long-term level by transitory shocks, so that there is a lot of fluctuation that is in some sense unnecessary—which would not be the case if real exchange rate movements reflected permanent changes in the long-run equilibrium. However, most economists would probably argue that the shocks that move the exchange rate are things like changes in monetary and fiscal policy; if these policies were stable, the exchange rate would be too. So next I need to make the case that a substantial share of the fluctuations in exchange rates may be attributed to speculative churning rather than movements in fundamentals such as economic policies.

2. The inefficiency of foreign exchange markets

When Milton Friedman argued against the possibility of destabilizing speculation, his starting point was an assertion that financial markets would tend to become efficient in the sense that no unexploited profit opportunities remained. Since that time efficient markets theory has grown into a massive structure, with a prestige that is based on its apparent empirical success in predicting that asset prices will follow random walks.

Now we already saw that tests of random walks may not have much power. In any case, however, it is important to know that the empirical evidence on foreign exchange markets provides *no* support to the idea that these markets are efficient.

The basic piece of evidence is on the forward exchange market. If risk were not too important, we would expect the forward discount on a currency to be the best available predictor of the actual depreciation of the currency over the contract period. A regression of the actual depreciation on the forward discount should yield a coefficient of one; adding other variables to the regression should not improve the explanatory power.

In fact, however, forward discounts have in general shown either near-zero or negative correlations with subsequent depreciation rates.[5] Notice that this is not a matter of a small technical failure of forecasting efficiency. The basic result is that the forward discount or premium on a currency has *no* usefulness as a predictor.

It is, of course, possible to reconcile this fact with an efficient markets model, just as it is possible to add epicycles to any model that fails to account for the data in its simplest version. By adding large and shifting risk premia, and arguing that forward discounts almost entirely reflect these risk premia rather than expected exchange rate changes, the basic result can be explained away.[6] But this is an *ex-post* rationalization, which incidentally requires risk premia that are larger and more volatile than anyone believed existed before the bad news on forward exchange rates came in. It also reflects an intellectual double standard: when random-walk-type results appeared to confirm simple efficient markets models, they were widely cited in support; now that they do not come out as expected, we are told that the theory doesn't really predict them anyway. At this point belief in the efficiency of the foreign exchange market is a matter of pure faith; there is not a shred of positive evidence that the market is efficient, and I think it is safe to say that an observer without an intellectual vested interest in the efficient markets theory would find the absence of any correlation between forward discounts and subsequent exchange rate movements a good piece of evidence that the market is not efficient.

It might be worth noting at this point that similar results obtain for other asset markets. That is, both the bond market and the stock market show behaviour that contradicts the simple random-walk-type models that we used to be told were borne out in the data, confirming the efficient markets theory. In the bond market, the rate differential between long-term and short-term bonds, which plays a role similar to that of the forward discount, shows a similar failure to predict future long-term rates; in the stock market, the volatility of prices

[5] There is by now an enormous literature on the forward discount's predictive power. An early paper was Tryon (1979); the paper widely regarded as decisive in demonstrating the failure of the simple speculative efficiency model was Hansen and Hodrick (1980).
[6] See Hodrick and Srivastava (1984, 1986).

is much greater than the volatility of a present value of future dividends. In each case, it turns out that a more complex efficient markets approach is consistent with the data; but these approaches are contrived and look somewhat implausible. The bottom line is that there is no *positive* evidence in favour of efficient markets, and if anything a presumption from the data that markets are not efficient.

From my point of view, the important conclusion from this discussion is that we are freed from Friedman's logic. Friedman's argument was that an efficient market could not exhibit destabilizing speculation; this seemed a fatal objection to the destabilizing speculator argument, because for a generation we have been told that overwhelming evidence supports the hypothesis of efficiency in financial markets. Now that we know that in fact no evidence supports this hypothesis —that it is one maintained purely on faith—we are free to examine whether the exchange markets really do seem to make big mistakes.

3. Evidence of destabilizing speculation

Destabilizing speculation has no precise definition, and I will not try to provide one here. What we mean in general, however, is that there are episodes in which markets get carried away, driving the exchange rate to levels that are not warranted by fundamentals, and that turn out to be unsustainable in the long run. There is an implication that such episodes also do real harm, but let us put that aside for the moment.

My own view that exchange markets are subject to destabilizing speculation has been strongly coloured by two episodes: the run-up of the dollar from the late summer of 1984 to its peak in early 1985, and the recent surge of the dollar from April to June of 1989. In both cases a huge exchange rate movement by historical standards took place with very little in the way of either current movements in underlying fundamentals or meaningful news about future fundamentals that could justify such a move. Furthermore, in both cases the resulting level of the exchange rate was one that appeared unsustainable.

What I mean by unsustainability is that the current level of the exchange rate is one that cannot be justified by any reasonable forecast about the future exchange rate. Or to put it another way, any current level of the exchange rate involves an implicit forecast on the part of the market about the future exchange rate. If we can show that this implicit forecast is unfeasible, then we can say that the current level is unsustainable.

I have applied this logic extensively to the 1985 dollar, both at the time and as a post-mortem (Krugman 1985,1989); so it is probably more interesting to use the same logic to argue that the level of the dollar as of early July 1989 was similarly unsustainable.

We first note that the exchange market is always in effect making a prediction about the future course of the exchange rate. Suppose that the interest rate on, say, ten-year bonds is 8 per cent in the United States and 5 per cent in Japan. Then the willingness of market participants to hold dollars is in effect a prediction that the dollar will decline by not more than 3 per cent per year over the next decade.

It is important to notice that in an efficient market it should not make any difference whether individual investors plan to hold assets for short or long time horizons. A short-term investor may be interested only in near term developments and in the price at which he can sell to another investor; however, this next investor will presumably be looking at prospective developments a little further ahead, and so on. So even if nobody invests for the long term, the market should be making implicit forecasts that are in fact feasible for the long term. The point of the sustainability analysis is that it appears that foreign exchange markets do not do this; that at times the exchange rate moves to levels that cannot be consistent with any reasonable long-term forecast.

What forecast was the market making in July 1989? Let us focus on the dollar–yen rate, where the case appears particularly clear cut. Long-term government bonds in Japan were yielding about 5 per cent, while in the US they yielded about 8 per cent. That is, the market was implicitly forecasting that over the next decade the dollar would decline against the yen no more than 3 per cent per year.

The sustainability argument is then that this was not a feasible forecast, for at least three reasons. The US had higher inflation than Japan, so that it needed a steadily falling dollar just to keep the real exchange rate constant; the US would have needed a substantially lower dollar even in 1989 just to stabilize its growing foreign debt; and the steady erosion of US technological superiority requires a real dollar that declines over time to compensate.

On the inflation issue, the underlying inflation rate in the United States is about 5 per cent, while in Japan the rate is more like 1 per cent. This alone means that if the dollar were to decline as slowly as the market apparently thought in mid-1989, the US would become steadily *less* competitive on costs and prices as time went by.

Meanwhile, almost all estimates suggest that the

dollar needed to fall substantially if the United States were not to have an explosive growth of foreign debt. Major international agencies like the IMF and the OECD were forecasting widening US deficits after next year even before the recent run-up in the value of the dollar. A recent study by William Cline (1989) at the Institute of International Economics, using middle-of-the-road numbers, estimated that simply to slow the rate of growth of US foreign debt to a sustainable rate—not to stop its growth, which would be much harder—the yen would have to rise 28 per cent from its late-1987 level in real terms, implying a dollar–yen rate in 1989 prices of not much more than 100. The longer that decline is delayed, the more the dollar must drop, because of the growing burden of foreign debt. So the prospect was that sooner or later the dollar would have to fall very sharply in real terms. Yet the financial markets seemed to believe that it could actually rise.

Finally, when looking at the dollar–yen rate it is impossible to avoid the broader issues of competitiveness. In 1980 many businessmen thought US firms could be competitive with an exchange rate that, at 1989 prices, would have been about 190 yen. In 1989 the Japanese obviously felt quite comfortable at 130 yen. This suggests a downward trend in the equilibrium real exchange rate. Such a downward trend is indeed visible in most empirical estimates of trade equations for the US, and a corresponding upward trend is visible in most trade equations estimated for Japan.

Putting these arguments together, it seems that on any sort of standard analysis it was impossible that the long-term path of the dollar would be as gentle a decline as financial markets seemed to expect in mid-1989. Admittedly, the market could simply hold a different view—believing, for example, that a huge US export surge would eventually result from the current level of the dollar. It is virtually certain, however, that alternative views about fundamentals were not at the heart of the willingness to hold dollars in 1989. Instead what was happening was that almost all market participants took a short-term view, and that they were attempting to forecast future prices without reference to future fundamentals. This is the essence of the destabilizing speculation that policy-makers worry about.

Many economists still have a visceral distrust of arguments that claim that markets are liable to make simple mistakes of the kind I am asserting. The point of the discussion of efficiency tests was to show that our tendency to believe in asset market efficiency is at this point entirely based on assumption rather than evidence. We may also point out that what we know about how the exchange markets function does not suggest a strong interest in checking to see whether current exchange rates make sense in a long-run context. It is noteworthy that even analytically-oriented financial journalists have little interest in long-run arguments about exchange rates. For example, the *Economist* of July 1, 1989, in a story about the dollar, explicitly dismissed sustainability arguments like the one made here on the basis that 'Financial markets are not worried about what happens to extrapolated lines for deficits or foreign debt ratios in five years' time—unless they think the authorities are worried.' This attitude puts us squarely in the world described by Keynes (1937), in which speculators, instead of trying to assess fundamentals, try to forecast the common opinion of common opinion—and thereby generate a propensity for sudden booms and crashes.

4. Verdict on the foreign exchange market

The purpose of this section has been to make three points. First, there is convincing evidence that floating exchange rates lead to much more volatile real exchange rates than occur under fixed rates. Second, there is no evidence that the foreign exchange market is efficient in the sense that prices move appropriately in response to changes in available information. Third, there are important episodes—including the situation at the time of writing—in which the exchange market appears to hold views about the future that are unfeasible under any reasonable forecast.

Taken together, these observations suggest that the old hypothesis of destabilizing speculation remains a viable one: that substantial excess volatility in real exchange rates results from the speculative character of markets under floating exchange rates. The next question is what harm this volatility does.

III. The costs of instability

1. Risk and international integration

What are the costs of fluctuating exchange rates? The most common answer is that they increase the risk of international trade and investment. Suppose, to make matters concrete, that the cost of goods shipped from Europe to the United States is fixed in ECU, but that the price at which they can be sold is fixed in dollars (and that the converse is true for goods shipped in the opposite direction). Suppose also that there is a substantial lag between ordering and sale. Then if the dollar–ecu rate fluctuates, exporters in each direction will face a new source of risk that should add to the costs of international trade. This loose story led some economists to expect that floating rates would have a substantial negative effect on international trade, and has led to a considerable literature on the question of whether some adverse effect on trade volumes has resulted from exchange rate instability.

The general conclusion of such tests has been that little or no adverse effect on trade volumes can be demonstrated, a conclusion that essentially reflects the fact that trade has continued to grow more rapidly than output throughout the floating rate period.

However, this result should not be a surprise. The extent of exchange rate fluctuations in the 1970s and 1980s should not have been expected to have very much effect via its risk aspects in discouraging trade, for two reasons: the risk is not very large, and much of it should be diversifiable.

The riskiness issue may be summarized briefly: since exchange rate volatility is small compared with that of other asset markets, even if exchange rate risk could not be diversified there should be little risk premium required to engage in trade. Suppose, for example, that the typical time from order to sale in trade were a full year, which seems fairly long. The average one-year standard deviation of the dollar–mark exchange rate since 1973 has been approximately 10 per cent. If this risk were completely undiversifiable, and if relative risk aversion were 2, the risk premium would be only one per cent. That is, the whole effect of the risk would be to add an implicit transportation cost of 1 per cent of the goods shipped. It should not be surprising that this effect cannot be identified in the data.

We may further add that much of the exchange risk should be diversifiable. A European exporter faces the risk of a capital loss if the dollar falls against the ecu; but a US exporter experiences a corresponding capital gain. By signing a forward contract with each other, both exporters can eliminate much of their exchange risk. This should therefore make the already small exchange risk premium on foreign trade even smaller.

The point of this discussion is not to suggest that there are no costs to exchange rate instability. Rather, the point is that increasing the riskiness of foreign trade is not the main effect, and looking at the volume of trade is not the place where one would expect to find the costs of fluctuating exchange rates.

2. Blurring the signal

Even if risk is not a major issue, exchange rate instability should change firms' behaviour. A recent theoretical literature originated by Avinash Dixit (1987) has pointed out that when future exchange rates are uncertain there is an incentive for firms to adopt a wait-and-see attitude toward investments which may reduce the rate at which trade patterns adjust to more fundamental factors.

The point may be made by considering the following example. Suppose that a firm is considering an investment in export capacity which will be difficult to reverse, and that this investment appears to have a small positive present value at the current exchange rate. If the firm were fairly sure that the current rate would last for a long time, it would make the investment. If, however, it views the future exchange rate as uncertain, *even if it regards appreciation and depreciation as equally likely*, it will have an incentive to delay the investment. To see this, suppose that the firm makes the investment and then finds that the home currency appreciates; it will then have made a costly long-term mistake. On the other hand, if it delays the investment and the home currency depreciates, it will have sacrificed some profits but can still choose to invest later. So the costs of being wrong in one direction are larger than the costs in the other, and the firm will therefore choose to err on the side of caution. It can be shown that this tendency to caution is a general result of exchange rate instability, and applies to disinvestment as well as investment (see Dixit, 1987, and Krugman, 1989).

This may now be argued to turn into a cost of instability. Suppose that there are real changes in the world economy that should be met by investment and disinvestment. Then firms faced with uncertain exchange rates will delay investment decisions that would raise world efficiency, and this delay will therefore degrade the effectiveness of world adjustment. The problem would not be one of too little trade; it would be one of too slow an adjustment of trade patterns to shifts in comparative advantage and other sources of change. In effect, exchange rate instability would reduce the gains from international trade by blurring the price signals that are supposed to regulate markets.

How would we see evidence of such a degradation of efficiency? One possibility would be that estimated lags in trade equations would appear longer. There is at present no evidence of such a lengthening of lags. However, this is a pretty crude test—and as argued below, there may be contrary effects. My own guess is that in fact the decision-delaying effects of uncertainty are substantial. This is only a guess; the basic point is simply that it is by this kind of microeconomic gumming up of the works, rather than by reducing the aggregate volume of trade, that adverse effects from exchange rate instability are likely to make themselves felt.

3. Excess capacity

It seems worth mentioning that another possible effect of exchange rate instability may be to encourage construction of excess capacity. Suppose that for other reasons a firm is considering installing capacity in more than one country, capable of serving several markets (or that several firms are installing capacity—one need not appeal to multinational firms for this argument). If the firm thinks that large exchange rate movements are possible in either direction, it will have an incentive to install some excess capacity, so as to be able to switch production to whichever location has the lower costs. This would, of course, impose a social cost because of the resources used to construct the extra capacity.

An interesting point about this argument is that it carries the opposite prediction about the response of trade flows to exchange rates from the previous argument: if excess capacity is installed in order to allow firms to respond quickly to exchange rates, then at least some trade adjustment will come *faster* under floating rates than under fixed. This is not really a contradiction of the earlier assertion about excess caution. What will happen is that firms become quick to respond to exchange rates with existing capacity, so that there is a fast short-run response of trade flows; but they become more cautious about adding or scrapping capacity, so that there is a delayed long-run response. We may pity the poor econometrician who attempts to sort all this out from limited time-series data. The point is, however, that it is in such subtle effects that the possibly sizeable costs of exchange rate fluctuations lie.

4. Political economy

Practical concerns about exchange rate instability are heavily influenced by a non-economic issue: the fear that countries with over-valued currencies will experience protectionist reactions. There is not much that this paper can say about this issue, except that it appears to be true that there is an asymmetry: the free-trade sentiments generated by a weak currency do not appear to offset fully the protectionist sentiments generated by a strong currency. Thus a system of large fluctuations in exchange rates tends, other things equal, to worsen the climate for maintenance of an open trading system. This is somewhat ironic, since one of the main arguments for floating rates was that they would reduce pressure for protectionist measures justified on balance-of-payments grounds (see, for example, Cooper, 1971).

5. Verdict on costs

The upshot of this discussion is inconclusive. The simplest hypothesis about costs from exchange rate fluctuations, that these fluctuations discourage trade, is not supported by the data—but we would not expect it to be so on an a priori grounds. The more likely sources of costs are probably hard to identify in practice.

IV. The other side: advantages of exchange rate flexibility

It is common for advocates of particular exchange rate regimes to belittle the advantage of other regimes, or to deny that there are any special problems associated with the proposed regime. This is clearly not appropriate in the present case, since there are well-known problems with a fixed rate system, problems that change character but do not go away if the system allows for occasional parity changes.

1. The adjustment problem

The basic problem with any proposal for exchange rate stabilization is that exchange rate changes are sometimes very useful and indeed nearly inevitable. The reason is that it is much easier to adjust relative national wage and price levels through exchange rate changes than through deflation and inflation.

The traditional example is still valid. Suppose that a country has for some reason managed to get its price level out of line with those of its trading partners— either because of an inflationary shock of some kind, or because some real shock has altered the equilibrium relative price. Then the country must either persuade its trading partners to inflate; engage in deflation itself; or devalue its currency. The first option is rarely available. Meanwhile, the experience of the 1980s has rejuvenated the Phillips curve: deflation or disinflation really is costly in terms of real output. So there is a clear advantage to being able to adjust relative prices through a devaluation rather than painful deflation.

The ability of countries to use exchange rate adjustment to avoid the need for deflation is a compelling argument against a system of unalterably fixed exchange rates. The experience of disinflation in the 1980s, while suggesting a short-run inflation-output trade-off a little more favourable than some pessimists had supposed, does not encourage a belief that the costs are minor. Comparing US output over the period from 1979–87 with a trend line suggests that output equal to at least 20 per cent of annual GNP was sacrificed in order to bring the underlying inflation rate down from about 10 per cent to about 5. This suggests that a country whose relative prices are, say, 10 per cent out of line would pay a huge price to preserve a fixed exchange rate, one that would be hard to justify in terms of any conceivable microeconomic gains. As James Tobin has pointed out, it takes a lot of Harberger triangles to fill an Okun gap.

The relevant comparison to make with the present system, however, is not with permanently fixed rates but with a system that allows occasional exchange rate adjustments—as was the case under Bretton Woods, and is the case within the EMS now.

2. Problems with an adjustable peg

Can occasional exchange rate changes, rather than continuously flexible rates, fulfil the need for an international adjustment mechanism? The answer is surely yes, except under two circumstances. The first is large differences in underlying inflation rates among the participating nations—a serious problem a decade ago, but much less of one now. The second is frequent sustained changes in equilibrium real rates. As pointed out in part II of this paper, while such real shocks are widely believed in among academic researchers, the evidence in fact does not support their importance. Thus a system under which nations devalue their currencies when they appear to be very far out of line on relative prices would probably not require such frequent realignments as to be meaningless.

The main problem with an adjustable peg system would be that speculators would try to anticipate devaluations, provoking occasional speculative attacks on currencies—either because speculators correctly perceive that an exchange rate change is imminent, or because of the same kind of destabilizing speculation that I have argued characterizes floating rates. The now extensive literature on balance-of-payments crises has confirmed the traditional concerns of central bankers that a fixed rate offers speculators a 'one-way option'. Some theorists, such as Obstfeld (1986), have argued that a fixed exchange rate system is just as vulnerable to destabilizing speculation as a floating rate.

This is a worrying possibility, especially if we consider that the Bretton Woods system and its short-lived successor, the Smithsonian system, both collapsed under the pressure of speculative attack. The formal result that the exchange regime offers no defence against speculative bubbles, however, applies only to markets where speculators behave rationally. As I have argued, the evidence is that markets are not rational in this sense. Since we have no general theory of irrational speculation, it is impossible to prove that a rate that is fixed but adjustable will give rise to less destabilizing speculation than one that is fully flexible. My own reading of the experience of both Bretton Woods and the EMS is that in spite of some serious problems with speculative attack, they were less troubled by destabilizing speculation than the floating rate system of the post-1973 period.

V. Exchange rate policy

The issue of the appropriate exchange regime is unusual in economics, in that it does not break down along the usual left–right or 'salt-water' vs. 'fresh-water' lines. The reason is that it involves a trade-off between two concerns. The main argument for stabilizing exchange rates in practice is fear of excess volatility and destabilizing speculation, possibilities that 'salt-water' economists take seriously but that their 'fresh-water' colleagues dismiss a priori. The main argument for flexible exchange rates is the need to facilitate adjustment of relative prices, which equilibrium macroeconomics dismisses as a problem. The result is that the field is full of strange bedfellows: advocates of flexible rates may be traditional monetarists who want freedom to target their favourite aggregates or Keynesians who want to be able to pursue active stabilization policy; advocates of fixed rates may be either global monetarists who want a monetary anchor or interventionist-minded economists who distrust financial markets.

My position is the 'salt-water', i.e. East Coast USA interventionist, one for fixed rates. I have argued that the old-fashioned view that exchange markets are prone to destabilizing speculation that disrupts the real side of the economy is nearer the truth than the Friedman view that exchange markets are as stable as the monetary policies of the countries whose currencies are exchanged. This means that the same arguments used to justify the Bretton Woods system remain valid today.

The problem is that the case for exchange rate adjustment as a tool of policy is also still valid. Indeed, the resurrection of the Phillips curve by the experience of the 1980s suggests costs to price adjustment large enough to make a fully fixed exchange rate system very difficult to justify even given the presumption that floating rates are subject to destabilizing speculation.

Thus the best that can be proposed here is an adjustable peg system, under which exchange rates may be altered when relative prices are clearly out of line, but are otherwise kept stable. Such a system will unfortunately be subject to speculative attacks. It is simply a judgement call that such attacks will be less virulent and do less harm than the speculation that takes place under floating rates.

The final point to address is about current events. Since early 1987 the G7 nations have attempted to maintain a system of target zones on exchange rates, motivated by pretty much the same concerns expressed in this paper. At the time of writing, unfortunately, this system appeared to be falling apart in the face of a speculative surge in the dollar. Why should this be happening?

The answer, I would argue, is that the diffident way in which the current system has been set up robs it of the stabilizing effect on speculation that is the main purpose of a return to more fixed rates. The Louvre system has deliberately avoided explicit statements about target zones, even though the existence of the targets is general knowledge, so that the governments can abandon their targets without too much political embarrassment. This is a peculiar set-up, which I would describe as a 'covert' target zone system, by analogy with US covert policies in such places as Central America and elsewhere—that is, the policy is not at all secret, but because it is not officially acknowledged the government is protected from the political costs of failure. The problem is that by offering themselves an easy path of retreat, the G7 nations also deny their target zones the credibility to prevent speculative runs of the kind that has just happened.

What is needed is a system in which the commitment to try to defend parities is explicit. This worked for two decades in the Bretton Woods era; it has worked better than anyone expected in the European Monetary System; it can work for the world monetary system as a whole—maybe.

References

Canzoneri, M. B., and Gray, J. A. (1985), 'Monetary Policy Games and the Consequences of Non-co-operative Behavior', International Economic Review, 26, 547–64.

Cline, W. (1989), US External Adjustment and the World Economy, Washington, Institute for International Economics.

Cooper, R. N. (1971), 'The Nexus among Foreign Trade, Investment, and Balance-of-Payments Adjustment', in Commission on International Trade and Investment Policy, United States International Economic Policy in an Interdependent World, Washington, US Government Printing Office.

Dixit, A. (1987), 'Entry and Exit Decisions of a Firm under Fluctuating Exchange Rates', mimeo, Princeton University.

Dornbusch, R. (1976), 'Expectations and Exchange Rate Dynamics', Journal of Political Economy, 84, 1161–74.

Frankel, J. (1988), 'International Capital Mobility and Exchange Rate Volatility', presented at Federal Reserve Bank of Boston conference at Bald Peak.

Frenkel, J., and Goldstein, M. (1989), 'Exchange Rate Volatility and Misalignment: Evaluating some Proposals for Reform', in Federal Reserve Bank of Kansas City, Financial Market Volatility, (proceedings of a symposium at Jackson Hole, Wyoming, August 17–19, 1988).

Friedman, M. (1953), 'The Case for Flexible Exchange Rates' in Essays in Positive Economics, Chicago, University of Chicago Press, 157–203.

Giavazzi, F., and Giovannini, A. (1989), Limiting Exchange Rate Flexibility: The European Monetary System, Cambridge, MIT Press.

Hansen, L., and Hodrick, R. (1980), 'Forward Exchange Rates as Optimal Predictors of Future Spot Rates: an Econometric Analysis', Journal of Political Economy, 88, 828–53.

Hodrick, R., and Srivastava, S. (1984), 'An Investigation of Risk and Return in Forward Foreign Exchange', Journal of International Money and Finance, 3, 5–29.

—— and —— (1986), 'The Covariation of Risk Premiums and Expected Future Spot Exchange Rates', Journal of International Money and Finance (Supplement), S5–22.

Kindleberger, C. (1981), International Money, London, George Allen and Unwin.

Krugman, P. (1985), 'Is the Strong Dollar Sustainable?' in The US Dollar: Prospects and Policy Options, Federal Reserve Bank of Kansas City.

—— (1989), Exchange Rate Instability, Cambridge, MIT Press.

Keynes, J. M. (1937), 'The General Theory of Employment', Quarterly Journal of Economics, 51, 209–23.

Marris, S. (1985), Deficits and the Dollar, Washington, Institute for International Economics.

Mussa, M. (1987), 'Nominal Exchange Rate Regimes and the Behavior of Real Exchange Rates', Carnegie-Rochester Conference Series on Public Policy.

Nurkse, R. (1937), International Currency Experience, Geneva, League of Nations.

Obstfeld, M. (1986), 'Rational and Self-fulfilling Balance-of-payments Crises', American Economic Review, 76, 72–81.

Stockman, A. (1987a), 'The Equilibrium Approach to Exchange Rates', Economic Review, Federal Reserve Bank of Richmond.

—— (1987b), 'Exchange Rate Systems and Relative Prices', Journal of Policy Modelling, 9, 245–56.

Tryon, R. (1979), 'Testing for Rational Expectations in Foreign Exchange Markets', Federal Reserve Board International Finance Discussion Papers, no. 139.

PART III

CONSUMPTION AND
THE BALANCE OF PAYMENTS

6

The balance of payments and international economic integration

CHRISTOPHER ALLSOPP

New College, Oxford

TIM JENKINSON

Keble College, Oxford

TERRY O'SHAUGHNESSY

St. Anne's College, Oxford[1]

I. Introduction

Through most of the post-war period, it was taken for granted that the balance of payments position of a country was a major and legitimate concern of governments and of international organizations. External balance, alongside full employment and the control of inflation, was seen as a principal objective or target of policy—though careful analysts distinguished between the ultimate objectives, concerning the growth and distribution of consumption possibilities, or welfare, and proximate targets such as the balance of payments or inflation control.

Since the 1970s, vast disturbances in external payments positions have taken place. Some of these are straightforward to explain, such as those following the oil price hikes and the implied changes in international income distribution and savings patterns. As can be seen in Table 6.1, a principal feature of the 1980s was the emergence of large imbalances within the industrial world; most notably, those between the G3 coun-

First published in *Oxford Review of Economic Policy*, vol. 6, no. 3 (1990). This version has been updated and revised to incorporate recent events.
[1] We are grateful to John Muellbauer and Andrea Boltho for helpful comments and suggestions. The usual disclaimer applies.

tries—the US, Japan, and Germany—but also, from a somewhat different perspective, within regional groupings such as the EC, where the imbalance between Germany and the rest has been similarly large. At an even more local level, there have been some dramatic swings such as the movement in the external position of the UK from surplus to a deficit in 1989 of 4.4 per cent of GDP before returning to balance in 1994.

The general questions that arise are, first, what accounts for these imbalances and, second, whether they matter, in the sense of calling for policy action. If they do matter, then there is a third question—what should be done about them? It is clear, however, that there are unlikely to be any general answers. Payments imbalances arise for different reasons. Whether they matter depends not only on how they arose, but also on circumstances, such as financing possibilities. The policies available depend on the regime in operation, such as whether the country has a separate currency or is part of a monetary union.

Nevertheless some general themes emerge. The first is the importance of economic integration in changing the nature of balance of payments problems and the terms of the debate. Imbalances, especially between the US, Germany, and Japan, which looked extremely

Table 6.1. Post-War Current Account Trends

	1956–60	1961–67	1968–73	1974–79	1980–84	1985–89	1990–95
US	0.4	0.8	0.1	0.0	−0.8	−3.0	−1.6
Japan	0.1	−0.2	1.3	0.3	0.9	3.2	2.5
Germany	2.1	0.4	1.1	1.0	0.2	4.1	−0.3
France	−0.1	0.4	−0.1	0.1	−0.9	−0.3	0.2
Italy	1.0	1.5	1.3	−0.2	−1.4	−0.6	−0.3
UK	0.5	−0.3	0.4	−1.0	1.4	−1.7	−1.4

Source: Annex table 51, 'Current balances as a percentage of GDP', *OECD Economic Outlook* 57, June 1995.

destabilizing when they emerged, seem to have caused less trouble than commonly anticipated. The fundamental change that has facilitated this is the increasing integration of international financial markets, which has allowed imbalances to be financed. Similarly, the British balance of payments deficit has, so far, not led to major financing or exchange rate difficulties. Moves towards monetary union in Europe raise, even more explicitly, the question of what happens to balance of payments problems and balance of payments policy as integration proceeds. Effectively, countries take on the characteristics of regions and what might have appeared as a payments problem, may, as the Delors Report suggests, reappear as a regional problem. At the same time, the menu of policy options is narrowed.

Financing problems can still exist, especially for countries with separate exchange rates where exchange rate risk must be taken into account. There is no doubt, however, that, for the industrialized countries at least, the financing constraint is less important than it was. This provides an important set of benefits. Macroeconomic consistency conditions applying to the domestic economy (for example, that savings must balance investment) are released—though they reappear at the international or global level and may imply the need for international co-ordination. The domestic economy has access to the international markets for investment finance and as a home for their accumulated savings. Policy problems, such as excess demand, appear in a different (and arguably more benign) form, as balance of payments deficits rather than inflationary pressure.

But with additional freedom comes additional responsibility. There are legitimate reasons why governments may be concerned with savings and investment and the pattern over time of the accumulation of capital and financial wealth within the economy. An aspect of these considerations may be concern over the balance of payments, which may appear as a symptom of more fundamental problems and difficulties. In principle, even a regional authority may want to affect the external payments position if there are reasons to suppose that the individuals and firms within the region are not behaving in a socially optimal way.

Another issue, traditionally associated with balance of payments problems, is competitiveness. Balance of payments problems have often been characterized as a conflict between internal and external balance, i.e. full employment and an acceptable current account. It is a moot point as to whether such a problem should be described as a balance of payments problem, or more simply as a competitiveness problem. An uncompetitive economy would show the symptom of an external deficit if, for example by policy action, domestic demand were high enough for full employment. Alternatively, the balance of payments problem may be suppressed at the cost of low demand and unemployment. In this latter case, there is a direct analogy between a country with balance of payments difficulties and an uncompetitive, high unemployment region within a country. Clearly, balance of payments difficulties associated with uncompetitiveness pose some of the most intractable policy problems.

The issues of finance, of capital accumulation and savings–investment balance, and of competitiveness interact. Nevertheless, it is extremely important in a given case to try to identify where the problems mainly lie. Thus Davies (1990), in his discussion of the UK's capital account, points to financing difficulties and high interest rates. Financing problems arise, however, partly from the short-term nature of the finance, and partly from the predominance of consumption demand in leading to the external imbalance. Holtham (1990), in his survey of the imbalances between the US, Germany, and Japan, notes a shift away from previous concerns over the financing of the imbalances to concern over the sustainability and desirability of the flows of financial capital. In Bliss's (1990) optimization framework, however, large imbalances may exist and persist which, by construction, pose none of the problems outlined above.

If a particular external situation can be diagnosed

as problematic—and this inevitably involves both empirical and normative judgements—then the question arises as to how the situation will change and/or be made to change by policy action. There are several aspects to this balance of payments adjustment problem. One, which is explicitly addressed by Venables (1990), concerns the response of firms to variations in the (real) exchange rate. He applies the techniques of modern industrial economics to look at the pricing and output decisions of firms in imperfect markets (with market segmentation) as the costs of one set of firms, those based abroad, changes. A feature of his model is that, in the presence of sunk costs, there may be ranges where the price in the home market does not vary at all. Moreover, hysteresis (or path dependence) is likely to be important. In practical terms, if the real exchange rate fluctuates, a return to the starting point will not leave industrial structure unchanged. Temporary changes in competitiveness will have long-term effects.

Clearly, a realistic view of the adjustment process is needed. Here the regional analogy is useful, and is directly addressed by Begg and Weale (1990) which looks at balance of payments issues in the context of an integrating Europe. Much work on the balance of payments makes the conventional assumption that factors are immobile between countries. However, economic integration will increase mobility, but at different rates in different markets. The extent to which integration will foster convergence in an automatic way is clearly of vital importance to European governments. There is a danger that pressure to narrow wage differentials could slow the process and lead to serious competitiveness problems for some regions or countries. This clearly happened in Germany following monetary union. It may also be a serious risk for the UK as it contemplates European Monetary Union.

Adjustment also needs to be looked at from the side of financial flows. Is it the case, for example, that private sector net asset decumulation leading to an external deficit can be safely left to the market—as has, for example, been argued for the UK. Clearly, there are automatic wealth effects to take into account, and adjustments will occur. But what policy problems could be raised and what are the possible responses, in the short run and in the longer run? What policies are available if it is desired to change the savings–investment position in the longer term?

It is evident that the number of issues that interact with, and impinge on, the analysis of the balance of payments is very large. It is wrong to think of the balance of payments as a problem in its own right: rather an adverse payments position *may* be a symptom and indicator of other more fundamental problems. These problems can exist, whether or not they express themselves in the balance of payments figures. There is much to be said for focussing on the problems themselves rather than on the external deficit or surplus *per se*.

The next section considers the fundamental question of what meaning can be given to the idea of a balance of payments problem, classifying the discussion into three areas—financing; savings, investment and capital accumulation; and competitiveness. Section III considers aspects of the adjustment problem, for countries or for regions within a common currency area. Section IV looks more generally at the interaction of the balance of payments with other aspects of economic policy, paying particular attention to how this might change as integration proceeds and as countries join common currency areas. Finally, section V looks more directly but highly selectively at two particular problems: the world imbalances, and the impact of European monetary integration, and the UK. The intention is to illustrate particular general problems in a practical context, as well as to throw light on the policy issues themselves.

II. What is a balance of payments problem?

An important strand of analysis considers balance of payments surpluses and deficits as part of intertemporal optimization by economic agents or countries. If we consider an individual agent, a standard life-cycle model would suggest periods of saving and dissaving, i.e. periods of current account surplus and deficit (which will equal savings if we neglect household investment). The same idea can be applied to a region or a country, and from the country's point of view it may be optimal to make use of international financial markets so that absorption (or expenditure) may take a different track from production. There are practical examples. For the UK, with an uneven profile of oil production, it was obviously advantageous to smooth the consumption profile via recourse to international financial markets; a closed economy would have to limit production to current needs. In the case of Japan, it is often suggested that an ageing population means that it is sensible to build up international assets which can be used to support consumption in future years; the alternative would be to suppress current savings, or to stimulate productive investment.

The general point that potential welfare is likely to

be improved if high savings areas make their savings available where returns are highest is surely not contentious. As Bliss (1990) notes, the possibility of swapping asset claims for current goods and services expands opportunities in much the same way as international trade and production according to comparative advantage expands opportunities. This is an aspect of resource allocation that would be taken for granted in a regional context.

This optimization framework is useful to counter the idea that payments imbalances are always a problem. Such models do not, of course, imply that observed imbalances arose for these reasons, or that they should be welcomed as necessarily contributing to world or national welfare. They do suggest, however, that it is important to examine carefully where problems might arise in an imperfectly functioning and highly uncertain world economy.

We consider, in turn, financing problems; problems concerned with savings, investment, and capital accumulation (which may include domestic and international co-ordination difficulties); and problems of competitiveness. Needless to say, the problems interact.

1. Financing

The institutional constraints of the Bretton Woods system meant that financing was almost *by definition* a potential problem for a deficit country. Continuing deficits would lead to a drain on reserves, which could not go on for ever, or borrowing on increasingly difficult terms. Actions, such as raising interest rates (to attract capital flows), deflation, or, at the end of the day, exchange rate depreciation, could effectively be forced. It is not surprising that the balance of payments, for deficit countries such as the UK, was often described as a constraint.

It is the reduction in the financing constraint in recent years that has most obviously altered the terms of the debate on balance of payments positions. With financial liberalization and integration, flows of international borrowing and lending are much easier. This does not mean, however, that finance and associated issues have simply ceased to be a potential worry. In a different context, the origins of the debt crisis lay in apparently easy finance, which dried up after 1982 (for a survey, see Allsopp and Joshi, 1986). International financial markets are subject to potential market failures (of over- or under-lending and of contagion effects) and these may be a reason for government concern and policy intervention.

Amongst industrialized countries, solvency and liquidity problems do not loom so large. But governments need to concern themselves with longer term issues, including future macroeconomic implications such as likely interest rate or exchange rate changes that may not, adequately, be taken into account by private sector agents. Most obviously, governments have an exchange rate to manage, a factor which may be ignored by private sector agents.

With the reduction in short-term financing difficulties, attention has naturally turned to the longer-term question of the sustainability of an external position. One criterion is whether the external debt position looks explosive, with debt ratios tending to rise without limit. There was a time, for example, with the dollar at its peak in 1984–5, when the US position did suggest unsustainability (see Bliss, 1986). However, forward-looking agents should foresee financing problems and the exchange rate should change accordingly. Alternatively, market operators might anticipate policy reactions from the authorities, again leading to changes in important variables such as the exchange rate. The ease of financing is unlikely to be independent of the reasons for the deficit. In particular, a deficit which reflects productive investment in the domestic economy will be viewed more favourably than one used to finance consumption or public expenditure.

The terms of the debate have, therefore, moved away from the external financing problem itself to a consideration of the longer-term and macroeconomic implications of the economic behaviour of agents within the domestic economy. These are best considered under the next sub-heading, although any potential problems, such as a trend towards excessive indebtedness, may clearly involve external financing difficulties as well.

2. Savings, investment, and accumulation

If we put aside the international financing problem for the moment, a concern with the current account of the balance of payments could reflect concern over the international position itself (for example, a concern over the stock of external debt, perhaps on welfare grounds) or more general concerns over capital accumulation and savings within the economy. In the latter case, the current account might be seen as a symptom of more general worries.

If we consider individual decisions to save and invest, the obvious question that arises is why these should not be left to market forces; generally a case for policy concern or intervention needs to be made. There are some standard types of problem that could constitute a case for policy intervention. Individual preferences, especially those involving intertemporal choices, may be myopic or irrational. Or individuals may act on incomplete or faulty information. Markets, such as those for finance, may be imperfect and subject to problems of market failure. Generally, incentives in imperfect markets, e.g. to innovate or invest, may not be consistent with social valuations.

One of the commonest explanations of the UK's move to deficit in the late 1980s was a reduction in previously binding borrowing constraints—especially against the equity in the housing stock—leading to a surge in consumption and a payments deficit (see Muellbauer and Murphy, 1990). Market failure and preference failure are appealed to by those worried by the phenomenon: thus individual borrowers are accused of irresponsibility, financial institutions of imprudence and over-enthusiastic lending policies, and governments of regulatory failure.

But whilst a case for intervention can be made on these grounds, care obviously needs to be exercised in the design of possible responses. Government intervention to improve the intertemporal allocation of resources has not been notably successful in the past and some types of policy could make things worse. Moreover, there is usually a case for trying to correct market failures as close to the source of the failure as possible. Thus, even if the case for concern were established, it is not at all obvious that it is helpful to see these issues as balance of payments issues. The external position is better seen as a symptom of a more general set of problems.

Closer to macroeconomic concerns over the external payments position are co-ordination problems—which underlie much of the case for macroeconomic policy. Individual actions may not be consistent with macroeconomic constraints. The issues come out most clearly in the case of a closed economy where, by definition, there can be no external borrowing or lending and domestic savings must balance domestic investment. The flow constraints mean that an increase (say) in investment would have to be balanced by a decrease in consumption.

In the closed economy some means must be found to equate savings and investment flows and to produce macroeconomic consistency. The means could be automatic, as in forward-looking new classical approaches; in practice governments usually give a hand, using the normal instruments of macroeconomic policy. Thus, for example, if savings were to increase

this could call for a reduction in interest rates to ensure that investment rose equivalently. Other policy responses would be possible: the increase in saving could be balanced by public deficits, so that the economy's total savings were unchanged, or more direct microeconomic policies could be directed towards lowering savings or stimulating investment.

If the UK were a closed economy, an increase in consumption would have to be balanced domestically. If it were not, excess demand and inflation would ensue. The rise in interest rates necessary to sufficiently stimulate savings or cut investment would be large and could conflict with longer-term objectives. A fiscal response would imply a large budget surplus whilst the financial stock adjustment was proceeding. Clearly, the freedom of the UK as an open economy to run an external deficit and to avoid much of the inflationary pressure is a major benefit to policy-makers and to the economy—so long as it does not disguise the underlying problems.

Along similar lines, one way of characterizing the Japanese situation is to argue that as the growth trend slowed, investment ratios fell whilst the savings propensity did not. In a closed economy this could cause major difficulties. Interest rates might have to be very low. Indeed, given that neither savings nor investment seem very sensitive to interest rate changes, a low enough interest rate might not be feasible. The alternative of large fiscal deficits (balancing the excess in private sector savings) might not appear attractive to the Japanese authorities. Direct action to lower savings, or to stimulate investment to the required extent, may also be difficult or conflict with other objectives. Clearly, from the Japanese point of view, it may be highly desirable to run an external surplus as the counterpart to the internal imbalance. The same appeared to be true of Germany, though the situation has markedly changed due to the needs of reconstruction in the East.

The potential benefits of international integration are clear. Not only does it ease the macroeconomic consistency conditions, it also means that investors have access to international rather than just domestic capital markets. None of this means, however, that governments no longer have legitimate cause for concern over trends in domestic savings and investment. If they do, it implies concern over the external position as well. Moreover, the easing of the macroeconomic problem may be more apparent than real. The need for consistency reappears at the international level. It is easy to imagine situations which simply do not add up, such as two 'Japans' both of which would like to run surpluses relative to each other. There may be a need for international co-ordination, not just on balance of payments issues but on other aspects of policy as well (see Artis, 1989, for a discussion).

3. Competitiveness

We have noted that the traditional diagnosis of a balance of payments problem was as a conflict between internal and external balance, involving a need to change competitiveness. It is still true that the most difficult policy problems arise when the underlying issue is a relative price or cost problem of this type.

The potential policy problems come out in a particularly clear way if we consider the regional analogy to a competitiveness problem. For a region within a common currency area, the external balance is not going to appear as a problem in its own right. Instead, an uncompetitive region (frequently industrial problem regions) would normally have high unemployment. Essentially, a high wage in relation to productivity feeds through automatically to low output and demand. Clearly in practice there may be adverse effects on the dynamics of regional development as well.

Such an uncompetitive region can be contrasted with regions which are simply poor and where low productivity is balanced by low wage costs as well. Here competitiveness is not the issue—though such regions may well pose policy problems if convergence of income levels is desired. Indeed, we can usefully put the two cases together and imagine, for the uncompetitive region, a response to high unemployment in terms of lower regional wages, which would improve competitiveness, stimulate exports and curtail imports, and generally have favourable effects on both the demand and supply sides of the picture. Such a response amounts, as we discuss in the next section, to a kind of regional devaluation.

Regions can become, and remain, uncompetitive in many ways. An adverse shock combined with real wage resistance is one possibility. One of the most important practical cases, however, is national wage scales and nationally centralized bargaining structures, which in the extreme case would lead to equalization of nominal wages between regions, despite variations in the underlying productivity position. One of the most important questions facing countries joining a common currency area is whether wage differentials would thereby be narrowed, hampering regional adjustment. This is a crucial question for Germany, where a large wage differential between East and West is essential to promote adjustment and convergence.

Clearly, a region could support living standards and wages in the short term by borrowing. This is the case for both uncompetitive and for poor regions; in the former case employment would be supported as well. If it is consumption that is being supported, however, the likely consequence is a worsening over the longer term. Borrowing for investment, assuming it is productive and profitable, looks, from a common sense point of view, quite different, and could be a route to regional development and convergence. Unfortunately, it is the less likely to happen, the more uncompetitive the regional economy.

III. The adjustment problem

This section looks at how balance of payments positions may adjust over time, or be made to adjust by policy action. The stock/flow identity between the current account flows and changes in the stock of external assets tells us that both sides of the picture have to change together. Nevertheless, more partial analyses are still useful so long as other adjustments and other policies are kept in mind. Thus, for example, if a change in competitiveness is to affect the external balance, domestic absorption in relation to production must change as well.

1. Competitiveness and exchange rate changes

The traditional way in which countries or regions adjust relative to each other is by changes in competitiveness, especially by changes in relative wage costs. The adjustment may come either domestically, by falls in wages and prices (and changes in the relationship between them), or it may be induced by exchange rate depreciation (if, that is, the depreciation sticks). In either case there is a real devaluation, which is part of the adjustment process.

Traditional approaches to this aspect of adjustment tend to assume perfect competition and analyse responses in terms of supply and demand elasticities. The assumption of perfectly competitive markets is highly unrealistic: forecasters know that the response elasticities built into their equations summarize a highly complex set of adjustments of prices and quantities in imperfect markets as well as involving longer-term adjustments, such as decisions over sourcing or location.

Venables (1990) uses a simple model to show just

how much difference a more realistic modelling strategy can make. Looking at a situation where markets are segmented between different countries, so that firms supplying imports can 'price to market', a number of real world phenomena are illuminated. The pass-through to prices from competitiveness changes will be less than under competitive assumptions. Sunk costs mean, for example, that the number of firms supplying imports may not vary over a range of values for relative costs. When uncertainty is introduced, hysteresis effects appear, even if firms are regarded as risk neutral. In practical terms, this means that fluctuations in relative competitiveness (e.g. due to exchange rate changes) will not leave the industrial structure (and hence prices and quantities) unchanged, even if competitiveness returns to the starting point. The path matters.

Informally, we can illustrate the potential importance of the kind of phenomena under discussion by referring to the experience of the United States and the UK, both of which went though periods when competitiveness deteriorated sharply due to exchange rate movements (the US in the mid-1980s and the UK in the early 1980s). One phenomenon, which is well attested, is that when the dollar fell back after mid-1985, the price response by firms supplying imports was attenuated: to remain in the market, margins were cut. One effect was that the inflationary effect of the dollar's fall was thereby weakened. But so too was the effect on trade volumes. It is also suggested that, for both the US and the UK, the periods of overvaluation led to firms leaving the market altogether or locating elsewhere. The costs of re-entry, the sunk costs, would mean that these decisions would not be reversed when competitiveness reverted to more normal levels. Again the history of the economy affects its state. Short-term impacts, such as temporary overvaluation, have long-term consequences.

Path dependence (or hysteresis) is usually discussed in terms of the labour market (Jenkinson, 1987). Morris and Sinclair (1985), discussing the origins and persistence of UK unemployment in the mid-1980s, argue that the exchange rate rise of the early 1980s (which led to a serious loss of competitiveness) interacted with sluggish responses in the labour market to raise unemployment. The main downward effect on inflation came, however, whilst unemployment was increasing and firms were under heavy pressure or going bankrupt. As unemployment stabilized, so did wage inflation. The high level of unemployment, especially long-term unemployment, had little effect on the wage bargaining process. If this is so, there is an unhappy implication. The forces tending, over time, to correct a situation where competitiveness is a problem may be rather weak. High unemployment may have rather little effect on wages and international wage differentials.

Turning to the exchange rate, the question arises as to the effects of exchange rate changes on competitiveness, and the potential use of the exchange rate as a policy weapon to affect the external balance of trade. One obvious point is that exchange rate changes, especially for the US and the UK, have in fact led to major changes in the real exchange rate, or competitiveness. The floating exchange rate system has not operated as the optimists hoped. Fluctuations in the real exchange rate may have adverse consequences for international resource allocation with risk-averse firms being unwilling to take advantage of cost differentials. The hysteresis effects referred to above also mean that fluctuations can be highly damaging. Arguments such as these account in large part for moves towards international co-ordination of exchange rates and for moves towards regimes such as the Exchange Rate Mechanism of the EMS, or monetary union.

The implied importance of nominal exchange rate changes is somewhat at odds with prevailing views about devaluation as a potential remedy for competitiveness problems. Here, the usual argument is that devaluation has no longer-term effect on competitiveness, and is simply inflationary. There are, in fact, two cases where devaluation would not be useful. One is where domestic wages and prices will adjust anyway—as in competitive models. The other is where real wage resistance is strong, in which case it would not work. Since exchange rate changes and domestic flexibility of wages and prices are substitutes for one another, the specific case where exchange rate changes are potentially useful is where they help to co-ordinate a change that would otherwise be resisted. The classic, but extreme, case is where nominal but not real wages are fixed, as in textbook Keynesian models.

In practice it may be easier to co-ordinate international wage changes by exchange rate movements rather than domestic adjustment. Concern over inflation, however, means that such a policy would need to be sparingly used if the credibility of the strategy were not to be lost. These points clearly bear on the issue of what degree of sovereignty is lost when countries join currency unions and give up the option of nominal exchange rate change. They are further discussed in the next section.

Regional adjustment

When people think of regional adjustment, the mechanisms of adjustment considered are broader than

Table 6.2. Hourly Compensation Costs for Production Workers in Manufacturing (US = 100 in each year[1])

	1975	1980	1985	1990	1994
Germany[2]	100	125	74	147	160
Sweden	113	127	74	140	110
Belgium	101	133	69	129	134
Netherlands	103	122	67	123	122
Italy	73	83	59	119	95
France	71	91	58	102	100
UK	53	77	48	85	80
Ireland	48	60	46	79	73[3]
Spain	41	60	36	76	67
Portugal	25	21	12	25	27
Japan	47	56	49	86	125
Asian NICs[4]	8	12	13	25	34

Notes: [1] All figures are converted into US $ before being expressed as index numbers.
[2] The former West Germany.
[3] 1993 Figures.
[4] Hong Kong, Korea, Singapore, Taiwan.
Source: US Department of Labor, *Monthly Review*, October 1995.

those usually discussed for inter-country adjustment. The basis of that discussion has typically been relative price and cost adjustments with immobile factors of production. For regional adjustment, the mobility of labour and of physical capital (the location decision of firms) are clearly crucial. They will become increasingly important at the international level as well, especially within an integrating Europe. Since these involve adjustment processes, much conventional economics is effectively useless.

Labour mobility as well as regional variations in pay can be fostered, but both can cause well known difficulties. For example, low pay can be associated with low regional demand for non-traded goods, poor infrastructure, and so on. Wage levels are, however, clearly important for the profitability of firms within the region and for the location decision of capital. In considering the latter factor, it is likely to be wage levels themselves that are important rather than the more usual indicators of competitiveness such as unit labour costs. The point is that mobile capital will normally bring with it techniques of production, management, and productivity levels which are comparable with international standards. A high productivity firm in a low average productivity/low wage area should be highly profitable.

Table 6.2 shows data for selected years on wages for production workers for major countries as well for selected newly industrializing countries (NICs). Some features stand out. Amongst the industrialized countries, there has been considerable convergence. In the mid-1980s, the high dollar clearly gave other countries

a considerable cost advantage; the decline since means that Northern Europe is a relatively expensive area in which to produce. The UK has substantially lower wages than, say, West Germany (balanced of course by lower average productivity). Parts of Southern Europe are still low wage areas. In terms of wage costs, the NICs still have an enormous locational advantage for high productivity firms—so long as transport costs are not too significant.

Adjustment issues are well illustrated by German Monetary Union. One of the contentious questions was over the conversion rate for the East German currency, which affected the initial wage differential between the two areas. At one-to-one, it was clear that much of East German industry would, given low productivity, be unable to compete, even though wages would be only some 40–50 per cent of West German levels. This was not, however, the only relevant factor. One issue was to mitigate somewhat the incentive to migration. The main longer-term question, however, was whether there would be sufficient incentive to invest and locate in the lower wage region. From this longer-term point of view, the wage differential could be argued to be adequate.

Consideration of such practical questions brings out the point that wage differentials, whilst important, are by no means the only factor to take into account, and policy towards wage differentials is certainly not the only (or even the preferred) policy that can affect the regional adjustment process. Whether a wage differential constitutes an incentive towards convergence depends upon a host of supply side factors—infrastructure, skills and training, cultural and political factors—as well, and policy towards these will be as important as policy towards the wage level.

2. Wealth and accumulation

The above has considered part of the adjustment process, that concerned with variations in competitiveness. As competitiveness varies, so too will other aspects of the economy, including the external balance. The simultaneities are important, and depend on circumstances. Thus, an improvement in competitiveness may lead to a change in the current account if the savings–investment balance changes—or is made to change. Or it could lead to a reduction in unemployment with little change in the external position. It is necessary to look briefly at the other side of the adjustment problem—the savings–investment side. How might that adjust?

In a closed economy, as noted, the flow constraint that savings must balance investment would, in many models, normally require co-ordination via the interest rate—the main intertemporal price. In practice, fiscal policy would usually also play a role. In a common currency area, the interest rate can no longer play that role: even for countries with a separate exchange rate, international financial integration means that interest rate policy is closely constrained. Does this mean that a country or region should simply accept whatever external position results from domestic savings and investment decisions? In section II it was suggested that there were a number of reasons, ranging from concern over international financing to microeconomic factors applying in the domestic economy, why a government or regional authority might still have grounds for concern. To focus on these concerns, it is useful to put aside the competitiveness problem—assuming, for example, that domestic wages and prices are sufficiently flexible for that not to be an issue.

One of the strands of the debate, which we consider first, is whether the external position should be regarded as self-adjusting. In turn, this bears on the important issue of whether there is a difference between public sector induced deficits and those that result from private sector behaviour. The basic argument as to why a private sector deficit should be self-adjusting is very simple: individuals and firms on the one side, and financial institutions on the other, can be relied on, it is asserted, to act in their own interests. Access to international financial markets should not make any difference. If the aggregate of individual decisions adds up to an external deficit, then so be it. If there is a large deficit at a particular time then, over time, wealth effects applying to individuals, as well as limits applied by financial institutions, would mean that the position would adjust, and adjust automatically, without policy intervention.

This argument, that private sector positions are self-adjusting is powerful. There is obviously a great deal in it. It is necessary to consider carefully, in terms of the potential problems outlined in section II where the grounds for intervention might lie.

At the international level, there are several. The first is international co-ordination. As in the closed economy, private sector positions may require co-ordination, which is particularly difficult at the international level since sovereign governments interact strategically. The second is financing. It is simply not true that private sector positions can never lead to international financing difficulties—as the debt crisis demonstrates. If they do, governments will normally be involved. Third, and closely related to financing difficulties, there may be consequences for the exchange rate. A feature of all these reasons for doubting

the efficacy of market adjustments is that there are domestic analogues. Financing difficulties can arise domestically—as occurred with the Savings and Loans institutions in the US. Excessive consumption would cause a problem in a closed economy just as much as in an open one—indeed, the problems posed would be the more acute without the international safety valve. The market failures, or the differences between social and private incentives and valuations, are general problems which may or may not manifest themselves in balance of payments or exchange rate difficulties.

Similar points apply to the public sector. There are no general reasons for supposing that public deficits are bad and private sector deficits good. Either could be a cause for concern and need correction, or neither. There are, however, reasons for singling out the public sector position for special treatment.

Public borrowing, whether domestic or international, takes place on different terms from private sector borrowing. The public sector can borrow, in effect, against the collateral of its powers to tax. It is thus not limited to the same extent as private sector agents. This means that the public sector *could*, if it wanted to, be particularly irresponsible, since normal commercial criteria do not apply. This seems to be what many commentators have in mind when they single out a public sector position as a cause for special concern in discussion of the balance of payments. A public deficit could be the cause of a payments imbalance, and a cause for concern if, on wider grounds, the macroeconomic consequences in the present and in the future were judged undesirable. It would be particularly problematic if, as in the US, the underlying budgetary position were, in political terms, very hard to change. The complaint is then, fundamentally, a complaint about the government's macroeconomic stance.

But it is precisely these kinds of arguments which, turned on their head, constitute the case *for* fiscal policy. If, due to the diagnosis of market failure, or a difference between private and social valuations, a private sector position is judged undesirable, then the government has a legitimate case for acting in an offsetting or correcting way. It is because the government has some advantage in being able to take macroeconomic considerations into account and some advantage in terms of its power to tax and to take expenditure decisions on other than purely commercial criteria that fiscal policy makes a difference. As monetary policy comes, more and more, to be determined by international considerations, fiscal policy will need to play a major role in balancing supply and demand and in controlling the external position.

IV. Macroeconomic policy and economic integration

This section is particularly concerned with what happens to balance of payments issues and the macroeconomic policy options available as countries change from having separate currencies to being part of a common currency area. These questions bear on the debate over European Monetary Union, including the question of whether the UK should enter.

Obviously, the main change that occurs with monetary union is that a country gives up the option of changing the exchange rate and, because of financial integration, it gives up the possibility of an independent monetary policy. Interest rates are given to the country from outside. Effectively, a country takes on the characteristics of a region—though a region with considerable political power. Perhaps a better analogy is with a federal system where the states retain a policy-making role.

The main loss of sovereignty is the loss of the freedom to devalue the currency and the loss of independent control over the interest rate or monetary policy. Clearly, the exchange rate and the interest rate are not independent instruments outside monetary union—though the relationships between the two are complex. Thus what is being given up is some control over the nexus of the interest rate and exchange rate. The question is whether this loss of economic sovereignty matters.

Before considering this, however, there are some obvious gains from a currency union. One is the reduction of transaction costs within the wider area. Beyond this, other aspects of economic integration may be fostered, and these may be beneficial. A third, which relates closely to the balance of payments issue, is that financing is likely to become easier, because of the absence within the area of exchange rate risk. This is related to a fourth: exchange rate fluctuations can themselves be damaging and shocks can emanate from the exchange markets themselves; within a currency area, this source of disturbance should be eliminated. These are important gains.

Turning back to the potential losses, these, in line with the discussion in previous sections, really divide into two areas. The first is the loss of an instrument of control over aggregate spending and the balance between savings and investment in the domestic economy. Since it has been argued that countries need to concern themselves with these issues even in the absence of financing problems, the loss of such an instrument is potentially serious. The second

is competitiveness. If exchange rate changes can help in adjusting competitiveness to appropriate levels, the loss of this possibility could also be serious.

In practice, as economies integrate internationally, both these aspects of sovereignty are more apparent than real. Moreover, interactions between the exchange rate and monetary policy mean that the use of these instruments to control the economy may be particularly unfavourable from a longer-term or intertemporal point of view. For convenience, the argument is couched initially in terms of an economy which suffers simultaneously from problems of excessive spending and a lack of international competitiveness (a situation, many would argue, that pertained in the UK in the late 1980s).

Taking the problem of competitiveness first, the limitations on the use of depreciation to improve competitiveness have already been discussed. More generally, exchange rate policy interacts with inflation policy: an exchange rate target, or, in the extreme, an irrevocably fixed exchange rate under monetary union, is a substitute for domestic 'nominal anchors' such as the money supply or nominal GDP. Seen in this way, the problem with a policy of exchange rate depreciation to improve competitiveness is not only that it might not work—e.g. because of real wage resistance—but also that counter inflation strategy, in the form of interest rates or the money supply, would have to be relaxed in order to induce the exchange rate depreciation in the first place. The credibility of the strategy would be lost. It is precisely because of the close interconnection between domestic monetary policy and the exchange rate that an exchange rate target is a possible substitute for domestic policy. In the extreme case of monetary union, there is no domestic monetary freedom at all.

Short of monetary union, in a world of uncertainty and shocks, the interconnections between the exchange rate and domestic monetary conditions will not be so neat. It is still true, however, that a policy of depreciation will normally require a relaxation of domestic policy to bring it about. Thus, given a domestic strategy against inflation, exchange rate options are closely constrained. They amount really to the freedom to allow exchange rate movements (e.g. in response to differential shocks) to occur, given domestic policy to control inflation. This may be important in some circumstances. For example, in the case of an oil shock which, within Europe, would affect the UK differentially, exchange rate flexibility could help adjustment of the real relative exchange rate. Against this, however, exchange rate flexibility introduces the possibility of shocks emanating from the exchange markets themselves, which experience suggests can be damaging. Thus, sovereignty in this area amounts in practice to little more than the freedom to allow exchange rate fluctuations given a strategy against inflation, a freedom which is useful in the presence of differential shocks, but potentially damaging if exchange markets themselves lead to instability.

A potentially much more serious loss of sovereignty with monetary union arises from the loss of the interest rate as a policy weapon for controlling domestic spending. Let us assume that, due to some domestic failure, consumption spending rises a good deal faster than seems prudent. In a closed economy, excess demand and inflation would ensue. Interest rate increases would be one way of checking the excess spending. It is already apparent that this response may not be an optimal response: in practice, an increase in interest rates would tend to check investment as well. The best response would be to do something about consumption itself by correcting the distortions that led to the problem arising in the first place. More formally, the macroeconomic response, whilst it improves the short term, may worsen the economy in the longer term.

In an open economy with a separate exchange rate, one of the main ways in which a higher interest rate would affect the economy would be through a rise in the nominal exchange rate, which would be helpful in pulling down inflation, but at the expense of, at least a temporary, deterioration in international competitiveness. Particularly for economies facing problems of inflationary pressure and competitiveness, the short-term responses could have extremely adverse longer-term consequences for the supply side and productivity, especially if hysteresis effects are important. Although the balance of payments would improve in the short term, the underlying situation would worsen.

If this is what the sovereignty gain due to having a separate exchange rate actually amounts to, it may seem that the loss of sovereignty due to entering a common currency area is not much of a loss. It would prevent policy responses which were undesirable anyway. That is, however, too simple a view. Mistakes and distortions will, from time to time, arise, and need correction. If the interest rates and exchange rates cannot be manipulated (as in a common currency area) what alternative policies are available to check a runaway inflationary boom?

There are two obvious candidates: quantitative influences on financial behaviour (such as credit restraints) and fiscal policy. We concentrate on the latter: differential credit controls would become increasingly difficult to operate as financial integration proceeded. Fiscal policy, however, might still be used to slow a consumer boom. But can fiscal policy work to slow consumer spending? It is necessary to take on directly the neo-Ricardian proposition that an increase in taxes, balanced by, in this case, a decrease in future tax liabilities due to the retirement of government debt, would have no effect on forward-looking, rational households. One obvious response is that there is little empirical evidence in favour of this invariance proposition: empirical studies continue to suggest that fiscal policy has powerful short-run effects.

A more theoretical argument would go as follows. In a situation where consumption spending rises excessively due to financial liberalization, the likely reason is the reduction of previously binding liquidity constraints. Assuming that constraints nevertheless continue, tax increases in the present (even if balanced by lower anticipated future taxes) re-introduce restraints in the short term and help to control the expansion of spending. The increased liquidity of consumers' wealth, due to financial liberalization, needs to be balanced, via budget surpluses, by a reduction in the stock of national debt, which can also be seen as providing liquidity. In the best known version of the neo-Ricardian story—Barro (1974)—it is explicitly assumed that the public sector has no advantage in the provision of liquidity. The above argument for fiscal effects explicitly assumes that it has.

Thus, a fiscal offset to a consumer boom can be seen as a response which is quite close to the underlying problem. If well designed, it might seem much closer and less damaging from a longer-term point of view than a monetary and exchange rate response. It is, of course, still second best and not as good as a response which tackles the distortions and failures directly.

The upshot of these arguments is that the effective loss of sovereignty due to joining a common currency area, in the situation described, is small. For an inflation-prone country, it would be extremely difficult to improve competitiveness by depreciation anyway. However, competitiveness may be a real problem requiring real solutions such as improvements in the labour market and other aspects of the supply side. The use of the exchange rate/interest rate mechanism to control excessive consumption looks highly damaging to the supply side in the longer term, and would be dominated by policies to deal with distortions directly, and by fiscal action.

V. Balance of payments issues

This section considers some current issues which relate directly and indirectly to the balance of payments.

Apart from the importance of the issues themselves, the intention is to illustrate some of the main points raised in previous sections.

1. World imbalances

The oil price rises of 1973–4 and of the late 1970s imposed an exogenous swing on the external balance of payments position of importing countries. The impacts were unfamiliar (or at least the first one was) in many respects. From a balance of payments point of view, they were unfamiliar in that the deterioration for importing countries (which amounted, *ex ante*, to about 2–3 per cent of their GDP) was effectively imposed from outside. Balance of payments deficits had, until then, been seen as domestic issues, arising from domestic causes.

The first issue raised is effectively one of international co-ordination. It was important that there should be no scramble to improve the external account arising from inconsistent national targets; there was a danger of trade wars and competitive deflations reminiscent of the inter-war years. Co-operation in this respect was only partially successful. One of the reasons for this was surely that the domestic implications of an imposed external swing were unpalatable. A move to external deficit implies by identity a move towards deficit by some domestic sector or sectors—households, companies, or government. Since it was unlikely in the circumstances (mid-1970s or early 1980s) that either households or companies would conveniently want to increase their borrowing, or lower their accumulation of assets (effectively by borrowing from OPEC) the tendency almost everywhere was for government deficits to increase. In fact, government deficits rose further than implied by the OPEC surplus or the external impact as financial accumulation by households increased, and borrowing by companies decreased.

As is well known, the imbalances of the 1970s were accommodated, not by industrial countries, but by large-scale borrowing by third world countries, especially in South America. There was no financing problem at the time. Some of the arguments used then to justify complacency were remarkably similar to those used more recently to justify complacency over the G3 imbalances. This is not to suggest that the situations are comparable, only that the arguments are. In the 1980s, the initial response was similar but rapidly changed with the onset of the debt crisis and a quick reduction in the OPEC surplus. As described by Holtham (1990), the feature of the 1980s was the rise in the internal and then external deficits in the United States, and the counterpart surpluses in Japan and Germany. Elsewhere, public deficit positions improved with the revival of investment in the late 1980s (in part a lagged response to oil price declines into 1986). In Japan and Germany private surpluses were balanced by external surpluses, rather than, as previously, by public deficits.

Much of the rise in the US external deficit can be explained by the policies followed and by relative growth rates. With a more balanced world economy in the last few years, the deficit has stabilized, declining as a proportion of US GDP. Naturally enough, this has focused attention on longer-term structural issues and consequences. It takes at least two to make an imbalance, and the explanations advanced to explain the US deficit and the Japanese surplus are strikingly different. In the US, the public sector deficit is the main counterpart, though low US private sector savings also contribute. In Japan, the main explanation is the high level of private savings, a feature of the economy for many decades.

For Japan, one set of explanations appeals to life-cycle phenomena, and an ageing population. This would suggest that accumulation of external assets makes sense for the Japanese and also that the surplus may be temporary. One problem with this is that other countries, including the US, face similar problems. What then accounts for the marked differences in response? An adequate explanation of differences in savings behaviour does not seem to exist, and whilst this is so, predictions of longer-term trends are extremely hazardous. By contrast with the savings side, explanations of the US budget deficit are all too easy, though prediction of the political response is not.

It is the uncertainty that needs to be stressed, and which gives grounds for continuing concern over the imbalances. Large changes are possible, indeed likely. Whatever they are, adjustments are bound to be painful and policy reactions uncertain. There are large systemic risks built into the system, which are increased by the degree of international financial intermediation that is necessary to support it. Experience of the debt crisis, as well as domestic financial problems (such as the fiasco over the Savings and Loans institutions in the US), suggests that financial markets are highly vulnerable to systemic risk.

2. European monetary integration

The balance of payments issues raised by moves towards monetary union in Europe are comprehensively discussed by Begg and Weale (1990). One of the most important features to bring out is that, for the foreseeable future, the central budget will remain small. At present the EU budget is of the order of 1 per cent of GDP, and structural funds amount to only one-third of a per cent of GDP. Thus an integrated Europe would be quite different from federal states, such as the US, Germany, or Australia.

This raises two issues which we concentrate on here. The first is, in effect, the question of regional convergence. The second is the question of regional stability, and the design of macroeconomic policy for individual countries within a common currency area.

Within countries, very substantial transfers between regions take place through the fiscal and public expenditure system: income disparities are reduced by some 40 per cent. It is the almost complete absence of such automatic transfers between countries that would distinguish the problems faced within a monetary union of EU countries from analogous problems within countries. Even within countries, it has proved very hard to bring about regional convergence. Thus, potentially, there could be a serious 'regional problem' between countries. The obvious counter to this kind of argument, which is often used as an argument against integration, is that sovereign countries face the problem of divergencies in incomes and productivity anyway. The question is then whether monetary union would make the process of adjustment easier or more difficult. The major reason why it might make it more difficult is the loss of the freedom to change relative wages via devaluation of the currency.

We have argued that the cases where wage rate adjustment would be promoted by exchange rate depreciation are limited. Indeed, the use of such policies is more and more constrained by the need to control inflation. In the particular case of inflation prone countries—the ones most likely to get into a position of uncompetitiveness—depreciation is particularly dangerous.

As countries integrate, growth in a particular region or country is likely to depend more and more on the supply side, and policies to promote growth will depend on policies to improve the supply side, including policies to foster needed wage adjustments. From this perspective, access to a wider capital market, is clearly beneficial. But though this suggests that the balance of payments would no longer be an issue, the underlying concerns remain. Thus no country or regional authority could be indifferent about the savings behaviour or the investment flows of the area they are concerned

with. It is for this reason that Begg and Weale (1990) suggest a savings target to replace the concern over the balance of payments as countries integrate and financing becomes easier.

One of the disturbing aspects of the present debate over European Monetary Union, is that the potential need for differential stabilization is barely discussed. One reason for this is, no doubt, that continental European countries did not, in the 1980s, experience widely differing trends. Policy design, however, is a matter of designing for contingencies. Consumer booms on the UK model, or for that matter, recessions, cannot be ruled out, especially as financial liberalization proceeds. This issue is more important than the longer-term co-ordination of fiscal positions, envisaged in the Delors Report, which we would see as secondary to a country by country concern over trends in savings and investment.

References

Allsopp, C. J., and Joshi, V. (1986), 'The International Debt Crisis', *Oxford Review of Economic Policy*, 2(1), i–xxxiii.

Artis, M. J. (1989), 'International Economic Policy Coordination: Theory and Practice', *Oxford Review of Economic Policy*, 5(3), 83–93.

Barro, R. (1974), 'Are Government Bonds Net Wealth?', *Journal of Political Economy*, 82(6).

Begg, I., and Weale, M. (1990), 'Monetary Integration and the Balance of Payments', *Oxford Review of Economic Policy*, 6(3), 68–81.

Bliss, C. (1986), 'The Rise and Fall of the Dollar', *Oxford Review of Economic Policy*, 2(1), 7–24.

—— (1990), 'Unbalanced Payments: Adjustment and Policy', *Oxford Review of Economic Policy*, 6(3), 40–8.

Davies, G. (1990), 'The Capital Account and the Sustainability of the UK Trade Deficit', *Oxford Review of Economic Policy*, 6(3), 28–39.

Holtham, G. (1990), 'World Current Account Balances', *Oxford Review of Economic Policy*, 6(3), 49–67.

Jenkinson, T. J. (1987), 'The Natural Rate of Unemployment: Does it Exist?', *Oxford Review of Economic Policy*, 3(3), 20–6.

Morris, D. J., and Sinclair, P. J. N. (1985), 'The Unemployment Problem in the 1980s', *Oxford Review of Economic Policy*, 1(2), 1–19.

Muellbauer, J., and Murphy, A. (1990), 'The UK Current Account Deficit', *Economic Policy*.

Venables, A. J. (1990), 'Microeconomic Implications of Exchange Rate Variations', *Oxford Review of Economic Policy*, 6(3), 18–27.

7

Consumer expenditure

JOHN MUELLBAUER

Nuffield College, Oxford[1]

I. Introduction

Consumer spending accounts typically for between half and two-thirds of total spending in the economy. The 1980s in the UK saw an almost continuous rise in the consumption-to-income ratio, peaking in 1987–8 at the highest levels seen in the last 40 years. The saving ratio of the personal sector, defined as one minus the ratio of consumer spending to personal disposable income, declined correspondingly.

The early 1990s saw a stunning reversal, with the personal-sector saving ratio in 1992, as measured by the latest data, at the highest level since 1980. The qualification 'as measured' is not an innocent one. As Hendry (1994) shows, there have been remarkably large revisions of data. In 1978, the Central Statistical Office (CSO) believed that, in the post-oil-shock recession, the 1975 saving ratio was 15.3 per cent, while in 1992 it told us firmly that the 1975 figure was 10.6 per cent. Nevertheless, it is incontrovertible that the 1980s saw an unprecedented consumer boom, while the early 1990s saw one of the sharpest spending reversals ever experienced in the UK.

Such behaviour was not confined to the UK. Berg (1994) shows how the Scandinavian countries went through a similar boom-to-bust cycle, with Denmark leading and Finland bringing up the rear. As in all the Scandinavian countries, the UK boom was accompanied by a massive rise in consumer debt relative to income, and by an asset-price boom in both the stock market and the housing market. Correspondingly, the slump in consumer expenditure was associated with a slump in asset values and a rise in housing-loan arrears, loan defaults, and repossessions to levels never previously experienced. In Norway, Sweden, and Finland, the banking system came close to collapse, having to be rescued by the respective governments. Even in the UK, as has subsequently been revealed, the Bank of England had to provide support of the order of £6 billion to keep a number of financial institutions afloat and to prevent knock-on effects destabilizing the whole system.

To their embarrassment, macroeconometric forecasters were all caught out by the consumer spending boom of the 1980s, and the Treasury, as most others, failed to predict the scale of the early 1990s slump (see Keith Church, Peter Smith, and Kenneth Wallis, 1994). Though real GNP is not a good welfare measure, there can be little doubt that the cumulative welfare losses resulting from policy failures from 1986 to 1992 need to be measured in tens of billions of pounds sterling.

Developing an understanding of this boom-to-bust cycle is part of the purpose of this survey. But there are other important puzzles concerning aggregate savings behaviour. From the mid-1970s to the mid-1980s, the US personal-sector saving rate underwent what appears to have been a secular decline. This has excited much controversy and many alternative hypotheses have been put forward to explain it. Indeed, the 1980s saw declining personal and national saving rates in many industrial countries, including Australia (see Lattimore, 1994) and Japan, though the relatively

First published in *Oxford Review of Economic Policy*, vol. 10, no. 2 (1994).

[1] I am grateful for helpful suggestions to Chris Allsopp, Janine Aron, Robin Burgess, Gavin Cameron, Angus Deaton, Matthew Ellman, David Hendry, Tim Jenkinson, and Lucy White and for longer-term research collaboration to Ralph Lattimore and, above all, to Anthony Murphy. I am grateful to the ESRC for research support under grants R00023 1184 and R00023 4954.

small scale of the Japanese decline in the face of an asset-price boom is a bit of a puzzle. This 'savings shortage' is widely suspected, alongside rising public-sector deficits, of having been a cause of the high levels reached by real interest rates in the 1980s.

'But I thought higher real interest rates make people save more,' you may be asking yourself. Here, it seems, causation went in the reverse direction. Indeed, the truth is almost certainly even more complex; the movement in the 1980s towards international and domestic financial liberalization in many countries raised both credit demand and interest rates, as credit rationing was used less and less to curtail demand.

The textbook idea that higher real interest rates raise the saving rate is, in any case, not very robustly borne out by empirical studies in a variety of countries over a variety of time periods. There is some tendency to find small negative influences of the real interest rate on consumption, but they are typically not very stable when re-estimated over different samples. This is another stylized fact that calls for an explanation.

Empirical models to explain the behaviour of consumption are called 'consumption functions', a term which once had an unambiguous meaning: relationships between consumer spending, income, assets, interest rates,[2] etc., such as those following pioneering work by Ando and Modigliani (1963), Ball and Drake (1964), Spiro (1962), and Stone (1964). At the micro level, such relationships were justified, at least loosely, as the solution to an optimization over an individual's life cycle, resulting in some version of the life-cycle (Modigliani and Brumberg, 1954, 1979) or permanent-income (Friedman, 1957) hypotheses—in other words a 'solved-out' consumption function.[3] Since 1978, another way of modelling consumption, the 'Euler equation' form, has come into use, as will be explained below.

The year 1978 proved to be a milestone for research on the aggregate consumption function. Empirical research of that time was, generally speaking, in a far from satisfactory state. The statistical problems of working with non-stationary time-series[4] data (such as consumption, income, and assets) were not really understood. In many countries, time-series data on household asset holdings were not readily available. Consistency with economic theory was problematic. And the treatment of expectations remained fairly *ad hoc*. Two papers were published in 1978 that proved to be key pointers for subsequent research. One was Davidson, Hendry, Srba, and Yeo (1978), DHSY. Though similar lag structures had earlier been used, e.g. by Stone (1964), this effectively introduced the notion of an error-correction model, setting the scene for subsequent work on co-integration[5] of non-stationary time series by Engle and Granger (1987). Hendry (1994) re-examines DHSY and its extension by Hendry and von Ungern-Sternberg (1981) to incorporate a liquid-asset effect.

The other paper of that year (Hall, 1978) on the face of it eliminated all the other problems listed above in one fell swoop: by effectively analysing *changes* in consumption, it made the series stationary; it eliminated the problem of paucity of asset data by making such data redundant; it claimed to solve the problem of theory consistency by estimating directly, from aggregate data, the first-order condition or 'Euler equation', for an optimal intertemporal consumption decision by a 'representative consumer'; and, building on the 'rational expectations' revolution following Muth (1960), it adopted the rational-expectations approach, assuming efficient information processing by consumers taken to be at least as well informed as the econometricians studying their behaviour. Hall showed that, under certain assumptions, this approach had the apparently revolutionary implication that the best forecast of next period's consumption was this period's consumption. Many researchers, no doubt, asked themselves what the point was of laborious and difficult research on the traditional solved-out consumption function when, supposedly, it had no forecasting value.

Research on Euler consumption equations was further stimulated by the relatively early discovery that empirical evidence did *not* after all, accord with Hall's proposition that consumption changes could not be forecast. One particular form of this forecastability is the empirical sensitivity of consumption changes to income changes predicted on the basis of lagged information. This is the 'excess sensitivity' puzzle. Much of

---

[2] However, real interest-rate effects were often found insignificant, as just noted, and omitted.

[3] Useful reviews of this literature on aggregate 'solved-out' consumption functions can be found in Evans (1969) and Modigliani (1975).

[4] A non-stationary variable is one which shows no tendency to return to some mean value or some well-defined trend path. Its forecast variance tends to infinity as the forecast horizon tends to infinity. The opposite is true for variables which are stationary or stationary around a deterministic trend (see Granger, 1986)

[5] Co-integration refers to the existence of a stationary linear combination of two or more non-stationary variables. Granger and Engle demonstrated that, if series are cointegrated, they can be consistently represented by an error-correction model, capturing the short-run dynamics of adjustment towards a long-run equilibrium relationship. See Banerjee *et al.* (1993) for comprehensive reviews of the issues.

consumption research has been concerned with relaxing one or more of the key assumptions that underlie Hall's result to explain why consumption growth is, in practice, forecastable. In the process, many important issues concerning consumer behaviour have been illuminated, which have implications for the specification of solved-out consumption functions and for developing better macroeconomics. Another puzzle which has been much discussed is the 'excess smoothness' or Deaton (1987) paradox: if income data are strongly persistent, then under the rational-expectations permanent-income hypothesis (REPIH), consumption should be as volatile as income. The fact that it is not is the paradox.

Partly because of this evidence from aggregate time-series data, and partly because of gathering evidence from household survey data, some reinterpretation has been taking place of one of the corner-stones of post-war economics, the life-cycle model of Modigliani and Brumberg (1954, 1979). The basic idea, based on intertemporal utility maximization,[6] is that households try to smooth consumption over time so that workers save in order to spend in retirement. One useful macro prediction from the theory is its emphasis on the role of asset or wealth effects on consumption; another is the association of high aggregate saving rates with population and income growth. Note that the latter result is due to aggregation rather than being implied by the behaviour of some representative agent. An economy with population growth will have more savers relative to dissavers than one with a static population, and so will have a higher aggregate saving rate. This is essentially an age structure effect. Similarly, one with economic growth will have dissavers living on assets accumulated out of lower incomes than current workers are earning. On the face of it, this should result in a higher aggregate saving rate. However, higher *expected* real *per capita* income growth for given individuals will, if borrowing is possible for the young, result in higher consumption for

them, which could partially offset, or even more than offset, for very high growth rates, the higher saving rate which comes from the bigger earning power of the working population.[7] In practice, credit constraints will eliminate the latter possibility: the young without collateral typically cannot borrow. Indeed, cross-country empirical evidence does show a positive correlation of saving rates with income growth and with population growth or a young age structure (Modigliani, 1990; Deaton, 1992, ch. 2.1).[8] This evidence is thus consistent with the life-cycle hypothesis holding for the majority, but not for a young, credit-constrained, minority.

Powerful evidence against the life-cycle theory, at least in its simplest form, has been building up from household survey data. Cross-section evidence suggests that consumption tends to follow the hump shape of income over the life cycle more closely than life-cycle theory implies and that the old do not dissave on the scale one might have expected. This suggests that simple life-cycle theory is in considerable need of modification.

This article uses a combination of theory and econometrics to explain the boom–bust cycles in the UK and Scandinavia and to illuminate the secular decline in international saving ratios and the more general consumption puzzles which have been discussed. Given the flowering of research in this area, it is hard to separate a discussion of the policy implications from a review of the theory and evidence. This article combines both.

The structure of the article builds up the economic theory of consumer expenditure piece by piece, integrating each element with relevant empirical evidence. The sequence begins in section II with an analysis of consumption decisions under point (i.e. non-probabilistic) expectations in a two-period model, extending to the multi-period model. In section III, probabilistic income expectations, in which households associate different probabilities with different possible future income levels, are introduced, first in a linear rational-expectations framework. Here Hall's Euler equation model is explained as the solved-out version of the linear REPIH. The 'excess sensitivity' and 'excess smoothness' puzzles are also discussed here.

---

[6] Intertemporal utility maximization also underlies Friedman's (1957) permanent-income hypothesis (PIH). Formally speaking, permanent income is defined as that income which, if sustained over the life cycle, has the same discounted present value as the actually expected income stream. Friedman focused not on wealth, retirement, and rate-of-growth effects, but on the measurement links between current and permanent income which explained why, in short-run data, the marginal propensity to consume out of income could be substantially less than one. It is often forgotten that Friedman's consumers have much shorter horizons than do those of Modigliani and Brumberg. Friedman (1963) claims that consumers behave in practice as if they applied 33 per cent per annum discount rates to future expected income, giving an effective horizon of only around three years.

[7] Note that growth which raises the incomes of cohorts born later but does not affect income profiles in individual life cycles, does not have such an expectations effect.

[8] However, it is not clear that, for income growth, the causal mechanism is unique: a higher saving rate may also generate more investment via lower interest rates, if international capital markets are imperfect, and hence more growth, as suggested by some endogenous growth theories.

The linear REPIH, as exemplified in the Hall model, rests on an extreme set of assumptions. Section IV discusses the consequences of relaxing these one by one. It begins by introducing precautionary behaviour in the face of risk. Credit constraints, interactions between consumption and leisure, and habits and durable goods are discussed next. The rationality of expectations is reconsidered and asset effects, including those of illiquid assets, on consumption are analysed. Aggregation, i.e. how one gets from micro theory to relationships that hold, at least approximately, for macro data, is considered. Throughout, the implications of the analysis for the forms both of the Euler equation and the solved-out consumption function are discussed.

After this consideration of all the important elements of a more general model, section V discusses data problems and gives an empirical example of an approximately theory-consistent solved-out aggregate consumption function. The principle of general to specific modelling, i.e. beginning with a general maintained hypothesis and testing a series of simplifications to reduce it to a parsimonious model consistent with the data, is fairly widely accepted as ideal econometric practice (Gilbert, 1986). Therefore it is desirable, whether under the Euler approach or the solved-out consumption function approach, to take a general model as the starting point for empirical work. Theory consistency is also a necessity for model building: there are never enough data and, unless many possibilities are excluded by prior theoretical considerations, confusion reigns. However, precise theory consistency and theoretical generality are mutually exclusive, whether in the context of the Euler or the solved-out consumption function approach. Creative approximations are essential to make progress.

The reasons for emphasizing the solved-out consumption function in section V are several. One is that the Euler approach has been given excellent coverage in Angus Deaton's 1992 book. Another is that the solved-out approach has been excessively dominated in macroeconometric research in the last 15 years, in North America if not in Europe, by the Euler approach and a small step is taken here to restore the balance. Each approach has merits and those claimed for the Euler approach have already been mentioned. Two important advantages of the solved-out over the Euler approach are as follows. First, the Euler, unlike the solved-out, approach involves quasi-differencing which eliminates from the data important long-run information, for example information on assets but also on consumption and income. DHSY (1978) and the

co-integration literature have taught us that this is often unwise. Second, consider how the Euler approach is to be useful for policy analysis, for example, for simulating the medium-term impact on consumption of a change in tax policy: one needs to solve a set of intertemporal efficiency (i.e. Euler) conditions forward, combining them with the period-to-period budget constraints, altered to take account of the change in tax policy and incorporating expectational assumptions. But this is just what the solved-out consumption function is designed to do. For policy analysis, both approaches need to make their expectational models explicit.

Each approach necessarily involves approximations, particularly when dealing with aggregation, uncertainty, and credit constraints, or other non-linearities in budget constraints. The simplifying assumptions usually made in the Euler approach are not necessarily superior, partly because, as we shall see, the aggregation problem has been neglected in much of the Euler literature.

Finally, section VI draws conclusions and considers key areas for future research.

II. Consumption over the life cycle

The context here is one of point expectations, i.e. the individual believes that next period's income will take a particular value rather than a range of values, each with an associated probability. The alternative of probabilistic or stochastic income expectations will be adopted in later sections. Part 1 of this section considers the simplest two-period consumption-choice problem. We begin with the period-to-period budget constraints which, given lack of money illusion, can be expressed in real terms. After reminding the reader of the distinction between real and nominal interest rates, we explain why conventionally defined saving rates are likely to rise with inflation even when the budget constraint in real terms is unaltered, an issue of some practical importance. Part 1 then considers both 'Euler equation' and 'solved-out' consumption functions resulting from utility maximization. The former is just the intertemporal efficiency condition linking consumption now with expected consumption next period. Research on Euler equation consumption functions has dominated the literature of the last 15 years, especially in North America. The solved-out

---

Box 7.1.   Intertemporal Budget Constraints

In nominal, i.e. current price, terms the period-to-period budget constraints are

$$p_1 c_1 + p_1 A_1 = p_1 y_1 + p_0 A_0 (1 + r_0{}^*) \qquad \text{and} \qquad p_2{}^e c_2 + p_2{}^e A_2 = p_2{}^e y_2{}^e + p_1 A_1 (1 + r_1{}^*) \qquad (1)$$

where the 'e' superscript indicates expectations.

This pair of nominal, period-to-period budget constraints is easily converted to real, i.e. constant price, terms by dividing throughout by p_1 and $p_2{}^e$ respectively. We then find

$$c_1 + A_1 = y_1 + A_0 (1 + r_0) \qquad \text{and} \qquad c_2 + A_2 = y_2{}^e + A_1 (1 + r_1) \qquad (2)$$

where r is the 'real' interest rate. The period 1 real interest rate is thus defined by

$$1 + r_1 = p_1 (1 + r_1{}^*)/p_2{}^e. \qquad (3)$$

The well-known conclusion that the real interest rate is approximately equal to the nominal rate minus expected inflation is easily established.[10] When end of period 2 assets, A_2, are zero because life is not expected to extend beyond, the full discounted present value form of the life-cycle budget constraint is derived from (2) by eliminating A_1:

$$c_1 + c_2/(1 + r_1) = A_0 (1 + r_0) + y_1 + y_2{}^e/(1 + r_1) \equiv W_1 \qquad (4)$$

which defines life-cycle wealth, W_1.

---

consumption function is the result of combining all the available intertemporal efficiency conditions with the life-time budget constraint to derive current consumption as a function of initial assets, current income, future income, and the real interest rate. This was the dominant paradigm for consumption research before the 1980s and also of the UK models surveyed by Church *et al.* (1994). Part 1 concludes by considering why the effect of the real interest rate on consumption is likely to be unstable and is even indeterminate in sign, an important issue for empirical work.

Part 2 considers the multi-period extension of the two-period model. We explain why the marginal propensity to consume (MPC) out of assets varies with the time horizon and so with age. Given the typical hump profile of income, the MPC out of income will be lowest just before income declines with retirement. Shifts in demography, as well as shifts in the distribution of income and assets over consumers of different ages, will then affect aggregate consumption, even where aggregate income and aggregate assets are held constant. We will also examine the spending implications of variations in needs over the life cycle and of the bequest motive.

1. The two-period case

We assume that consumers do not suffer from money illusion and are concerned with consumption, c, non-property income, y, and assets, A, all measured in constant prices, i.e. in real terms. To convert to current prices, we multiply by the price index p.[9] We can then write down the period-to-period budget constraints in nominal terms. These constraints just say that the net change in assets, $(p_1 A_1 - p_0 A_0)$, is the difference between income (including property income coming from asset ownership), $(p_1 y_1 + r_0{}^* p_0 A_0)$, and consumption expenditure, $p_1 c_1$, where $r_0{}^*$ is the nominal rate of return on assets held at the end of the previous period. This nominal rate of return includes any dividend or interest income from these assets as well as capital appreciation.

Box 7.1, on intertemporal budget constraints, shows the relationship between period-to-period budget constraints in current and in constant prices.

---

[9] See Deaton and Muellbauer (1980, chs 5 and 7), for discussion of the index number and aggregation issues involved in treating a composite bundle of goods and the corresponding price vector as scalar indices.

[10] Take logs of (3) so that $\ln(1 + r_1) = \ln(1 + r_1{}^*) - \ln(p_2{}^e/p_1)$. Then $r_1 \approx r_1{}^* - \Delta \ln p_2{}^e$, using the approximation $\ln(1 + x) \approx x$, for small x.

It explains how to define the real rate of interest (which is approximately the nominal rate minus the expected inflation rate) and the full life-cycle budget constraint. This says that (assuming no bequests) the discounted present value of consumption equals the discounted present value of income plus the initial asset endowment from last period, augmented by the asset return.

The budget constraint (2) implicitly defines the Haig Simons-Hicks (see Haig, 1921; Simons, 1938; Hicks, 1946) concept of real income as that level of consumption which can be had while keeping the real asset level constant. Setting $A_1 = A_0$ in (2), this is $y_1 + r_0 A_0$, i.e. real disposable non-property income plus the real rate of interest times the real asset level. In contrast, real disposable income under national accounting conventions is defined by real disposable non-property income, y, plus a distorted measure of real property income, defined as the nominal return excluding capital gains on illiquid assets (but including imputed rental income from owner-occupied housing), all deflated by the current price level, p. Higher inflation for a given real rate of return implies a higher nominal rate. Since the rise in the nominal rate will be much sharper than the rise in the price level, p, it is clear that the deflated nominal return will rise and measured real disposable income will rise, even though, in real terms, nothing has altered. If, indeed, real consumption is unaltered, the conventional saving ratio defined as disposable income minus consumption divided by disposable income will rise. (See Lattimore, 1994, for an empirical illustration for Australia.) The practical implication is to take care in interpreting conventionally measured saving ratios and, wherever possible, to use real disposable non-property income in preference to conventional real disposable income in modelling consumption.

Having considered the budget constraints, let us now turn to consumption decisions. Most of the consumption literature assumes intertemporally additive preferences in which life-time utility, U, is the sum of the sub-utilities of consumption in each period, discounted using the subjective discount rate, δ. The term δ reflects impatience, which means attaching a lower weight to the utility of future consumption. The intertemporal optimization is set out in Box 7.2. This derives the Euler equation, i.e. the first-order condition which says that marginal utility now equals next period's marginal utility, weighted by the ratio of market and subjective discount terms. As we will see in section III, under probabilistic expectations we simply take the mathematical expectation of the same

expressions. Box 7.2 also shows how the solved-out consumption function is obtained.

The two functional forms for preferences which dominate the literature are also considered in the box. The first is quadratic preferences which results in marginal utility being a linear function of consumption and is the whole basis of the linear rational-expectations approach discussed in section III. The further, widely adopted, simplifying assumption of a subjective discount rate equal to the market real interest rate results in the specially simple result of complete stabilization of consumption so that the optimal consumption plan is for no changes. Then, as for the second form of preferences, consumption is proportional to life-cycle wealth defined in (4).

The second form of preferences results from the combination of homothetic[11] and additive preferences which gives the constant elasticity of substitution (CES) form. The elasticity of substitution, σ, measures how responsive is the ratio of consumption in the two periods to relative prices i.e. to $1/(1 + r_1)$. Indeed, taking logs of the Euler equation (12) in this case shows that the planned log-change in consumption is approximately equal to $\sigma(r_1 - \delta)$. Thus, the greater the elasticity of substitution, the greater the reduction in current consumption in order to substitute into higher future consumption when the real interest rate rises.

There are good reasons to believe that intertemporal substitutability is low: in other words, consumers prefer consumption to be very steady over time. This would suggest an elasticity of substitution certainly below unity and probably under one-half.

The special case where the elasticity of substitution is zero yields important insights into the effects one might expect the real interest rate to have in a solved-out consumption function. Consider the case where, in addition, future non-property income is expected to be the same as the present, resulting in the solved-out consumption function (15). Here it is clear that an increase in the real interest rate, r_1, *increases* current consumption: effectively, initial assets yield higher returns and consumption rises with the real rate of return. The bigger initial assets are, the stronger will be the positive effect of the interest-rate increase on consumption. If the elasticity of substitution is positive and assets zero, then the interest-rate increase has an unambiguously negative effect on

---

[11] Homotheticity has the implication that the ratios of consumption in different periods are independent of income or wealth under linear budget constraints.

Box 7.2. Intertemporal Optimization and Consumption Decisions

$$U = u(c_1) + \{1/(1+\delta)\}\, u(c_2) \qquad \text{where } u' > 0, u'' < 0, \text{ and } \delta \geq 0. \qquad (5)$$

We can optimize by substituting the period-to-period budget constraints (2) into the utility function (5), optimizing with respect to A_1. At the optimum we find that

$$\partial U/\partial A_1 = (\partial u/\partial c_1)(\partial c_1/\partial A_1) + (1/(1+\delta))(\partial u/\partial c_2)(\partial c_2/\partial A_1) = 0. \qquad (6)$$

We can also maximize (5) subject to (4), using the Lagrangian technique to obtain the same relationship between current marginal utility and next period's planned marginal utility as implied by (6):

$$\partial u/\partial c_1 = \{(1 + r_1)/(1+\delta)\}\, \partial u/\partial c_2. \qquad (7)$$

This is the Euler equation. Under more general probabilistic expectations, the Euler equation is identical to (7) but with the expectations operator, E, in front of the right-hand side.

The Case of Quadratic Preferences	The Case of CES Preferences
Here $$U = -[(1/2)(\beta - c_1)^2 + \{1/(1+\delta)\}(\beta - c_2)^2] \quad (8)$$ where β is the bliss point: $c < \beta$ for non-satiation. In this case, the Euler equation (7) takes the form $$\beta - c_1 = \{(1+r_1)/(1+\delta)\}(\beta - c_2). \quad (9)$$ Much of the literature takes the special case $r_1 = \delta$ when $$c_1 = c_2. \quad (10)$$ To derive the solved-out consumption function, substitute (10) into the life-cycle budget constraint (4) to find $$c_1 = W_1/\kappa_1 \quad (11)$$ where the inverse MPC out of assets, $\kappa_1 = 1 + 1/(1 + r_1)$. The more general case comes from substituting (9) into (4).	Here $U^{-\rho} = c_1^{-\rho} + \{1/(1+\delta)\}c_2^{-\rho}$ where the elasticity of substitution, $\sigma = 1/(1 + \rho)$, $\rho > -1$. In this case, the Euler equation (7) takes the form $$c_1^{-1/\sigma} = \{(1+r_1)/(1+\delta)\}\, c_2^{-1/\sigma}. \quad (12)$$ To derive the solved-out consumption function, take the $-\sigma$ power of (12) and combine with the life-cycle budget constraint (4) to find that (11) holds again, but where $$\kappa_1 = 1 + \{1/(1+\delta)\}^\sigma \{1/(1+r_1)\}^{1-\sigma}. \quad (13)$$ For small values of δ and r_1, $$\kappa_1 \approx 1 + [1/\{1 + \sigma\delta + (1-\sigma)r_1\}]. \quad (14)$$ Here the inverse MPC out of assets, κ_1, depends on the weighted average of the subjective discount rate, δ, and the market rate, r_1. In the special case where $\sigma = 0$ and $y_2^e = y_1$, the solved-out consumption function is[12] $$c_1 = [\{A_0(1 + r_0)\}/\{1 + \{1/(1 + r_1)\}\}] + y_1. \quad (15)$$

consumption.[13] With positive assets the overall effect is ambiguous if the elasticity of substitution is below

unity, but more likely to be negative the higher the elasticity. It is less likely to be positive with a low ratio of assets to income. Since the ratio of assets to income varies over the business cycle, one should not expect a stable aggregate real interest-rate effect. An elasticity of 0.5 or less would be consistent with a relatively small and unstable real interest-rate effect in aggregate consumption functions—which accords with most,

---

[12] Note that $\kappa_1 = 1 + \{1/(1 + r_1)\}$, which cancels with a similar term in the discounted present value of income, $y_1[1 + \{1/(1 + r_1)\}]$.

[13] It is also worth noting that when income is expected to grow, the negative effect of the real interest rate on consumption, via the discount factor applied to future income, is greater than when income is expected to be static or to fall.

though not all, empirical evidence, see Deaton (1992, ch. 2.2).

2. Age and consumption: the multi-period case

In this section, we consider the multi-period extension of the two-period model. We explain why the MPC out of assets varies with the time horizon and so with age. The implication is that demographic and distributional changes affect aggregate consumption. We will also examine the spending implications of variations in needs over the life cycle and of the bequest motive.

Maintaining the assumption of point (i.e. not probabilistic) expectations, the multi-period extension of (4) simply extends the utility function (5) and the discounted present value of non-property income, often termed 'human capital', to the horizon, T, instead of just to the next period. The Euler equations are the same as in the two-period case, while the solved-out form of the consumption function maintains the generic form (11). It is often assumed that the same real interest rate will persist across periods.[14]

We can now illustrate the effects of age on the marginal propensity to consume out of assets. For simplicity, suppose the subjective and the market discount rate coincide. Then, in the special case of static real income expectations (i.e. $y_2^e = y_1$),

$$c_1 = A_0(1 + r_0)/\kappa_1 + y_1. \qquad (16)$$

Under these assumptions, the effect of age on consumption will operate entirely through κ_1, the inverse MPC out of assets. To illustrate this, take a discount rate of 0.05. Then, by the previous footnote, with $a = 1/1.05$, making explicit the dependence of κ on the horizon, T, $\kappa(10) = 8.1$, $\kappa(20) = 13.1$, $\kappa(30) = 16.1$, $\kappa(40) = 18$, and, for an infinitely lived consumer, $\kappa(\infty) = 21$.

This has the immediate implication that older people, with their shorter time horizons, have a larger MPC out of assets than younger people. Note that the difference between the old and young MPCs rises as the average discount rate falls. For the aggregate consumption function the old/young difference is important because the age distribution of assets will then affect consumption. This issue is central to the question of the effects of *per capita* real income growth on aggregate savings behaviour. An economy with a higher

growth rate will, generally speaking, have a bigger share of wealth held by the young—though there is the proviso discussed above that, if the young were free of credit constraints, then with optimistic expectations regarding their future income they would borrow to smooth life-cycle consumption. Another qualification arises if the old are heavily invested in illiquid assets, such as equities and land, that may appreciate with economic growth, especially if this growth is partly unexpected.

Let us now relax the assumption of static real income expectations and replace it by the hump profile of earnings which is widely observed. The smooth nature of the aggregate profile reflects, among other things, variations over individuals in the age of retirement. Someone at the earnings peak in middle age can only expect lower incomes ahead and will have the lowest MPC out of current earnings. Deaton (1992) notes that the age profile of saving varies a good deal across countries. The single most common pattern appears to be raised saving rates in the years just before retirement. Thus, the expected drop in income at retirement clearly has an important influence on saving in the years immediately preceding.

Based on the above, we might expect something like the following pattern for the MPCs out of assets and income over the age profile:

	MPC out of assets	MPC out of non-property income
young	small	large
pre-retirement	medium	small
retired	large	medium

where the MPCs out of income exceed those out of assets, assuming expected incomes are positively related to current income.

Empirically, consumption follows income rather more closely over the life cycle than predicted by the simplest life-cycle models where consumers try to keep planned consumption constant, see Carroll and Summers (1991) and Deaton (1992, ch. 2.1). As we saw in section II.1, discounting future income provides part of the reason. As we will see in section IV, credit constraints and uncertainty are further important reasons for this empirical result. But there is another explanation which fits into the above model, without incorporating either credit restrictions or uncertainty: the variation of needs over the life cycle, especially related to child rearing. The reason why many young couples,

---

[14] The inverse MPC out of assets, for the CES case, is the straightforward generalization of (14), $\kappa_1 = 1 + a + a^2 + \ldots a^{i-1} = (1 - a^T)/(1 - a)$, where $a = 1/(1 + \sigma\delta + (1 - \sigma)r)$.

who expect higher earning opportunities later, may not want to borrow (even if they could), is that they also expect higher expenses, particularly if one of them stops or reduces work for several years to bring up children. Needs variations can be readily introduced into the CES utility function by dividing consumption in period s by the equivalence scale, m_s, where m_s increases with household needs such as the number and type of dependants present or expected to be present.[15] To some degree, given planned parenthood, these equivalence scales will reflect income expectations. The elasticity of substitution, σ, needs to be less than unity for this specification to give the intuitively plausible results one expects of higher needs at period s leading to higher consumption at s. Empirical estimates for a model of this kind have been obtained by Banks *et al.* (1993).

Planned parenthood should thus have some fairly sharp consequences for the savings behaviour of households. There is strong empirical evidence for the effect of the presence of young children on labour-force participation and hours of work by mothers. Lower income is typically associated with the higher needs following the arrival of young children. One would therefore expect to see low saving rates associated with large numbers of preschool children as households run down assets in the expectation, in most cases, of work being resumed when the children reach school age. Brenner *et al.* (1992) find evidence for this.

Depending upon the system of finance for higher education,[16] one may expect saving for college education to become important as children enter the teenage years, followed by a certain amount of running down of assets when the children reach college age. For many households this may leave a relatively short gap before retirement age in which to top up saving for retirement. Over the individual life cycle, this suggests quite sharp fluctuations in the saving rate. To pick up these effects in cross-section data, it would be necessary to condition quite carefully on the age and number of children as well as on the age and expected income profiles of the adults. On aggregate time-series data alone, the effects of such demographic variations are very difficult to estimate robustly.

The context of parenthood is the natural one in which to raise the issue of bequests. Economists usually assume that bequests are made because the utility

of the children's consumption enters the parents' utility function. The marginal utility to the parents of the assets bequeathed can be low for two reasons: one is that the parents apply a bigger subjective discount factor to the utility from consumption of the children, and the other is that the receiving household may have its own assets and income prospects. It would, therefore, be far from appropriate to treat the parents' behaviour as that of an infinitely lived household with geometrically declining subjective discount factors as assumed by Barro (1974). This view is consistent with empirical evidence against Barro's proposition of Ricardian equivalence, under which households do not respond to tax cuts now on the grounds that they or their descendants will have to pay higher taxes in the future. However, there are a number of other reasons to explain the failure of complete Ricardian equivalence, including uncertainty and credit constraints (see Tobin, 1980).

We examined above the effect of the size of the remaining horizon on the marginal propensity to consume out of wealth, $1/\kappa_1$. The bequest motive is likely to reduce this MPC for retired households, narrowing the gap between retired and younger households. The size of this gap is critical for the size of demographic, population, and income growth effects on aggregate consumption or saving.

III. The consumption function under stochastic income expectations

More plausible than point expectations about future income is the association of different probabilities with different income levels. In the last two decades, the Muthian rational expectations assumption has been the most popular way of treating stochastic expectations about such variables as income. For tractability, this has been largely used in the framework of linear models. We develop the implications of linear rational expectations models both for Euler equation and for solved-out consumption functions. In section III.1, we explain the Hall (1978) Euler equation consumption model. This implies that consumption changes are not forecastable. The result is derived with a quadratic utility function. This section also discusses the relationship between news about income and the consumption innovation or surprise. Section III.2 introduces the literature devoted to testing the Hall model and develops one particularly important aspect in more detail. This is the 'excess sensitivity', found in practice, of

---

[15] The equivalence scale thus says that, for higher needs, more consumption is needed to give the same utility (see Deaton and Muellbauer, 1980, ch. 8).

[16] Feldstein (1994) argues that, in the US, rules which reduce college scholarships if parents have significant assets act as a major disincentive to save.

consumption changes to anticipated income changes based on lagged information, when, under the Hall hypothesis, consumption changes should be independent of any lagged information. Section III.3 turns to the solved-out consumption function, in which consumption depends on initial assets and on the mathematical expectation of permanent non-property income: hence the 'rational-expectations permanent-income hypothesis'. We consider the issue of the 'persistence' of income. The more persistent is income, the more permanent are income innovations and the more responsive, other things being equal, is consumption to current income. If the stochastic process for income has a 'unit root', so that first differencing is required to make the process stationary, then (subject to a qualification discussed below) persistence is complete and the consumption innovation should be as large as the income innovation, making the variance of consumption changes at least as large as that of income changes. Aggregate time series evidence appears to support the unit root hypothesis. Yet the variance of consumption change (on seasonally adjusted data) in most countries is one-half or less of that of income changes: consumption is 'excessively smooth'. This is the excessive smoothness or Deaton paradox, see Deaton (1987), and is the subject of section III.4.

1. The Hall Euler equation model

Let us turn to the Hall (1978) result. Making expectations explicit in the Euler condition which holds in the multi-period as well as the two-period case, gives the result, see (7) in Box 7.2, that marginal utility in period 1 equals the mathematical expectation of the product of the ratio of discount factors $(1 + r_1)/(1 + \delta)$ and of the marginal utility in period 2.

Two problems arise in deriving an exact analytical solution for consumption, c_1. One is that, in general, the real interest rate is stochastic (i.e. imperfectly predictable) and the expectation of the product of two stochastic variables is not the product of the expectations. The other one is that, in general, the marginal utility is non-linear in consumption and then the expectation of next period's marginal utility of consumption is different from the marginal utility of expected consumption.

One of the few cases where an easy exact solution is possible is when preferences are quadratic, see (8) in Box 7.2, which implies that marginal utility is linear in c. The Euler equation is then given by the expectation of the linear equation (9). If the market interest rate

and the subjective discount rate coincide, current consumption equals the expectation of next period's consumption:

$$c_t = E_t(c_{t+1}) \tag{17}$$

where E_t denotes expectations given the information set at t. Since $c_{t+1} = E_t(c_{t+1}) + \varepsilon_{t+1}$ where ε_{t+1} is a surprise or innovation error, i.e. is non-forecastable, and shifting time back one period, we obtain

$$c_t = c_{t-1} + \varepsilon_t. \tag{18}$$

This is the Hall (1978) type stochastic Euler equation.[17] Equation (18) is an example of a 'martingale process'. If, in addition, the variance of the innovation, ε, is constant, (18) is called a 'random walk'.

This result has the apparently revolutionary implication that, under certain assumptions, the best forecast of next period's consumption is this period's. No other currently available information is of any use. The nine assumptions are:

  (i) no credit restrictions or other non-linearities in the budget constraint;
 (ii) a quadratic utility function additive over time;
(iii) no habits or adjustment costs;
(iv) non-durable goods;
 (v) the subjective discount rate, δ, the same across consumers and, for (17), equal to the market real interest rate, r;
(vi) no measurement errors or transitory shocks to consumption;
(vii) the coincidence of the frequency of consumers' decision making with the frequency of the data;
(viii) a constant real interest rate;
(ix) rational expectations.

Box 7.3 explains an important insight into the nature of the consumption innovation: it is equal to the news about permanent non-property income, defined as that income level which, sustained over the life cycle, has the same present value as the actually expected income stream. The restriction imposed on (18) by rational expectations is the orthogonality condition on the disturbance term. Income revisions must be orthogonal to (independent of) lagged available information, with the implication that consumption growth should not respond to variables anticipated last period.

---

[17] Hall himself argued for (18) as an approximation which holds if the subjective and market discount rates are close together and if consumption shocks are small relative to the level of consumption (see Hall, 1978, p. 987).

Box 7.3.   The Solved-out Consumption Function under the Rational Expectations Income Hypothesis

---

Consider the quadratic preferences case in Box 7.2, and assume that r is constant. Allowing stochastic income expectations simply puts the expectations operator, E, in front of the right-hand part of each of the expressions (9)–(11). Thus (11) implies

$$c_1 = E_1 W_1/\kappa_1 = [A_0(1+r) + y_1 + E_1\{ y_2/(1+r)\}] / [1 + \{1/(1+r)\}]. \tag{19}$$

The definition of permanent non-property income, y^P, given in the text implies, for the two-period case,

$$y_1^P[1 + \{1/(1+r)\}] = y_1 + E_1\{ y_2/(1+r)\}. \tag{20}$$

Hence

$$c_1 = E_1 W_1/\kappa_1 = A_0(1+r)/\kappa_1 + E_1 y_1^P. \tag{21}$$

Thus, under the REPIH, consumption equals permanent non-property income, y^P, plus a component proportional to assets, A_0. Under the T-period generalization, the outcome (21) is identical and the extensions of (19) and (20) are obvious. In the T-period case, a quasi-difference transformation[18] can be used to transform $c_t = E_t W_t/\kappa_t$ and establish that the error term, ε_t, in (18) is the news about permanent income.

---

2. Testing the Hall model and the excess sensitivity of consumption

Typically the orthogonality assumption has been tested by adding variables (dated $t-1$ and earlier) to a specification akin to (18) and testing for their significance. For example, Hall (1978) provided some empirical evidence that appeared to support the theory when he found that the change in consumption was independent of lagged income using standard significance levels, but not of lagged stock prices. Despite his formal rejection of the model, he noted that the additional information contained in stock prices contributed little to consumption growth and that, therefore, the REPIH was a reasonable approximation.

Most tests across different periods and various countries reveal violation of the orthogonality condition of the REPIH model. In the UK, Davidson and Hendry (1981) and Daly and Hadjimatheou (1981) found lagged income, consumption, and liquidity measures had significant explanatory power; Johnson (1983) found lagged unemployment significant in Australia. Part of the interest of the Davidson and Hendry study was the demonstration, via a Monte Carlo study, that a random walk in consumption could be quite hard to distinguish from an error-correction form of the

consumption–income relationship. In other words, the Hall model might well be accepted when, in fact, it did not hold.

Another strand in the testing of the Hall hypothesis (exemplified by the studies of Bilson, 1980; Flavin, 1981; Muellbauer, 1983; Blinder and Deaton, 1985) aimed to identify separately the consumer's reaction to anticipated and unanticipated income (and other) shocks by modelling both the income and consumption processes. Muellbauer (1983) found that lagged income did have explanatory power once the regressions took account of a break in the series in the early 1970s. Flavin (1981) also found that lagged income had explanatory power on US data.[19] Such findings led to the stylized fact that changes in consumption appeared to exhibit excess sensitivity to anticipated income (and other variables). It is important to emphasize that excess sensitivity is not to the innovation in income, but to the *anticipated* component of income, which should under rational expectations have already been incorporated into lagged consumption.

The modelling of joint income/consumption processes and the growing interest in integrated processes led to two other major insights. First, as observed by Mankiw and Shapiro (1985, 1986) some tests of the REPIH assumed income was stationary around a

---

[18] The transformation is $E_{t+1}\{W_{t+1}/(1+r)\} - E_t W_t$, which, because of the structure of the discounted income stream, with its geometrically declining weights, eliminates the predictable part of future incomes, see Muellbauer and Lattimore (1994) and Deaton (1992, pp. 81–3), but note that Deaton uses beginning of period rather than end of period assets.

[19] Blinder and Deaton (1985), who incorporate a richer set of surprises from non-income variables (such as wealth, inflation, and relative prices), however, found no role for unanticipated income in the US after all. But they find that anticipated relative prices and inflation matter for current consumption, which still violates the orthogonality assumption of the REPIH.

deterministic trend (for example, Flavin, 1981). Consider a regression of the change in consumption, Δc, on lagged levels of income, y. If, in fact, y is a nonstationary variable while Δc is stationary, the different sides of the equation have different orders of integration. Then the problems in making inferences about the coefficient on lagged income are essentially the same as the problems that occur in discerning the existence of a unit root in a univariate time series, a point first noted by Mankiw and Shapiro (1985).[20] Use of the standard normal tables at usual significance levels would result in over-rejection of the Hall model.

However, not all excess sensitivity tests have non-standard distributions. An example arises in the regression of Δc on lags in Δy and on lagged saving, which is stationary even if y is not (see Campbell, 1987). Most studies using such an approach confirm the rejection of the Hall model.

3. The solved-out consumption function and the persistence of income

An important issue in macroeconomics is the size of the response of consumption to current income, the MPC out of current income. Under the permanent-income hypothesis, the critical issue here is how responsive is expected permanent (non-property) income to current (non-property) income.[21] Under the further hypothesis of rational expectations, this responsiveness hinges on the degree of *persistence* in the income process. This is well illustrated by the following example. Suppose income follows the process

$$y_t = \lambda_0 + \phi t + \lambda y_{t-1} + \varepsilon_t, \tag{22}$$

where ε_t is a random income shock with mean zero and ϕt represents a deterministic trend effect.

When $\lambda = 1$, there is complete persistence: all income shocks, ε_t, translate one-for-one into changes in permanent income. The best estimate of permanent income, y_t^p, is then current income, y_t, plus the discounted present value of any drift term, λ_0, which then reflects trend-like growth in income.[22]

When $\lambda = 0$, persistence is zero. The best estimate of permanent income only gives any weight to current income to the extent that, in a finite sample, the current observation is included to derive estimates of λ_0 and the trend coefficient, ϕ. In general, the response of permanent income to current income depends on λ, the interest rate, r, and the length of the horizon.

Small differences in the degree of persistence, λ, can alter remarkably the response of y^p to y. In the case of an infinite horizon, the response of y^p to y can be shown to be $r/(1 + r - \lambda)$. To illustrate with r at 3 per cent, the response of y^p to y varies from 13 per cent when λ is 0.8, 23 per cent when λ is 0.9, 38 per cent when λ is 0.95, 75 per cent when λ is 0.99, and to 100 per cent when λ is 1.[23] In practice it is remarkably difficult to distinguish empirically the hypothesis $\lambda = 1$ from $\lambda = 0.95$, and hence to establish the degree of responsiveness, under rational expectations, of permanent to current income.[24]

4. The excess smoothness debate

Statistical tests find it hard to reject the hypothesis that income has a unit root; that is, first differencing appears to be required to make aggregate *per capita* real income stationary. If income has a persistent unit root process (see footnote 24), then permanent income responds strongly to shocks in current income.

Under the Euler equation (18), in the absence of transitory consumption shocks or measurement errors, the consumption surprise equals the surprise in permanent income. Thus, the variance of the consumption innovation should be at least as great as that of the current income innovation, if income has a persistent unit root process.

In fact, in the US (and other countries) seasonally adjusted consumption growth has a variance of one-half or less of income. Consumption thus appears to be excessively smooth, i.e. insufficiently volatile, for the assumption that income is generated by a difference stationary (DS) process. Since one of the stylized facts

---

[20] The Dickey–Fuller tests (Fuller, 1976; Dickey and Fuller, 1979) for the presence of a unit root, are then appropriate. The asymptotic statistical theory underlying these problems of inference is described in detail in Phillips (1986, 1987) and Phillips and Durlauf (1986), while a good summary exposition is provided by Banerjee and Dolado (1988).

[21] The persistence issue was rather neglected by Friedman (1957) in favour of merely the randomness of income, which is a rather different matter. Farrell (1959) made more of the issue through the 'elasticity of expectations', measuring the sensitivity of income expectations to current income.

[22] Note that with $\lambda = 1$, ϕ would have to be zero to prevent income from accelerating permanently ($\phi > 0$) or decelerating permanently ($\phi < 0$).

[23] The above examples assumed an infinite horizon. For shorter horizons there is greater responsiveness of y^p with respect to y when λ is less than one.

[24] These simple insights generalize to more complicated stochastic processes for y (see Hansen and Sargent, 1981). The key issue is whether the income process has a 'unit root' that is, whether it can be written in the form

$$\Delta y_t = \Sigma_{i=1}^{k} \beta_i \Delta y_{t-i} + \varepsilon_t,$$

where ε_t is a stationary random error. In this case, there will be more than complete persistence if $\Sigma \beta_i/(1 + r)^i$ is positive, and less than complete persistence if it is negative (see Deaton, 1992, ch. 4.2).

that the PIH was intended to explain was the smoothness of consumption, this gives rise to a paradox. This paradox was brought to light in Deaton (1987, also see 1992) and Campbell and Deaton (1989).

However, Deaton's paradox is fragile to many amendments in the simple underlying model. As noted, it is far from trivial empirically to distinguish unit-root processes from processes that are stationary about a deterministic trend (see Deaton, 1992, pp. 110–17; Stock, 1990, 1993), or stationary about a smooth, but still stochastic trend.

The notion that the income dynamics can be described by a single univariate process over many years is also highly contestable. Ignoring structural breaks owing to regime shifts in the series (such as a shift to floating exchange rates, different monetary regimes, and idiosyncratic shocks such as the oil shock in 1973), biases tests towards finding higher orders of integration, and therefore greater persistence than actually holds. This has spurred work using segmented trends, where occasional large shocks perturb the slope and/ or constant in what are otherwise deterministic trends (Perron, 1989; Banerjee et al., 1992). An alternative is to assume that there is a slowly evolving underlying trend in real income, say θ_t, that reflects primarily productivity growth and demography, and perhaps long-term trends in the world economy.[25]

Muellbauer and Murphy (1993a) applied both the split-trend approach, introducing a shift in the time trend in 1974, and a stochastic trend to US data on real personal disposable non-property income. The income forecasting model also incorporated lagged unemployment and the change in a short-term interest rate, both of which are stationary series for the US. The split trend is very significant, indicating a slow-down in US growth in the 1970s. The stochastic trend approach gives quite similar results, though a marginally worse fit. Either way these results suggest that income is not a difference-stationary variable at the macroeconomic level. This can be interpreted as follows: there are ceilings to output levels in the short run, given by a slowly evolving supply side, and there are forces for recovery if income drops far below trend. Macro-policy makers change policy when unemployment reaches levels perceived to be too high. There are also spontaneous

recovery forces via downward pressure on wages and commodity prices, and the wearing out of durable goods, which create private-sector demands for replacement investment, helping the upturn. In such a world, rational consumers would become more pessimistic about growth prospects if growth had been very strong lately, and more optimistic if growth had been weak for some time. Permanent income then would be smoother than current income: a large part of income innovations would not be permanent and consumption would be smoother than current income. There is then no Deaton paradox.

One of the problems that arises in investigating income processes in order to investigate whether consumption is excessively smooth, is the possibility that private agents may have more information than the econometrician studying the behaviour of aggregate income. As Deaton (1992, p. 122) puts it, 'a one period windfall to the analyst may be known by the agents to be the first instalment of a multi-period gain, with quite different implications for consumption'.

Campbell (1987) suggested an ingenious way of controlling for the superior information of private agents over econometricians. Campbell argued that under the null of REPIH, saving should encapsulate the superior information of the agent: saving should help to forecast income. Campbell set up a model for the joint consumption/income process consistent with this idea but the REPIH is still rejected. Similar results were found by Attfield et al. (1990) for the UK.

However, Muellbauer and Murphy (1993a,b) find, both in the UK and the US, in the context of an only moderately sophisticated income-forecasting process, that lagged saving does not have a significant negative effect on subsequent income.[26] This throws doubt on whether private agents have superior information, at least about the macroeconomy. The issue of excess smoothness can be resolved in other ways. As we shall see in section IV, precautionary behaviour and liquidity constraints can explain the phenomenon, as can habits of consumers.

We end this subsection with a note of caution which applies with equal force to all of section III. Much of the literature referred to here which works with linear rational-expectations models in aggregate income, consumption, or saving, claims to rest on rigorous microfoundations. Linearity requires a quadratic utility function. Working with linear aggregates requires that age is irrelevant for individual behaviour, i.e. that

---

[25] Such a trend would not be I(1) but I(2), i.e. twice differencing is required to make it stationary. Following Harvey (1989) and Harvey and Jaeger (1993), we represent such a trend, θ_t, by a varying coefficient model:

$$\Delta\theta_t = \beta_t + \beta_0$$
$$\Delta\beta_t = \varepsilon_t$$

where ε_t is white noise. As Harvey and Jaeger explain, such a model is close to the Hodrick–Prescott filter for generating a smooth but time-varying underlying trend for income.

[26] Alessie and Lusardi (1993) also find no evidence in favour of the Campbell effect on Dutch panel data for individual households, although, as in many panel data studies, the conclusions may be sensitive to the treatment of measurement errors.

behaviour is equivalent to that of infinitely lived agents, or at least of agents whose survival probability is independent of age (see Blanchard, 1985). Often, the simplifying assumption is added that the market discount rate is constant and equal to the subjective discount rate, as in Campbell (1987). It is one thing to derive some stylized parables about behaviour, but it is quite another to take such assumptions literally, for they have quite absurd implications. If individuals can expect income to grow and are interested only in the expectation of income, and not in its riskiness, then aggregate saving would have to be negative since, as Campbell noted, positive saving forecasts future income *declines*. Apart from the fact that, around the world, aggregate household saving is positive, how could systematically negative saving be financed?

If one stays within the framework of the linear life-cycle/permanent-income rational-expectations framework, the fact of positive aggregate savings demands the assumption of finite lives, as discussed in section II. This raises issues of aggregation over households varying by age and by demographic composition. These issues are discussed in section IV.8, where we see that aggregation can explain, at least in part, both the excess sensitivity of consumption growth to anticipated income and the excess-smoothness phenomenon.

There are two further problems stemming from implications of quadratic utility. The first is the bounded bliss point in (8): eventually satiation is reached. The problem can be postponed by making the bliss point a function of time, or of past consumption, but is not removed unless extreme assumptions are made. More subtly, the linear-invariables framework implied by quadratic utility does not cope well with exponentially growing data, which is likely to lead to problems of heteroskedasticity.[27] The second, even more serious problem, is that income uncertainty is irrelevant for behaviour under quadratic utility. Amending this implausible implication is the first concern of the next section.

IV. Beyond the linear REPIH

In this section we relax a number of the key assumptions that underlie the simple surprise version of the Hall Euler equation for consumption. In turn, we introduce precautionary saving in the face of uncertainty,

credit constraints, interactions between consumption and leisure, lags owing to habits, adjustment costs, or durability of consumption goods, alternatives to rational expectations, and assets that vary in their liquidity characteristics or which enter the utility function directly, and consider the problem of aggregation posed by moving from micro theory to macro data.

1. Uncertainty and precautionary saving

It seems hard to believe that income uncertainty is irrelevant for consumption decisions. Given that consumers spend a great deal on other types of insurance, it seems almost incontrovertible that there is a precautionary element in savings decisions. As noted in Blanchard and Fischer (1989, pp. 288–90), under constant absolute risk-aversion and normally distributed income, the effect of income uncertainty on saving can be derived analytically. Exact analytical results are otherwise rare, though the implications of uncertainty about a different issue, namely length of life, are straightforward. This section considers both types of uncertainty and uses an approximation argument to analyse the implications of income uncertainty. Applications of risk-averse behaviour to explaining part of the decline in the US saving rate are also discussed.

One can retain the attractive features of the CES utility function and its implied constant relative risk-aversion to obtain approximate analytical results of great intuitive appeal for the solved-out consumption function. One can show that the savings decision under income uncertainty is, to a near approximation, equivalent to a certainty-equivalent problem, in which expected income is reduced by a discount factor reflecting uncertainty.

This can be explained, in a two-period context, using the approach of Skinner (1988) and Kimball (1990). Certainty-equivalent income, y_2^*, is defined as that certain income level which makes second-period marginal utility equal to expected marginal utility, given the probability distribution of y_2. One can show that $y_2^* < Ey_2$, see Kimball (1990). With the CES utility function, it can be shown that, with $A_1 = 0$, $y_2^* \approx Ey_2/(1+ \sigma z)$, where z depends on σ (and hence on the degree of risk-aversion) and on the coefficient of variation of income in period 2. For example, with a 10 per cent coefficient of variation and $\sigma = 1/3$, the discount on expected income is 2 per cent. With a 20 per cent coefficient of variation, the discount rises to 8 per cent. For a consumer with debts, i.e. $A_1 < 0$, the discount is bigger, while for a consumer with an asset cushion, the discount is smaller. Thus an asset cushion provides

---

[27] However, there may by other ways round this, for example by scaling the change in consumption by lagged income.

insurance against risk so that consumers can spend more now, while debts will make consumers even more cautious in the face of uncertain income. This is what Carroll (1992) and Deaton (1992, ch. 6) call a 'buffer-stock theory of saving', in that a major motive for holding assets is to shield consumption against unpredicatable fluctuations in income.[28] It can also be shown that unpredictable variations in consumption needs have a similar effect to income uncertainty in inducing cautious behaviour.

One important implication of these considerations is that if future income is more heavily discounted because of uncertainty, then current income must necessarily play a bigger role in determining current consumption. And this must be particularly so for the young without a cushion of assets. This helps to explain why consumption follows income more closely over the life cycle than simple life-cycle theory would suggest.

In the context of section II's analysis of intertemporal choice with the CES utility function, the implication of the analysis is that a higher discount factor, i.e. the real interest rate plus a risk premium for income and needs uncertainty, is applied to future income than with point expectations. However, the MPC out of assets (the inverse of κ_1 in (11)) depends, as before, only on the real and subjective discount factors. In the many period extension, it will not, in general, be true that the same geometrically decaying discount factor applies to future incomes into the more distant future. Nevertheless, this may be a reasonable approximation for empirical work.

Hayashi (1982) used the assumption of constant discount factor and geometrically decaying weights to suggest a method for estimating solved-out consumption functions, which avoids having to generate explicit income forecasts. This method uses a quasi-difference transformation, exploiting the geometrically declining weights associated with future incomes. However, because the discount factor including the risk premium is different from the real interest rate, the resulting equation relates c_{t+1} to c_t, A_t, A_{t-1}, and y_t, instead of just to c_t, so that lagged information other than consumption matters. Hayashi, using annual US data, finds the real interest rate to be around 3.4 per cent and the discount rate for income to be 13.2 per cent. In other words, the data suggest that, indeed, US households discount future income at a much higher rate than would be implied by the real interest rate. Such a possibility was earlier raised by Friedman (1957, 1963), who suggested the income discount rate might be as high as 33 per cent. Similar results to Hayashi's for UK data are reported by Weale (1990) and Darby and Ireland (1994). We should not think of Hayashi's equation as an Euler equation because it is not the first-order condition for an intertemporal optimization. It shares all the characteristics of a solved-out consumption function, except that income expectations are eliminated.[29]

The effects of uncertainty in the Euler equation when consumers are risk averse can also be examined directly. Higher uncertainty operates like a higher real interest rate here, too: it depresses consumption in the first period, raising the planned growth rate, other things being equal.[30]

Let us now turn to three general implications of precautionary behaviour in the face of income and consumption needs uncertainty. First, precautionary behaviour may enhance the stability of consumption and hence help to explain the excess-smoothness paradox. A large positive income shock is likely to be taken as an indicator that there is more uncertainty about the new, higher level of income and so lead to a smaller rise in consumption than in the absence of the precautionary motive. Second, as far as testing the Hall model and 'excess sensitivity' is concerned, precautionary behaviour as modelled by Hayashi (1982) clearly implies that lagged income and other information matters. The same implication follows from the approximate Euler equation which incorporates uncertainty.

Third, as far as the aggregate saving rate is concerned, the classic life-cycle model without uncertainty implies a relationship between it and pension and social security systems. Other things being equal, a higher level of pension provision lowers saving out of after-tax income (see Feldstein, 1977, 1980). Risk-averse behaviour under uncertainty suggests additional reasons for such a relationship. Unemployment benefits insure against income shortfalls, as does an income support floor present in the welfare states of advanced industrial countries. Also, progressive taxation reduces the upper tails of the distribution of agents' uncertain future

---

[28] Lattimore (1993) shows that skewness of the probability distribution of future income in the form of a small probability of a large drop, for example, owing to becoming unemployed, implies a higher discount factor than a symmetric probability distribution with the same coefficient of variation. Such a probability distribution, together with the assumption that consumers cannot hold net debt at death, drives Carroll's theory.

[29] Blanchard (1985) derives the continuous time equivalent of Hayashi's equation for aggregate data from a set of microeconomic assumptions without any approximation arguments. These include a survival probability independent of age and point expectations about income.

[30] See Hansen and Singleton (1982). The result can also be derived as an approximation without their special assumptions.

incomes. By making income less risky, these institutions reduce precautionary saving. One reason for secular declines in saving rates in industrial countries may lie in improved income and health safety-nets which have reduced uncertainty.

Recently, Hubbard *et al.* (1994) have suggested that features of the US income-support system additionally discourage saving and may help to account for the decline in the US saving rate in the 1970s and 1980s. The US system makes payment of most welfare benefits conditional on the household owning very little in the way of assets. For households with some probability of income falling below the support level, this strongly discourages the accumulation of assets.[31]

Another kind of uncertainty concerns the date of death. Even in the absence of a bequest motive this implies that many individuals die leaving substantial levels of assets. Let us take a simple case in which there is no uncertainty about income, given survival—only survival itself is uncertain. The expected utility function is then identical to (5) except that the subjective discount rate, δ, also incorporates the survival probability of the individual. This therefore lowers the MPC out of life-cycle wealth, $1/\kappa_1$.

Thus, uncertainty about date of death, as with the bequest motive, modifies the conclusions of section III.2 above: the MPC out of assets for the retired will be rather smaller than the theory predicts when there is no uncertainty. Both considerations help to explain why, according to survey evidence,[32] the retired do not dissave on anything like the scale predicted by simple life-cycle theory. The uncertainty about consumption needs, discussed above, is likely to be particularly acute as individuals age into their 70s and 80s, when health becomes frail, and the risk of large medical or other support expenses looms. These considerations imply that the MPCs out of assets vary much less over age than implied by simple life-cycle theory. This result has several important practical implications. One is to reduce the importance of aggregation in explaining variations in saving rates across economies that differ in rates of population growth, and hence demographic structure, and that differ in rates of *per capita* income growth.[33] Another important implication of uncertainty in old age is that it helps to account for large-scale bequests. This provides some rationale for Kotlikoff and Summers (1981), who have argued, as against Modigliani,

that most of US wealth is the result of bequests rather than life-cycle accumulation for later expenditure. Meanwhile, the controversy continues (Modigliani, 1988; Kotlikoff, 1988).

Uncertainty affects not only income, consumption needs, and date of death, but the rate of return on saving is, in general, uncertain. Much of the literature on Euler equations and the consumption capital asset pricing model (Lucas, 1978; Breeden, 1979; Grossman and Shiller, 1981; Hansen and Singleton, 1983; Singleton, 1990), assumes that a safe asset exists. In the UK, this may not be a bad approximation, given the indexed gilt market. But in most countries, inflation-indexed securities do not exist. If the rate of return on saving is uncertain, the above analysis of certainty equivalence applied to income can be used to analyse the implications of an increase in rate-of-return uncertainty. Realistically, rate-of-return uncertainty and income uncertainty have often risen together, with inflation and inflation uncertainty. Empirically, the two effects may be hard to distinguish, especially as we have seen that, for aggregate consumption, interest-rate effects are likely to be small and time-varying.

2. Credit restrictions

In the US and the UK, 70–80 per cent of household debt is backed with housing collateral. A substantial fraction of the remainder finances purchases of cars and other consumer durables, which serve as collateral for the loans. Unsecured borrowing is only available in relatively small amounts and to a restricted set of people. The young on low incomes, but who expect to earn more in the future, appear to be obvious candidates for the category of persons who would like to borrow but are denied credit. Credit restrictions are an obvious reason for the tendency, observed in household surveys, for consumption to follow the hump-shaped profile of life-cycle income more closely than is predicted by the life-cycle model.

In this section, we ask under what conditions households are most likely to want to borrow and hence most likely to run into credit restrictions. We go on to discuss the effects of credit constraints on aggregate solved-out consumption functions and on aggregate Euler equations. As we shall see, credit constraints offer an explanation for the 'excess sensitivity' of consumption changes to predictable income changes. We suggest that credit constraints also offer a potential explanation of an 'error correction' form of the consumption function.

To derive the conditions under which households

---

[31] Feldstein (1994) has pointed to similar consequences of assets tests associated with college scholarships.

[32] See for example, Börsch-Supan and Stahl (1991).

[33] One incidental benefit is to reduce the aggregation biases in aggregate time series consumption functions conditioned on average incomes and average assets.

are most likely to want to borrow, we take the consumption function (11) and incorporate uncertainty by applying the extra uncertainty discount to next period's income. We can then derive the implications for the desired end-of-period asset level, A_t. If A_t is negative, so that borrowing is desired, the consumer may run into credit constraints. It can be shown that this is more likely if initial income and assets are low (or debts are high) relative to expected income, and if the real interest rate is low. Note that increased uncertainty about future income makes it less likely that the consumer would want to borrow and so run into the credit constraint.

In practice, for the majority of credit-constrained households, consumption equals income, though some of such households will have small levels of initial assets, and some may have debts. In the Euler equation literature, when credit constraints are considered, it is usually assumed that consumption is just equal to income for the credit-constrained (see Hall and Mishkin, 1982; Campbell and Mankiw, 1989, 1991).

If the fraction of credit-constrained households is π, aggregate consumption is given by

$$c = (1 - \pi)c^u + \pi c^c. \qquad (23)$$

The consumption of the credit-unconstrained, c^u, is given by (11) modified for uncertainty of their future income, while c^c for the credit-constrained is given by income, y^c. To make the resulting expression for aggregate consumption empirically tractable, one would typically assume that the incomes for the two types of household move in parallel with observed aggregate non-property income, and that π is constant, as in the comparable Euler-equation literature which we discuss just below. Then we can see that the implications of credit constraints for the solved-out consumption function are, in aggregate, to reduce the MPC out of assets and to increase the MPC out of current income.

More generally, there is an interaction of income uncertainty with the possibility of future credit constraints. The consequence is to increase precautionary savings compared with the situation where no future possibility of credit rationing arises, see Deaton (1991, 1992, ch. 6.2).

Let us return to the Euler equation literature. Much of the debate since Hall's famous 1978 paper has been about why the simple model (18) fails. Given the nine assumptions, listed below equation (18), for that equation to hold, there are nine reasons why the error term may not be an innovation but be correlated with earlier information. Papers have been written about all nine reasons, but credit restrictions,

suggested by Hall himself in Hall and Mishkin (1982), have proved the most popular.

The basic idea is very simple. The martingale condition (18) holds for the non-credit-constrained: $c_t^u = c_{t-1}^u + \varepsilon_t$, while it is assumed that the credit-constrained just consume income: $c_t^c = y_t^c$. If the latter holds without error, first-differencing of (23) gives

$$\Delta c_t = (1 - \pi)\Delta c_t^u + \pi \Delta c_t^c = (1 - \pi)\varepsilon_t + \pi \Delta y_t^c. \qquad (24)$$

Generally speaking, the change in income for the credit-constrained is proxied by the change in average non-property income, which is partly predictable and so provides an explanation of the excess sensitivity of consumption changes to anticipated income changes.

Deaton (1992) suggests a direct link between excess sensitivity and excess smoothness, the latter being a direct result of the fact that changes in consumption are not unpredictable. Equation (24) implies that the consumption innovation is the weighted average of ε_t, the innovation in permanent income for those whose consumption follows permanent income, and of the current income innovation for the credit-constrained. Generally speaking, if ε_t and the current income innovation are not very highly correlated, and if the variance of ε_t is no bigger than the variance of the income innovation, the variance of the weighted average, i.e. the innovation in total consumption, is likely to be less than the variance of the income innovation (see Campbell and Mankiw, 1991, p. 729). However, this is by no means inevitable, particularly if income had a unit root, when the current income innovation and the permanent income innovation would be very highly correlated. However, Deaton's case would be enhanced if some other lagged information entered (24).

One such possibility can arise with the existence of transitory consumption for both household types. Muellbauer and Lattimore (1994) show that this implies an error-correction form close to that of DHSY (1978), but in an expectational formulation,[34] i.e.

$$\Delta c_t \approx \beta_0 + \beta_1 E_{t-1}\Delta y_t + \beta_2(y_{t-1} - c_{t-1}) + \varepsilon_t. \qquad (25)$$

The basic idea behind the derivation is to note that the presence of transitory consumption implies that the change in transitory consumption is added to equation (24). But, at time t, last period's transitory consumption should be known, and can be estimated from observed c_{t-1} and y_{t-1} (and, indeed, lagged assets,

[34] An alternative way of expressing the link between DHSY and credit constraints was considered by Muellbauer (1983) who showed that the shadow price of the credit constraint entered the Euler equation and that the shadow price was positively related to $(E_{t-1}y_t - c_{t-1})$.

A_{t-2} which, strictly speaking, should be included in equation (25)).[35]

Campbell and Mankiw (1989, 1991) deal with transitory errors by taking expectations at $t-2$ of the relationship between consumption growth and income growth. If transitory errors are serially uncorrelated, taking expectations at $t-2$ eliminates transitory errors at $t-1$ as well as at t. They assume that observable *per capita* real disposable income growth is a good proxy for the income growth of households who are credit-constrained, or who simply spend current income because of myopia. Estimates of the consumption share of such households for six countries lie in a 0.2–0.6 range, except in Japan where the income forecasts are too poor. They suggest that the cross-country differences may be related to how developed the credit markets are: France has a relatively high value of the share of income-constrained households, while Sweden, Canada, and the US have lower values.

The authors also examine whether there are trends in the share of income-constrained households which might be associated with financial liberalizations. Unfortunately, for the UK they find that this share has a significant *upward* trend, implying, apparently, that credit rationing has tightened over the period 1957–88, a result also obtained by Jappelli and Pagano (1989). In the UK, for much of the 1980s, consumption growth exceeded income growth. If consumption growth is to be explained only by expected income growth, allowing a rising value of the share of income-constrained households is the only way the Euler equation framework can explain the data. Incorporating uncertainty explicitly works in the wrong direction, since income uncertainty almost certainly fell in the 1980s, while lagged uncertainty has a *positive* effect on consumption growth. Introducing a variable real rate of return may be helpful. However, the timing is not ideal, since the consumption boom was at its strongest several years after the biggest rises in the real interest rate.[36] Apart from problems in measuring the real interest rate, the UK evidence raises doubts about the rational expectations assumptions in the Euler equation—see section IV.5 below.

Here we have focused on the macroeconometric literature on the Euler equations with credit constraints. Many of the applications, following Hall and Mishkin (1982), have been on micro panel data, raising additional problems such as income measurement errors. For surveys of this literature, see Deaton (1992, pp. 140–63) and Hayashi (1987).

3. Saving and leisure

So far, we have conditioned consumption decisions on non-property income. This is valid if consumption decisions are separable from leisure choices. If this is not so, consumption decisions depend also on leisure and leisure expectations. As we shall see, this gives yet another reason for the failure of the random-walk consumption model.

For some households at least, income is a choice variable to the extent that decisions on hours, labour participation, and human capital accumulation are made by individuals in households. For many others, hours of work, or even having a job, are constrained. If intertemporal consumption decisions are separable from leisure decisions, conditioning consumption on labour income incurs only a potential endogeneity bias. If separability does not hold, and consumption and leisure were specific substitutes, one might expect high earnings parts of the life cycle to exhibit high levels of hours at work and high consumption to compensate. On the other hand, if they were complements, because leisure is needed for consumption of goods, consumption would tend to be lower in high earnings phases of the life cycle.[37] Browning *et al.* (1985) argue that neither assumption can explain the way hours, wages, and consumption move together in the UK in terms of voluntary choice by households. Ando and Kennickell (1987) report similar negative results for the US.

If separability is violated, we would have, in the additive context, each of the period utility functions depending both on consumption, c_s, and on leisure, l_s. With l_s constrained to $\bar{l}_s$, this is like introducing a variable taste parameter into each of the period utility functions. If these remain homothetic in c_s, the generic form of the consumption function is as before, but with additional conditioning variables, $\bar{l}_1$, and expectations of $\bar{l}_2, \ldots, \bar{l}_T$. In the aggregate, the unemployment rate and changes in it are likely to be reasonable proxies for the labour market opportunities and expectations of

[35] This derivation is also consistent with another feature of the full version of DHSY: a negative inflation effect. We know that life-cycle wealth depends on assets as well as on current and expected non-property income. Empirically, inflation over the last 40 years has been associated with falling real asset values and thus declines in life-cycle wealth.

[36] Patterson and Pesaran (1992) incorporate a variable real interest rate in the Euler equation following Wickens and Molana (1984). Testing parameter stability for the UK in the 1980s, they find π to be constant, but tendencies for lower coefficients on the real interest rate in the 1980s. As in many aggregate Euler equation studies, however, their results are quite sensitive to the instruments chosen.

[37] See Deaton and Muellbauer (1980, pp. 310–13) for more discussion and Blundell (1988) and Blundell *et al.* (1994) for evidence on separability between consumption and leisure.

opportunities reflected in $\bar{I}_1, \bar{I}_2$, etc. It can be shown that the corresponding Euler equation implies a negative relation between consumption growth and growth in unemployment, implying an alternative explanation of excess sensitivity.[38] Unfortunately the change in the unemployment rate is also likely to be a good proxy for uncertainty about labour income; while, among other things, the unemployment rate will be correlated with early retirement, and thus with shifts in the distribution of consumption between the retired and the pre-retirement population of households. This is bound to make unique interpretations difficult.

4. The role of lagged consumption and durability

A classic conundrum in applied econometrics is whether lagged dependent variables represent agents' expectations or adjustment costs, habits, or the durability of goods. In the Hall (1978) type of Euler equation, lagged consumption has a pure expectational interpretation: it reflects the consumer's best prediction last period of future circumstances, particularly permanent income.

Habits, like convex adjustment costs, imply partial adjustment of consumption to life-cycle wealth, defined by,

$$c_t = \beta(W_t/\kappa_t) + (1 - \beta)c_{t-1}, \qquad (26)$$

where β is a parameter reflecting the size of adjustment costs (see Muellbauer, 1988). This offers a partial explanation for 'excess smoothness', because only part of any period's shock is adjusted to within the period. It also gives an alternative link between aggregate saving rates and aggregate growth rates because consumption lags behind an increase in the growth rate of income. In the Euler equation context, habits can explain why changes in consumption are not innovations: in the simplest case,

$$\Delta c_t = a\Delta c_{t-1} + \varepsilon_t, \qquad (27)$$

where a is the habits parameter (see Muellbauer, 1988). However, the balance of empirical evidence does not favour equation (27) as the complete explanation of the failure of the Hall model (18).[39]

Habits or convex costs of adjustment also imply partial adjustment in stocks of durables. This would imply a relationship similar to (26), not for purchases, but for stocks. The stock-flow relationship is

$$S_t = (1 - d)S_{t-1} + cd_t \qquad (28)$$

where S is the stock, d is the rate at which the stock wears out and cd is the flow of purchases. In the Euler equation context, without habits or adjustment costs, the Hall model (18) applies to stocks. Because of (28) there should then be negative first-order serial correlation of residuals in the equation for the change in purchases. One of the points made by Caballero (1994) is that habits or adjustment costs would push this serial correlation further into the past. He also explores the implications of lumpy adjustment costs. If adjustment costs are non-convex or lumpy, there are discrete jumps in individual behaviour, which are smoothed out in the aggregate, as economic conditions vary. And the history of recent shocks to the economic environment matters in a way not allowed for by the usual partial adjustment story. Indeed, there is empirical evidence against the latter.

5. Alternatives to rational expectations

The principal focus of the last 15 years of US consumption research has been in finding out why the REPIH is typically rejected at the aggregate level. In most cases, rejections have been of the *joint hypotheses* of rational expectations and of the assumptions underlying the formulation of the intertemporal optimization problem. This has led to a re-interpretation of the basic Euler approach, relaxing its restrictive assumptions to examine the effects of variable interest rates, finite lifetimes and retirement, temporal and cross-sectional aggregation bias, abandonment of certainty equivalence (i.e. relinquishing quadratic utility functions), credit constraints, relaxing intertemporal separability of preferences (i.e. habits allowed), and relaxing separability between consumption and leisure or other determinants of utility.

---

[38] Bean (1986) argues for an explanation on these lines, but his results for the US are not robust to small alterations in the sample period, as Muellbauer and Bover (1986) showed.

[39] The empirical issues are not trivial: in high-frequency data, e.g. monthly, measurement errors are likely to be important, while in low-frequency data, time-aggregation problems arise. The problem of time aggregation refers to a situation in which decision times do not coincide with the units of time over which the data are measured. It arises in the attempt to distinguish hypotheses about habits and durability from simpler alternatives. See Muellbauer (1988), Heaton (1993), and Ermini (1994) for a more comprehensive discussion. But even in simpler contexts, time aggregation can be important, see, for example, Hall's (1988) discussion of time aggregation and the measurement of real interest-rate effects in Euler equations. Also, relative to a span of a month, many consumption items e.g. a can of baked beans, appear as effectively durable when they would be classified as non-durable over the span of a year. For purchases of durable goods not subject to habits or adjustment costs, $a < 0$ in equation (27).

Rather less attention has been given to the assumption of rational expectations. Pesaran (1987) recommends the terminology 'Muthian', rather than the value-laden 'rational', to refer to the assumption that consumers (like the econometrician!) have costless knowledge of the true model. There are two main types of alternatives to rational expectations. First, alternative rules of thumb for forecasting can be asserted without providing any well-defined theoretical basis for why agents might behave this way. Thus, in Friedman's original formulation of the PIH, income expectations were modelled as a distributed lag of past income. Second, different micro theories about consumer behaviour can be used to infer expectations processes. For example, the optimizing problem for consumers can be reformulated to take account of the costs of information acquisition relative to the benefits of forecasting accuracy. One instance of this is Pischke's (1991) model, in which consumers do not have the expensive macro information to distinguish micro and macro elements in the shocks they observe. As Deaton (1992, p. 174) observes, this has predictable consequences for aggregate behaviour quite similar to those of habits. Using a model calibrated to the average US consumer, Cochrane (1989) found very small wealth costs from near-rational alternatives to full-information intertemporal maximization. Once information acquisition and processing costs are introduced, the benefits of full information intertemporal optimization may be too small to warrant the costs—simple rules of thumb are likely be optimal (see Simon, 1978).[40]

One plausible example of rule-of-thumb behaviour is the use by consumers of an error-correction mechanism (ECM), such as (25), interpreted as a servo-mechanism. A consumer following (25), with the βs between 0 and 1, would partially adjust consumption to current income changes, but always, in the long run, aim to keep consumption approximately in line with income. If last period's consumption was higher than income, this year's would tend to be lower, other things being equal. A possible analogy to the decision problem of a consumer has been suggested by Lattimore (1993) as follows. Imagine a car travelling on the left-hand side of a road with a line clearly marked in the centre. Heavy trucks are travelling in the opposite direction on the right-hand side at high speed, so that a cautious driver might wish to avoid them. The verge on the left-hand side is muddy and rocky: a car can only travel here very slowly without ruination. The windscreen has been smeared with dirt so that it is almost impossible to look forwards: the future direction of the road is therefore perilously uncertain. One thing is clear for the driver: the rear view. What should a driver wishing to get to her destination as quickly as possible do? The answer is to look backwards at the road, keeping the centre line well to the right, and correcting the steering as the car deviates from the line. In this metaphor for the consumer's dilemma, the heavy lorries are bad income states to be avoided because of convex marginal utility, the muddy verge is the cost of excess saving, while the dirt-encrusted windscreen represents the difficulty of forecasting future states.

The assumption of rational expectations underlies the Lucas (1976) critique of traditional econometric models. For example, if there is a change in the underlying process generating income, then the model of consumption conditional on income (and indirectly on expected income) should also change. This provides a vehicle for testing whether expectations are primarily backward- or forward-looking (see Hendry, 1988).

An alternative approach is to make use of consumer survey information on consumer sentiment, in particular responses to questions concerning future economic conditions (see Evans, 1969; Carroll *et al.*, 1993). More work on empirical models of expectations formation is needed.

6. The role of assets and asset prices

As we have seen, assets are accumulated to provide for retirement and bequests, and for rainy days that can arise earlier because of income falls resulting, for example, from unemployment or ill health, or consumption needs (e.g. paying for a child's education or medical bills). Assets are also accumulated to purchase 'big ticket' durables such as a car or furniture.

In this section, we suggest that marginal propensities to spend are less for illiquid assets than for liquid ones. We also discuss the special role of house prices for expenditure decisions where a positive wealth effect for some is offset, in part, by a negative price effect.

Households typically hold a balance of liquid assets, which can easily be converted into expenditure when needed, and illiquid assets, that typically yield a higher rate of return. Illiquidity has the following dimensions:

(a) Capital uncertainty: for example, equity or property prices might be low just as cash is needed for spending.

(b) Transactions costs: stockbrokers charge commission; in some countries, transactions costs, including taxes, of selling a house can be over 10 per cent.

---

[40] If more radical changes to the economic framework are visualized, some of the basic axioms of consumer behaviour may be discarded (see, for example, Tversky and Kahneman, 1974; Shefrin and Thaler, 1988; Thaler, 1990).

(c) Transactions restrictions: pensions are usually only accessible at retirement; tax benefits from many saving schemes are lost if they are cashed in early.

(d) Indivisibilities: houses, yachts or paintings can only be sold as units.

Different assets have different liquidity characteristics. Housing, pension funds, and life insurance funds are at the illiquid end of the spectrum. However, note that a general easing of credit eases these constraints to a degree: if housing or a painting can be used as collateral for a loan, the market value can, to a degree, be accessed. However, lenders, bearing in mind default risk and the volatility of asset prices, are unlikely to offer 100 per cent collateral-backed loans—though the UK and Scandinavian housing markets seem to have been a temporary exception in this respect after financial deregulation in the 1980s.

Differences in liquidity suggest, as far as spending decisions are concerned, associating different spendability weights with different types of assets and debt. The characteristics of illiquidity are complex and multidimensional and it is no easy matter to formulate the budget constraint to reflect points (a)–(d) above. If one were to do so, one could associate shadow prices or Lagrange multipliers with each of the constraints. One could then measure the marginal utility of increasing each asset by £1 relative to the marginal utility of increasing current income by £1. Cash then has a relative shadow price or spendability weight of 1, and less liquid assets would have lower spendability weights.[41] One would expect the assets with the highest long-term after-tax returns to have the lowest spendability weights. This is analogous to the index-number approach to measuring monetary aggregates (see Barnett, 1981). There, liquidity is indirectly measured by potential return foregone. However, one would expect spendability weights on illiquid assets to increase with financial deregulation—even though, in the short term, returns on such assets increased after deregulation. Potentially, as we shall see, the increased spendability of illiquid assets could be the most important consequence of financial liberalization.

The complex nature of illiquidity has made difficult the empirical integration of the consumption function with the modelling of portfolio decisions. This was one of the great hopes of the consumption capital asset pricing model (CAPM), a natural spin-off from the Euler equation (see Breeden, 1979; Singleton, 1990). The consumption CAPM examines the set of intertemporal efficiency conditions holding for each of a vector of assets. Unfortunately, as noted above, realistic specifications of the budget constraints entail non-convex transactions costs and other trading restrictions on illiquid assets. These destroy the simple structure of the consumption CAPM and are not only difficult to articulate, particularly for aggregate data, but are also likely to alter as credit conditions alter.

It is important to note that owner-occupied housing wealth, which is the most important single asset for the majority of households, has spending consequences which differ from those of illiquid financial assets in one significant respect. This is because housing services also appear in the utility function. Housing is the consumer durable par excellence. It is distinguished by the low rate of wearing out and also by its different treatment in national accounts statistics compared with other durables. Purchases of houses are not regarded as spending to be included in total consumer expenditure. Instead, the imputed value of owner-occupied housing services is included both in consumption and in income in the national accounts. This recognizes the asset acquisition aspect of house purchase.

In general, the relative price of durables enters the consumption function for non-durables, as well as its own expenditure equation. Similarly, the demand for non-housing consumption will be affected by the price of housing. To see how, consider a permanent change in the price of housing after which the relative prices of non-housing consumption and housing are expected to remain fixed. It can be shown that there are two effects.[42] One is a positive wealth effect for those with housing wealth and the other is a negative combined income and substitution effect from facing a higher house price. For those without housing wealth, there is no wealth effect so that the negative effect remains. The smaller the proportion of people who are not owner-occupiers but have aspirations in that direction, the more the wealth effect dominates at the aggregate level. One might expect non-owner-occupiers to save more when real house prices rise. Their saving would also depend on the size of the required deposit as a proportion of the price of a house when obtaining a mortgage. In the UK and in Scandinavia in the 1980s, financial deregulation reduced these deposit-to-value ratios. Further, those living in a deregulated private rented sector may anticipate having to pay higher rents in the future when they see real house prices rise and

---

[41] There are previous studies which have allowed different weights on liquid and illiquid assets (e.g. Patterson, 1984), and there are a number which have included liquid asset effects (e.g. Zellner et al., 1965; Hendry and von Ungern Sternberg, 1981).

[42] See Miles (1993) and Muellbauer and Lattimore (1994) for a simple derivation.

hence save more. In the UK, the deregulated private rented sector is under 7 per cent of households.

It is important to note the difference between housing wealth increases and house price increases. The accumulation of owner-occupied housing capital through investment and the transfer of publicly-owned housing into private hands at discounts of around 50 per cent experienced in the UK in the 1980s, are household wealth increases that do not rest on increases in real prices. Such increases have clearly positive expenditure implications.

Increases in real house prices tend to redistribute wealth between young households and older households, since the young have typically accumulated less housing wealth. To the extent that older households may have higher MPCs to spend out of housing wealth, this redistribution adds to the aggregate spending effect.

Evidence is accumulating that house prices have these dual effects discussed above: a positive wealth effect, which depends on the degree of liquidity of houses, and a negative relative price effect. Indeed, it is impossible to obtain sensible wealth effects for the aggregate Japanese consumption function without a negative relative price of land[43] effect (see Murata, 1994). Lattimore (1993) finds a similar effect for Australia. For the UK, it is hard to pin down a negative relative price effect in aggregate data, but on regional UK consumption data, where there is more price variation, Muellbauer and Murphy (1994) find a negative relative price effect alongside a positive wealth effect. A recent survey on the consumption effects of house price increases in a range of OECD countries finds mixed effects,[44] consistent with the dual role of house prices (Kennedy and Andersen, 1994).

7. Financial liberalization

Financial deregulation has been an important phenomenon in many countries, particularly in the 1980s. Moves in this direction started a little earlier in the US with the phasing out after the mid-1970s of regulatory ceilings on interest rates paid by savings and loans institutions. The move to targeting monetary aggregates made interest rates increasingly volatile at this time and interest-rate regulation increasingly out of place both in the US and the UK. Furthermore, the UK abolished exchange controls in 1979, which opened up

its domestic credit markets to international capital movements.

To understand the financial deregulation that occurred in the UK and the Scandinavian countries in the 1980s, it is helpful to understand the preceding regime of credit controls. The UK and Scandinavian countries in the 1970s had very progressive income-tax systems and tax deductibility of interest payments on debt, though in some cases, as in the UK, restricted to mortgage debt. After the first oil shock of 1973–4, inflation rose sharply in these countries, while nominal interest rates rose only moderately. Real interest rates became hugely negative given tax deductibility of interest payments. To illustrate, with inflation at 15 per cent and the nominal borrowing rate at 13 per cent, a marginal tax rate of 50 per cent implies an after-tax real interest rate of $6.5 - 15 = -8.5$ per cent. At a marginal tax rate of 80 per cent, common for higher incomes at this time, the real interest rate was $2.6 - 15 = -12.4$ per cent. This created strong incentives to borrow and a demand for credit which was held in check by rationing.

In the UK, from 1979, nominal interest rates rose, and foreign-exchange controls were abandoned. The 'corset' which had restricted bank lending was removed in 1980, and the banks, suffering losses from third-world lending in the 1970s, were anxious to enter domestic mortgage markets. These markets became much more competitive. Restrictions on building societies were progressively relaxed, culminating in the 1986 Building Societies Act, and a new breed of mortgage lenders, often financed by overseas banks, entered the market. Credit became so easily available that, by 1986–8, many first-time buyers were offered 100 per cent loan-to-value ratios. From 1980 to 1989, household debt-to-income ratios in the UK more than doubled, becoming one of the highest ratios in the world, and there was a boom in house prices in which real prices in the UK doubled over the same period. These developments were not solely the result of financial deregulation, since there was sustained economic growth and falling unemployment from 1986 to 1990. But it is hard to deny the connection.

Similar developments occurred in Scandinavia, with Denmark followed by Norway, then Sweden, and finally Finland (see Englund, 1990; Lehmussaari, 1990; Koskela and Viren, 1992; and Berg, 1994). In all four countries, debt-to-income ratios grew strongly, real house prices boomed, and household saving ratios fell sharply.

As far as a consumption function is concerned, financial liberalization can have several effects on the parameters. Most obviously, it could reduce the share of the credit-constrained. However, as noted in section

---

[43] In the absence of good house-price indices, the price of land is the best proxy.

[44] However, these results are not based on formal econometric work or a comprehensive model of consumption.

IV.2, Euler equation evidence for the UK appears to suggest that the proportion of credit-constrained households *rose* in the 1980s UK. One explanation is a shift in the spendability coefficients associated with illiquid assets and debt. Financial liberalization, by making asset-backed credit more easily available, made these illiquid assets more spendable. This makes good sense in the context of the shadow price interpretation of these weights in section IV.6. It is also plausible that it would have increased the spendability weight on debt, bringing it close to that on liquid assets, but with the sign reversed. During the credit-rationing regime of the 1970s, a household with big debts could count itself lucky, even though, with *given* assets and income, lower expenditure would still be associated with a bigger debt.

There are other possible effects. During credit rationing, intertemporal substitution, which is partly represented by the real interest effect, is less likely to be operative.[45] Thus, we would expect more powerful real interest effects after financial liberalization. More subtly, since part of the uncertainty discount factor applied to future expected income growth rests on the possibility of future credit rationing, it is possible that the coefficients on expected income growth would increase with financial liberalization.

It is too much to ask aggregate time-series models to pick up all these possibilities, though there is some empirical support in the UK for both of the last two. Distinguishing such effects is especially hard because *international* financial liberalizations reduced the balance-of-payments constraints on UK and Scandinavian economic growth. In the UK this gave rise to the Burns–Lawson doctrine[46] which claimed that, as long as the government did not run a deficit, the balance-of-payments deficit was self-correcting as the private sector made the necessary adjustments. Indeed, in forecasting income growth in the UK, Muellbauer and Murphy (1993*b*) found that the coefficient on the lagged balance-of-payments deficit fell in the 1980s, reducing the balance of payments constraint on growth. Since the rise of the Burns–Lawson doctrine occurred at about the same time as domestic financial liberalization, it is hard to separate precisely the direct growth effect of international financial liberalization from the effect via the higher spendability of illiquid

assets or a lower weight on credit-constrained households. The evidence we have so far is consistent with some effect on the spendability weights but not a very precisely determined one.

8. Some aggregation issues

One of the key insights of Modigliani and Brumberg's (1954) analysis of aggregate savings behaviour is that aggregate behaviour may be very different from that of any individual, even from that of a hypothetical 'representative' consumer. For example, each individual household may plan to spend in retirement everything saved during its working life. Yet in the aggregate, with population growth or productivity growth, saving will be positive. Another example arises both for Euler equations and for the solved-out consumption function in aggregation over credit-constrained households and those not so constrained. As we saw in section IV.2, aggregate behaviour is unlike the behaviour of either type of household. The problem of aggregation, how one goes from models of individual behaviour to a model of aggregate behaviour, has been particularly neglected in the Euler equation literature. This may in part be because it calls into question the link between theory and modelling the data, on which this approach has so prided itself.

First, we discuss some aggregation problems in Euler equations. We begin by illustrating the lack of one-to-one connection between individual and aggregate behaviour. Consider the Euler equation for a household with quadratic preferences, but where the real interest exceeds the subjective discount rate. By (9) in expectational form this implies that planned consumption grows steadily over each household's life cycle. Since the Euler equation is linear and, by assumption, the parameters β, r, and δ are the same across households,[47] it may seem that the same relationship should hold for aggregate consumption data, which should also grow. This is not so. Indeed, in the context of a stationary population, in which each generation has the same permanent income and in which there are no bequests, average consumption will be constant and the aggregate saving rate would be zero.

What has gone wrong with the attempt to go from (9) in expectational form to its equivalent for average consumption is the neglect of the distinctive behaviour of exiting households. In a stationary population, exiting households are replaced by entering households.

---

[45] However, when inflation and interest rates both rise and credit is rationed, a 'front-end loading problem' arises for borrowers whose cash debt-service payment rises, as a bigger fraction of the long-term burden of interest charges and debt repayment is loaded on to the current period. This can give rise to a negative nominal interest-rate effect on aggregate spending.

[46] Named after the then Chief Economic Adviser and the Chancellor of the Exchequer, respectively (see Muellbauer and Murphy, 1990).

[47] This is a questionable assumption where δ incorporates the survival probability of households.

Table 7.1. Successive Revisions of CSO Estimates of 1974–5 Consumer Expenditure and Saving Rates

Data vintage	1978	1980	1983	1986	1989	1992
1974 nominal consumer expenditure*	52.0	52.1	52.6	53.1	53.2	53.7
1975 nominal consumer expenditure*	63.6	63.7	64.7	65.2	65.5	66.1
1974 saving ratio %	14.1	14.2	12.2	11.9	11.1	10.0
1975 saving ratio %	15.3	14.7	12.6	12.8	12.0	10.6

Note. * In £10,000m.
Source: Table 5, Economic Trends Annual Supplements.

At that point in the chain linking the household consumption level to that of the following year, there is a sudden drop from the high level in the last year of life to the low level in the first, compensating for the marginal growth implied by (9) for all the other years of the life cycle, and leaving average consumption constant. Of course, with productivity growth, the permanent income of each cohort would be higher, resulting in growth of average consumption. But this has everything to do with the growth of average income and nothing to do with the amount by which the market interest rate, r, exceeds the discount rate.[48] Indeed, the aggregation phenomenon could explain some of the aggregate time-series evidence against the Hall Euler equation model.

Turning to the aggregation of solved-out consumption functions, consider the aggregation problem in terms of a simplified model. Suppose that on a *per capita* basis there is a linear relationship for household, h, between consumption and assets and income, $c_h = A_h + \beta_h y_h$. Then for aggregate *per capita* data a similar relationship, $c = \alpha A + \beta y$, holds, where α and β are parameters that vary with the population shares and relative assets and incomes of the different subgroups in the population. These will depend both on the rates of population and income growth and distributional changes. *A priori*, one might have expected substantial consequences for the aggregate saving ratio. There are few countries where there are likely to have been bigger changes than in Japan. However, Murata (1994) considers some numerical illustrations for Japan which suggest that in the last 30 years it is unlikely that more

than a 3 percentage point fall in the saving ratio can be explained by these demographic and age-distributional changes. Similarly, US estimates of the effect of demography on the aggregate saving rate also suggest they are quite small, see Bosworth *et al.* (1991) and Kennickell (1990).

V. Empirical evidence

1. Data issues

No discussion of the consumption function can be complete without mentioning data issues. Let us illustrate some of the measurement issues of consumer expenditure and the personal sector by examining some data revisions in the UK that are of particular concern. Table 7.1 shows figures for 1974 and 1975 seen from different later stand-points.

These figures show that consumer expenditure in current prices for 1974 was revised up from 1978 to 1992 by 3.3 per cent and for 1975 by 3.9 per cent. Thus, most of the downward revision in the saving ratios, from 14.1 per cent to 10 per cent and from 15.3 per cent to 10.6 per cent, is the result of revisions in consumer expenditure rather than in income. Figures for 1976–80 have been revised in the same direction.

The sources of data for consumer expenditure include the Family Expenditure Survey (FES), the Food Survey, retail sales inquiries and tax collection data from Customs and Excise, used particularly for drink and tobacco, which are understated in the FES. One of the major consistency checks in the national accounts is to compare expenditure- and income-based estimates of output. In the 1960s and 1970s the income-based estimates were typically 2 or 2.5 per cent below the expenditure-based estimates. This was thought reasonable in the light of the tendency to under-record income in data based on tax records. But in the late 1970s, the expenditure-based figures for output began to be

---

[48] Furthermore, note that at the point in the chain where exiting households are replaced by entering households, the change in consumption depends in particular on the income and expenditure history of the exiting households, thus violating the irrelevance of lagged information to explaining the period-to-period evolution of consumption. Gali (1990, 1991) and Clarida (1991) develop these points and also show how aggregation of households with finite lives can provide yet another explanation for the 'excess smoothness' of consumption. Deaton (1992, pp. 37–43, 65 75, and 167–76) has an excellent discussion of these and other aspects of aggregation in the context of Euler equations.

exceeded by the income-based ones. This led to a search for ways in which expenditure might have been underrecorded[49] and upward revisions took place in consumer expenditure figures back to the early 1950s, though the biggest revisions of data occurred for the 1970s (particularly for 1974–5). As the table suggests, one of the biggest upward revisions took place as recently as post-1989.

These major revisions reduce the estimated increase in the saving ratio, previously thought to have occurred in the great inflation that followed the 1973 oil and raw material price shock. If they are correct, they reduce estimated asset effects on consumption: the fall in real asset values in the 1970s *should* have led to a substantial increase in the saving rate. If it did not, asset effects must be weaker than would have been thought on the old data. This also reduces the negative inflation effect in the DHSY model of consumption, which is probably capturing the same phenomenon (see Hendry, 1994). They must necessarily affect the stability of econometric models re-estimated over the same sample dates but taking a later vintage of data. Sefton and Weale (1994) have undertaken the major task of rebalancing the UK national accounts from 1920 to 1990 using a development of methods put forward by Stone *et al.* (1942). Their work suggests a substantial downward bias in official estimates of mid-1970s saving rates.

Another significant source of revisions in data results from the rebasing of the national accounts in constant prices of different vintages. Price relativities can change considerably. Real energy and raw material prices have fluctuated a great deal. Technological advances have brought down quality-adjusted prices of anything applying micro-chip technology: computers, television, hi-fi, washing machines. Durable goods have tended to fall in price relative to non-durable goods. Thus, 1990 prices will give such goods a smaller weight than will 1980 and 1970 prices. Since volume increases in such goods have been greater than for, say, food, aggregate constant price consumption growth will be lower at 1990 prices than at 1970 prices. Economists have long known that Divisia or chain indices, which use annual reweighting, give a more sensible picture of the evolution of price and volume indices. In the US they are coming into more widespread use in the national accounts, alongside the conventional fixed-weight measures.

In most national accounts, consumption and income data are classified as among those having the lowest sampling errors. Flow of funds of assets and liabilities and the corresponding stock data are regarded as rather less reliable. Pesaran and Evans (1984) argue that large errors of cumulation can occur in building up estimates of asset levels. This is a major reason for their (quasi-) differencing of a solved-out consumption function: they argue that differencing reduces the errors in asset data to white noise. On the other hand, as originally argued in DHSY (1978), differencing can also remove valuable long-run information in the data. This is typical of the kind of trade-offs which must be faced in empirical work.

At the micro level, the problems are no less serious. There is evidence that asset data are widely understated in surveys, especially by the wealthy. Such a conclusion comes from Ferber (1965, 1966) where information from bank and insurance companies on asset holdings was obtained first and these records compared with the subsequent survey responses.

Unless a household is observed over time, it is hard to take appropriate account of the effect of past on present behaviour and to model expectations. Variations in tastes affect not only consumption but, if they are long-lived differences, also asset holdings and income. A lower subjective discount rate will result in more thrifty behaviour and, other things being equal, higher accumulation of assets and human capital. Preferences between goods and leisure may affect labour supply and so income. Again, panel data on households are essential properly to control for such correlations. Synthetic panels of aggregate cohorts of households surveyed in repeated cross-sections, such as the UK FES, have been used as a partial substitute (see Deaton, 1985; Attanasio and Weber, 1992, 1994). Because the same households are not sampled repeatedly, population heterogeneity and the shifting response rate in the survey generate noise that, for example, would tend to underestimate the effect of lagged dependent variables. In the UK, major discrepancies developed in the 1980s between FES estimates of consumption and income growth and the national accounts estimates: for example, the national accounts show a big fall in the saving rate in 1986–8, absent in the FES. These discrepancies probably reflect some combination of a change in the sampling frame, of shifts in the FES response rate, changes in the way some transactions, particularly involving credit and durable goods, were recorded in the FES, and continuing systematic biases such as the under-recording of purchases of alcohol and tobacco and the under-representation of the self-employed in the

---

[49] One of the areas known to be relatively badly measured is the service sector. However, it is puzzling that under-recording of expenditure on services was not accompanied by a similar under-recording of incomes from services. If both had been revised up, little change would have been recorded in the saving ratio.

survey.[50] In this period, there can be little doubt that the aggregates implied by the FES give a much less convincing picture of consumption and income trends than do the national accounts. Further, these surveys typically contain little direct asset information.

As we have seen, aggregation is potentially a serious issue for the aggregate time-series consumption function. The appropriate solution would be to pool micro and macro information, where this is available in acceptable quality.

2. Some empirical illustrations

The following equation, estimated by Muellbauer and Murphy (1993a) for the US, incorporates, in an empirical approximation, many of the theoretical features discussed in section IV. These are lags, some households experiencing credit constraints, real interest rate and uncertainty effects, explicit income expectations generated by a separate econometric model, liquid and illiquid asset effects with different weights, and demographic and distributional effects. The equation has an error-correction form and is formulated for log-consumption[51]

$$\Delta \ln c_t = \text{constant} + 0.51 \, (\ln y_t - \ln c_{t-1})$$
$$(0.12)$$

$$+ 0.17 \, \Delta \ln y_t - 0.22 \, r_t - 0.44 \, \Delta u_t - 0.20 \, ad_t$$
$$(0.08) \qquad (0.11) \quad (0.012) \qquad (0.070)$$

$$+ 0.066 \, LA_{t-1}/y_t + 0.024 \, IA_{t-1}/y_t + 0.14 \, y2h_t$$
$$(0.042) \qquad\quad (0.0085) \qquad\quad (0.07)$$

$$+ \text{'demographic and distributional effects'}$$

$$\text{s.e.} = 0.0036, \; \overline{R}^2 = 0.950, \, DW = 1.93.$$

The estimation period on annual data is 1956–88. The dependent variable, c, is aggregate consumer expenditure (including durable goods) in constant prices; y is real personal disposable non-property income; income uncertainty is proxied by Δu_t, the change in the unemployment rate, and by ad_t, the absolute deviation between current income growth and average income growth over the last 5 years; and $y2h_t$ is the fitted value of expected income growth, defined as

$$E_t \, [0.5 \ln y_{t+1} + 0.5 \ln y_{t+2} - \ln y_t].$$

Liquid assets, LA, is defined in net terms by subtracting debt, including consumer credit and mortgage debt, from gross liquid assets. Illiquid assets, IA, include equities, bonds, and housing wealth.[52]

From this equation we can identify the parameters of an underlying structural equation. The lag parameter β as defined in (26) is estimated at 0.51. Comparable estimates range from 0.46 for Australia (see Lattimore, 1994) and 0.48 for UK regions (see Muellbauer and Murphy, 1994), to 0.66 for Japan (see Murata, 1994), and close to 0.7 for the UK aggregate (see Muellbauer and Murphy, 1993b). The coefficient on $\Delta \ln y_t$, which is interpreted as the product of β and the consumption share of credit-constrained households, is estimated for the US at 0.17. This implies an estimate for the consumption share of the credit-constrained of 35 per cent. For the other countries, the estimates range from around 20 per cent for Japan to 48 per cent for Australia. Apart from that for Australia, none of these estimates is at all precisely determined. However, the Australian estimate may be biased up by the omission of income-growth expectations, which may be correlated with current income growth.

The real interest-rate effect is estimated as negative for the US, negative but somewhat smaller in importance[53] in the UK, omitted because of its insignificance in Australia, and significantly positive in Japan, an intriguing result. As we saw in section II.1, when σ, the elasticity of intertemporal substitution, is low (or consumers are very risk-averse, which is equivalent, given a CES utility function) and the asset-to-income ratio is high and the income growth rate is low, then consumption is likely to rise with the real interest rate. Since the early 1970s, when Japanese growth moderated, these conditions seem to have applied in Japan. Asset-to-income ratios are higher than in the UK or the US, yet, despite improvements in the welfare state, the Japanese saving rate remains high. This is at least consistent with a low value of σ. See Horioka (1990) and Hayashi (1986) for an exhaustive analysis of other potential reasons for the high Japanese saving rate. The positive real interest rate effect in Japan thus seems quite consistent with theory.[54]

---

[50] Slesnick (1992) discusses the consistency of the US National Income and Product Accounts with household survey data. He reports widening unexplained discrepancies between the two in the 1980s, a feature also of UK data, see below. Measurement problems from a US perspective are further discussed by Wilcox (1991) and Bosworth et al. (1991).

[51] This helps to avoid heteroskedastic disturbances and has the implication that asset effects appear in ratio form to income. The log-asset form used in the UK studies surveyed by Church et al. (1994) is less satisfactory.

[52] The figures are an average $0.4IAh_t + 0.6IA_{t-1}$ to reflect the fact that decisions are made over the course of a year rather than only at the beginning of the year. IAh_t is the fitted value of IA_t to overcome the endogeneity of end-of-year asset holdings, and similarly for liquid assets.

[53] However, there is evidence that the coefficient drifted up in the 1980s.

[54] The evidence in Murata appears to contradict two other possible explanations. First, the positive real interest-rate effect is not merely a disguised negative inflation effect. The hypothesis of equal and opposite signs on the nominal interest rate and the inflation rate can be accepted.

The coefficient on the uncertainty proxies have similar long-run coefficients in the US and the UK, while in Japan, where unemployment is low and has not varied greatly, income volatility as measured by *ad* has a marginally higher coefficient than in the US and the UK. Lattimore (1994) finds neither to be significant, though he does find a role for the level of the unemployment rate, and his credit acceleration measure may, in part, reflect uncertainty.

The two wealth variables have fairly similar coefficients in the US and the UK, rather higher in Australia and rather lower in Japan. To illustrate for the US, the coefficient on the illiquid asset/income ratio is 0.024. This suggests a long-run coefficient for households who are not credit-constrained of 0.072 and 0.2 for liquid assets, reflecting the greater spendability of liquid assets. The latter is everywhere less accurately determined than the former.

For Australia and Japan there is strong evidence for the negative real house-price effect discussed in section IV.6, which partly offsets the positive wealth effect of illiquid assets including housing. Indeed, both Lattimore and Murata report that it is otherwise hard to get sensible wealth effects for these countries. This negative real land-price effect, proxying a house-price effect, is the major reason why the Japanese household saving rate did not fall substantially at the time of the 1980s asset price boom. For UK regions there is evidence for a weaker, though significant, effect of this type (Muellbauer and Murphy, 1994). For UK aggregate data, the effect appears not be significant, probably because of collinearity of data; and for the US, house price data before 1963 are of poor quality which limits testing for such an effect.

Expected income growth has a roughly comparable long-run coefficient in the US and the UK and a somewhat higher coefficient in Japan, while Lattimore finds no effect in Australia[55] which corresponds to the case where no households are forward-looking. The model can, in principle, handle any mixture of myopic and forward-looking behaviour. The 'demographic' effects estimated for the US are implausibly large in magnitude and are almost certainly proxying other reasons for the gradual decline in the saving rate between the mid-1970s and mid-1980s.

The decline in the US saving rate from the mid-1970s to the mid-1980s remains only partly explained by wealth effects, a slow-down in growth, and other conventional macroeconomic variables. There is widespread agreement that changing demography can account for only very small changes (Kennickell, 1990; Bosworth *et al.*, 1991).[56] In section IV.1 we reviewed a hypothesis connected with aspects of risk-averse behaviour suggested by Hubbard *et al.* (1994), who emphasize changes in the US welfare benefit system. Brenner *et al.* (1992) suggest that female human capital accumulation in place of asset accumulation in response to higher divorce probabilities is a partial explanation. Bosworth *et al.* (1991) and their discussants, Poterba and Summers, point to more widespread ownership of pension and insurance plans as well as improved regulation of pension schemes, which raised the probability that pension promises are fulfilled. Poverty among the old fell in the 1970s and 1980s, suggesting that expected income security in old age improved, thus reducing saving. Participation rates for medical insurance rose too, thus reducing the need for precautionary saving.

The hypothesis advanced by Boskin and Lau (1988) of a cohort effect whereby those born after 1930 have a lower savings propensity than those born in the Depression years appears not to be strongly supported by the micro-evidence discussed by Bosworth *et al.* (1991) but may nevertheless have played a part.

Another hypothesis, advanced by Bernheim (1991) and Hatsopolous *et al.* (1989), is that corporate restructuring, encouraged by the tax system, which retired much corporate equity and replaced it by debt, put a great deal of cash into the pockets of US owners of equity in the 1980s. Poterba (1991) argues that the marginal propensity to spend such cash receipts is around 50 per cent, much higher than typical marginal propensities to spend out of wealth.

A final hypothesis is the one of financial liberalization, which improved access to credit, particularly for home loans, and reduced the ratio of down payment to house price for first-time buyers. This has already been discussed in the UK context. Though there is considerable circumstantial evidence of changes in US credit markets, it is difficult to distinguish changes in credit-market conditions that stem from supply-side changes from those on the demand side.

Second, this does not appear to be a case of reverse causation or endogeneity bias: if the authorities raise interest rates when consumption growth is too rapid, one might find a positive association between the real interest rate and consumption. However, Japanese short-term interest rates are driven mainly by inflation, US interest rates, and lagged growth. There remains one other possibility. The real interest rate, which was low in the 1970s and rose in the 1980s, may be a proxy for financial liberalization in Japan. This could induce a positive association with consumption.

[55] Paradoxically, this may be the result of attributing to households an overly sophisticated econometric model for forecasting income.

[56] Though it may be that conditioning not on whether someone is over or under 65, but on retirement status, would yield somewhat bigger estimates. Earlier retirement and lower participation by the over 65s have been a feature of the 1970s and 1980s.

Overall, Carroll's (1992) persuasive review of the evidence argues for a combination of easier access to credit and reduced uncertainty as the main factors that explain the decline in the US personal saving rate.

As far as financial liberalization in the UK is concerned, interaction effects with illiquid wealth and debt are fairly imprecisely estimated in Muellbauer and Murphy (1993b). Some of the reasons for this were discussed in section IV.7. The implication is that a relatively simple equation, such as that estimated for the US, would have tracked consumer expenditure quite well in the 1980s, without the massive breakdown experienced by all the UK macroeconometric models extant in the 1980s. Our US equation and its UK equivalent[57] give valuable insights into the explanations for the boom–bust cycle of the last decade and parallel developments in Scandinavia outlined by Berg (1994). For the UK, we have decomposed fluctuations in the ratio of consumer expenditure to non-property income into the various estimated effects we are able to isolate. The rise in the consumption-to-income ratio of around 10 percentage points between 1980 and 1988 was unprecedented in recorded UK history. The contribution of forecast income growth to this rise was just over 2 percentage points between 1980 and 1988. But the rise in the real interest rate reduced the consumption-to-income ratio by around 1 percentage point over the same period.

The estimated contribution of the fall in the change in the unemployment rate was around 2.5 percentage points from 1980 to 1988, with a little over 0.5 percentage points from the fall in income volatility measured by ad. Altogether, this suggests reduced income uncertainty contributing around 3 percentage points to the upturn in the consumption-to-income ratio, c/y.

The higher growth rate of current income, which enters the equation if some households are credit rationed, explains around 0.5 percentage points of the rise in c/y. While this is small, it should be noted that the effect of current income on consumption is already built in.[58]

The biggest single contributor to the rise in the c/y ratio was the rise in the spendability-weighted net asset-to-income ratio, which explains 5 percentage points (i.e. half) of the rise. This large positive effect was partly offset by around 1 per cent from the rise of inequality as measured by the ratio of non-manual to manual earnings. This is incorporated on the argument that higher income households tend to have higher saving rates. The reason is that households with a greater preference for thrift and human capital accumulation will, on average, have higher incomes.

Interestingly, the consumption-to-income ratio peaks before the asset-to-income ratio: the down-turn in c/y, while assets to income are still rising, is explained by a combination of factors. These include a sharp fall in expected income growth, a sharp rise in the rate of increase of unemployment, an increase in income volatility, a rise in the real interest rate, and the down-turn in current income growth.

By 1991 the lagged asset-to-income ratio was falling too, mainly because of the direct and indirect consequences of the sharp increases in short-term interest rates, which doubled between 1988 and 1990. This is all consistent with the dramatic fall that occurred in the consumption-to-income ratio between 1989 and 1991, and beyond. The lower values of real share prices and later, real house prices, and the high levels of debt—which has perhaps three times the spendability weight of illiquid assets, but a *negative* effect on consumption—help to explain the depth of the UK consumer recession in 1992 and the sluggishness of recovery in 1993, despite much lower interest rates.

Similar patterns of influences plausibly explain the analogous boom and bust experienced in consumer expenditure in Scandinavian economies in the 1980s and 1990s. For the US as a whole, fluctuations were much less pronounced, though individual regions were as volatile as the UK and Scandinavia. Our estimates in Muellbauer and Murphy (1993a) show US asset coefficients of similar magnitude to those in the UK. But they explain less of the variation in c/y because they change by much less than in the UK. Undoubtedly, the debt over-hang and employment insecurity contributed to the sluggish pace of recovery in the US until 1993, despite the very low levels reached by interest rates. The rapid increases in prices of financial assets experienced in 1993, together with falling unemployment and improved growth expectations, explain much of the recovery in the US seen in 1993 and early 1994.

[57] In addition to the variables in the US equation, the UK equation contains an interaction effect between illiquid wealth and a financial liberalization indicator; the change in a measure of the tightness of credit controls on consumer durables finance which was an important policy instrument before 1980; and a measure of strike incidence.

[58] Another variable linked with credit rationing had no long-run effect. Before 1979 the government used consumer credit controls to regulate consumer expenditure. These took the form of legal restrictions on down-payments and length of contract when buying durable goods by 'hire purchase'. It is impossible successfully to model expenditure on durables (which is part of total consumer expenditure) in the UK before 1979 without taking variations in these restrictions into account. Since

their effect is only transitory and they were abolished in 1982 they had no effect on the 1980–8 comparison.

VI. Conclusions

These conclusions begin with methodological issues, then discuss gaps in our knowledge and future research needs, and end with current policy issues. As we have seen, research on the Euler equation approach to aggregate consumption has flourished in the last 15 years, largely displacing work on the more traditional 'solved-out' consumption function. The Euler approach assumes rational expectations and focuses on the intertemporal efficiency condition linking consumption in adjacent periods. The solved-out consumption function, in its most basic form, solves the full set of these efficiency conditions stretching to the horizon to derive a solution for consumption in terms of initial assets and human capital, i.e. the present value of current and expected non-property income. Most of the Euler equation literature on aggregate consumption treats it as generated by a rational, infinitely lived, representative consumer operating in efficient financial markets without credit restrictions or transactions costs and restrictions. There is a widespread presumption that this is more rigorous than trying to develop more comprehensive, realistic, but necessarily approximate, models.

Apart from a desire to follow fashion, there are several reasons why a researcher might prefer modelling Euler equations to modelling solved-out consumption functions. The first is that, for individual households under rational expectations, taking expectations of the intertemporal efficiency condition eliminates the need to model expectations explicitly and to introduce asset data, which may not always be accurate or easily available. The second is that under uncertainty, with risk-averse behaviour, the individual household Euler equation can, in principle, be estimated directly using the generalized method-of-moments estimator, see Hansen and Singleton (1981). In contrast, as we have seen in section IV.1, there does not exist an exact analytical expression for the solved-out consumption function for this case, though there are convenient approximations.

However, there are five significant reasons for questioning the alleged superiority of the Euler approach. First, except at the microeconomic level in the context of panel data, the advantage that there is a one-to-one connection under the Euler approach between theory and estimation is something of an illusion. If a particular Euler equation held at the micro level, it would not hold for aggregate data as we saw in section IV.8, except under incredible assumptions about households behaving as if life were infinite. Fundamentally, this is because when households die and are replaced (or more than replaced in a growing population) by new households, the information link connecting consumption in adjacent periods is broken. Further, non-linearities and household heterogeneity will complicate the mapping from micro to macro, except under special restrictions.

Second, when some households are credit-constrained, simplifying approximations for modelling aggregate behaviour are needed in both the Euler and solved-out approaches. This is because the fraction of credit-constrained households is not observable and is unlikely to be constant, and because not all the credit-constrained spend only current income.

Third, knowledge of the parameters of an Euler equation consistent with risk-averse behaviour (and hence non-linear in form) would be of limited usefulness for forecasting or policy analysis. To work out the effects of a tax cut or an increase in asset prices on consumption over the next few years, necessitates obtaining an explicit solution for consumption of just the type that the solved-out consumption function is designed to provide. Inevitably, because no analytic solution exists, and because measures of uncertainty and expectations have to be given empirical content, this entails making the kinds of approximations discussed in section IV, and illustrated in section V. The usefulness of the solved-out approach in decomposing cyclical and secular movements in the saving ratios into different elements, such as income growth effects, asset effects, demographic effects, uncertainty effects, and financial liberalization effects, was amply demonstrated in section V, despite the inevitable empirical controversies surrounding such interpretations.[59] At least it offers a framework for analysing current developments.

Fourth, when a significant fraction of households do not have rational expectations, the martingale or 'surprise' formulation of the Euler equation breaks down. Admittedly, in some of the Euler equation literature on excess sensitivity (e.g. Campbell and Mankiw, 1989, 1991), the parameter π in (23) is interpreted as the proportion of credit-constrained or myopic consumers. However, there is no reason why myopic consumers

---

[59] A North American critic might well accuse such solved-out models of being over-parameterized (i.e. having more parameters than could robustly be estimated from the limited number of observations) and of suffering from simultaneity bias, which can occur when consumption shocks feed back within the period of observation on to the explanatory variables. A response would be that such models become more credible if similar estimates are obtained for a range of countries, if they hold up well to predictive tests, and if quarterly models, where there is less scope for within-period feedbacks to occur, give similar results to those estimated on annual data.

do not save or dissave out of accumulated assets and so no reason for their consumption to be constrained by income.

Finally, because the Euler approach involved something close to first-differencing of the data, it throws away long-run information on the levels of consumption, income, assets, and demography. As emphasized in the literature on co-integration and error correction, exploiting such information can convey substantial statistical advantages, providing that measurement errors in such data are stationary.

Thus, we conclude that the enthusiasm for the Euler approach and the large-scale abandonment, particularly in North America, of the solved-out approach for modelling aggregate consumption was somewhat misplaced. This is not to deny that many useful insights have been gleaned from the Euler equation controversies of the last 15 years—in conjunction with a great deal of analysis of survey data. The latter has suggested that consumption follows income more closely over the life cycle than the simple rational expectations permanent income hypothesis or its life-cycle equivalent would suggest. The explanation lies largely in a combination of credit constraints, precautionary behaviour under uncertainty, consumption habits depending on previous consumption levels, and, to some extent, in the variation of needs over the life cycle being correlated with income. Survey data also suggest that the old do not dissave on the text-book scale. The explanation, again, is in terms of uncertainty about income, needs, and length of life, and, to some extent, the bequest motive.

Many, though not all, econometricians favour general to specific modelling, i.e. begin with a general maintained hypothesis and apply a sequential testing procedure of simplifying restrictions to allow the data to speak in arriving at a more parsimonious model. As we have seen, both Euler equations and solved-out consumption functions need to incorporate the effects of age (and perhaps other household characteristics), and a reasonable treatment of aggregation over households, precautionary behaviour in the face of uncertainty (which rules out the quadratic utility function/ linear rational expectations approach popular in much of the literature), credit constraints, habits or costs of adjustment, and, where relevant, the durability of goods. To obtain tractable models, approximations are essential. We have demonstrated that there is scope for a more imaginative use of theoretically based proxies, particularly for the solved-out consumption function, but also for Euler equations, where consumption is modelled directly rather than indirectly via marginal utility.

Let us turn now to gaps in our knowledge and future research. The pooling of micro and macro information to deal better with aggregation issues, and hence with the effect of demographic and distributional changes on consumption, is a high priority, despite the problems discussed in section V.1. The development of good proxies for uncertainty and for income expectations, perhaps drawing on regular survey data, and comparative studies on regions or countries to test different specifications on a wide set of sample information are needed. More work is needed on the spendability of assets with different liquidity characteristics and on relating differences in the spendability of housing wealth across countries to the tax systems, transaction costs, and other institutional features. For example, the lower weights on life insurance and pension fund wealth predicted by theory need empirical investigation. Potentially, the resulting weighted wealth aggregates could be of major importance in tuning monetary and fiscal policy and in macroeconomic forecasting.

One of the biggest lacunae, in the consumption literature and more generally in empirical macroeconomics, both at the level of theory and of data, concerns variations in credit conditions. Particularly in the 1970s, when negative after-tax interest rates were common, credit-rationing was rife, though often frustrating portfolio investment more than consumption: borrowing at negative interest rates and investing in illiquid assets such as houses, paintings, or equities with more favourable after-tax return prospects than liquid assets, was a highly desirable prospect. The freeing of interest rates, financial deregulation, the fall of inflation, and tax reform altered these conditions, but asymmetric information between borrowers and lenders ensures that, generally, credit contracts do not depend only on price. However, few countries have regular surveys of credit conditions in which both financial institutions and households (and firms) are questioned about their perceptions of credit conditions, i.e. what credit contracts are on offer and whether, relative to the past, conditions are tighter or looser. This has made it difficult to distinguish precisely the effects of financial liberalization on the fall in saving ratios, which was such a feature of the 1980s.

Missing data on variations in credit conditions has contributed to the problems of integrating the modelling of consumption and portfolio decisions, see section IV.6. This remains an important area for further research. Progress is likely to come from an approach that eschews, at least on macro data, the two extremes of overly demanding theoretical rigour and empiricism. One needs to avoid, on the one hand, the theoretical rigour that demands precise analytical solutions which are only derivable under extreme, unrealistic assumptions, and, on the other hand, empiricism, which fails to take on board the many modelling insights which theory has to offer. It is possible, though by no means easy, to learn both from theory and data in somewhat more flexible ways than have often been prevalent in the research of the last 15 years.

To turn to some policy conclusions, it is now clear that a major failure of economic policy in the 1980s, both in Scandinavia and the UK, was to deregulate financial markets without at the same time reforming the tax system to make borrowing less tax-advantaged. Tax reform has lagged far behind, exacerbating the early 1990s' downturn, particularly in Scandinavia, as noted by Berg (1994). The models for the US, UK, Japan, and Australia which were discussed in the previous section offer important insights into fluctuations in consumer spending in these economies and in Scandinavia. The UK model tracks well both the boom in consumer spending in the 1980s and the bust in the 1990s.

It is worth highlighting some of the differences from the UK models reviewed by Church *et al.* (1994) in terms of implications for modelling the 1990s downturn. One important difference is that the Muellbauer and Murphy (1993*b*) model gives debt a larger negative role, and debt-to-income ratios have recently been at an all-time high. The other UK models aggregate all wealth into a single aggregate while we allow spendability of gross liquid assets minus debt to be two or three times as large as the spendability of illiquid assets. A second important difference is the explicit model for income expectations which incorporates the negative effect on future growth of increases in interest rates and of unemployment changes. This brings a significant downturn following the interest-rate rises of 1988–90. A third difference is the explicit income uncertainty proxy which rose sharply in 1990–2, dampening consumption.

As far as developments in 1994 are concerned, it is hard to believe that the tax rises from April 1994 and the falls in asset values since February will not be reflected in dampened consumption growth in the remainder of 1994, reducing pressure on the balance of payments and on interest rates. Consumption function research confirms the role of asset values in contributing to the determination of consumer expenditure in the UK. The Bank of England is right to monitor asset prices in evaluating the stance of macroeconomic policy in the UK.

References

Acemoglu, D., and Scott, A. (1994), 'Consumer Confidence and Rational Expectations: Are Agents' Beliefs Consistent with the Theory?', *Economic Journal*, **104**, 1–19.

Alessie, R., and Lusardi, A. (1993), 'Saving and Income Smoothing: Evidence from Panel Data', Center, Tilburg University.

Ando, A., and Kennickell, A. B. (1987), 'How Much (or Little) Life Cycle is there in Micro Data? The Cases of the United States and Japan', in R. Dornbusch, S. Fischer, and J. Bossons (eds), *Macroeconomics and Finance: Essays in Honor of Franco Modigliani*, Cambridge MA, MIT Press, 159–223.

Ando, A., and Modigliani, F. (1963), 'The Life-cycle Hypothesis of Saving: Aggregate Implications and Tests', *American Economic Review*, **53**, 55–84.

Attanasio, O. P., and Weber, G. (1992), 'Consumption Growth and Excess Sensitivity to Income: Evidence from US Micro Data', Stanford University and University College, London (April), mimeo.

—— (1994), 'Consumption Growth, the Interest Rate and Aggregation', *Review of Economic Studies*, **61**, 19–30.

Attfield, C. L. F., Demery, D., and Duck, N. W. (1990), 'Saving and Rational Expectations: Evidence for the UK', *Economic Journal*, **100** (December), 1269–76.

Ball, R. J., and Drake, P. S. (1964), 'The Relationship Between Aggregate Consumption and Wealth', *International Economic Review*, **5**, 63–81.

Banerjee, A., and Dolado, J. (1988), 'Tests of the Life-cycle Permanent Income Hypothesis in the Presence of Random Walks', *Oxford Economic Papers*, **40**, 610–33.

—— Galbraith, J. W., and Hendry, D. F. (1993), *Co-integration, Error-Correction and the Econometric Analysis of Non-Stationary Data*, Oxford, Oxford University Press.

—— Lumsdaine, R. L., and Stock, J. H. (1992), 'Recursive and Sequential Tests of the Unit Root and Trend Break Hypothesis: Theory and International Evidence', *Journal of Business and Economic Statistics*, **10**.

Banks, J., Blundell, R., and Preston, I. (1993), 'Life-cycle Allocations and the Consumption Costs of Children', London, Institute for Fiscal Studies.

Barnett, W. A. (1981), *Consumer Demand and Labor: Supply: Goods Monetary Assets and Time*, Amsterdam, North-Holland.

Barro, R. J. (1974), 'Are Government Bonds Net Wealth?', *Journal of Political Economy*, **82**, 1095–117.

Bean, C. R. (1986), 'The Estimation of "Surprise" Models and the "Surprise" Consumption Function', *Review of Economic Studies*, **49**, 497–516.

Berg, L. (1994), 'Household Savings and Debts: The Experience of the Nordic Countries', *Oxford Review of Economic Policy*, **10**(2), 42–53.

—— and Bergström, R. (1993), 'Consumption, Income Wealth and Household Debt: An Econometric Analysis of the Swedish Experience 1970–1992', Working Paper 1993. 12, Dept of Economics, Uppsala University.

Bernheim, B. D. (1991), *The Vanishing Nest Egg: Reflections on Saving in America*, New York, Twentieth Century Fund, Priority Press.

Bilson, J. F. O. (1980), 'The Rational Expectations Approach to the Consumption Function', *European Economic Review*, 13, 273–99.

Blanchard, O. J. (1985), 'Debts, Deficits, and Finite Horizons', *Journal of Political Economy*, 93, 1045–76.

—— Fischer, S. (1989), *Lectures on Macroeconomics*, Cambridge, MA, MIT Press.

Blinder, A. S., and Deaton, A. S. (1985), 'The Time Series Consumption Function Revisited', *Brookings Papers on Economic Activity*, 2, 456–521 (includes discussion).

Blundell, R. (1988), 'Consumer Behaviour: Theory and Empirical Evidence—A Survey', *Economic Journal*, 98, 16–65.

—— Browning, M., and Meghir, C. (1994), 'Consumer Demand and the Life-cycle Allocation of Household Expenditures', *Review of Economic Studies*, 61, 57–80.

Börsch-Supan, A., and Stahl, K. (1991), 'Life-cycle Savings and Consumption Constraints. Theory, Empirical Evidence, and Fiscal Implications', *Journal of Population Economics*, 4, 233–55.

Boskin, M. J., and Lau, L. J. (1988), 'An Analysis of Postwar US Consumption and Saving: Part 1 and Part 2', NBER Working Papers 2605, 2606.

Bosworth, B., Burtless, G., and Sabelhaus, J. (1991), 'The Decline in Saving: Evidence from Household Surveys', *Brookings Papers on Economic Activity*, 1, 183–256 (with discussion).

Breeden, D. T. (1979), 'An Intertemporal Asset Pricing Model with Stochastic Consumption and Investment Opportunities', *Journal of Financial Economics*, 7, 265–96.

Brenner, R., Dagenais, M. G., and Montemarquette, C. (1992), 'The Declining Saving Rate: An Overlooked Explanation', CRDE Papers, 0792, University of Montreal, Centre of Research and Development in Economics.

Brodin, P. A., and Nymoen, R. (1992), 'Wealth Effects and Exogeneity: The Norwegian Consumption Function', *Oxford Bulletin of Economics and Statistics*, 54(3), 431–54.

Browning, M. J., Deaton, A. S., and Irish, M. (1985), 'A Profitable Approach to Labor Supply and Commodity Demands over the Life Cycle', *Econometrica*, 53, 503–44.

Caballero, R. (1994), 'Notes on the Theory and Evidence on Aggregate Purchases of Durable Goods', *Oxford Review of Economic Policy*, 10(2), 107–17.

Campbell, J. Y. (1987), 'Does Saving Anticipate Declining Labour Income? An Alternative Test of the Permanent Income Hypothesis', *Econometrica*, 55, 1249–74.

—— Deaton, A. S. (1989), 'Why is Consumption so Smooth?', *Review of Economic Studies*, 56, 357–74.

Campbell, J. Y., and Mankiw, N. G. (1989), 'Consumption, Income and Interest Rates: Reinterpreting the Time Series Evidence', in O. J. Blanchard and S. Fischer (eds.), *NBER Macroeconomic Annual 1989*, Cambridge, MA, MIT Press, 185–216.

—— (1991), 'The Response of Consumption to Income: A Cross-Country Investigation', *European Economic Review*, 35, 715–21.

Carroll, C. D. (1992), {The Buffer-Stop Theory of Saving: Some Macroeconomic Evidence', *Brookings Papers on Economic Activity*, 2, 61–156 (with discussion).

—— Fuhrer, J. C., and Wilcox, W. (1993), 'Does Consumer Sentiment Forecast Household Spending?', *American Economic Review*.

—— Summers, L. J. (1991), 'Consumption Growth Parallels Income Growth: Some New Evidence', in B. D. Bernheim and J. B. Shoven (eds), *National Saving and Economic Performance*, Chicago, Chicago University Press for NBER, 305–43.

Church, K., Smith, P., and Wallis, K. (1994), 'Economic Evaluation of Consumers' Expenditure Equations', *Oxford Review of Economic Policy*, 10(2), 71–85.

Clarida, R. H. (1991), 'Aggregate Stochastic Implications of the Life-cycle Hypothesis', *Quarterly Journal of Economics*, 106, 851–67.

Cochrane, J. H. (1989), 'The Sensitivity of Tests of the Intertemporal Allocation of Consumption to Near-rational Alternatives', *American Economic Review*, 79(3), 319–37.

Daly, V., and Hadjimatheou, G. (1981), 'Stochastic Implications of the Life Cycle–Permanent Income Hypothesis: Evidence for the UK Economy', *Journal of Political Economy*, 89, 596–9.

Darby, J., and Ireland, I. (1994), 'Consumption, Forward Looking Behaviour and Financial Deregulation', University of Strathclyde, mimeo.

Davidson, J. E. H., Hendry, D. F., Srba, F., and Yeo, S. (1978), 'Econometric Modelling of the Aggregate Time-series Relationship Between Consumers' Expenditure and Income in the United Kingdom', *Economic Journal*, 88, 661–92.

Davidson, J. E. H., and Hendry, D. F. (1981), 'Interpreting Econometric Evidence: The Behaviour of Consumer Expenditure in the UK', *European Economic Review*, 16, 177–92.

Deaton, A. S. (1985), 'Panel Data from a Time-series of Cross-sections', *Journal of Econometrics*, 30, 109–26.

—— (1987), 'Life-cycle Models of Consumption: Is the Evidence Consistent with the Theory?' in T. F. Bewley (ed.), *Advances in Econometrics, Fifth World Congress*, Vol. 2, Cambridge and New York, Cambridge University Press, 121–48.

—— (1991), 'Saving and Liquidity Constraints', *Econometrica*, 59(5), 1221–48.

—— (1992), *Understanding Consumption*, Oxford, Clarendon Press.

—— Muellbauer J. N. J. (1980), *Economics and Consumer Behaviour*, Cambridge, Cambridge University Press.

Dickey, D. A., and Fuller, W. A. (1979), 'Distribution of the Estimates for Autogressive Time Series with a Unit Root', *Journal of the American Statistical Association*, 74(366), 427–31.

Engle, R. F., and Granger, C. W. J. (1987), 'Cointegration and Error Correction: Representation, Estimation and Testing', *Econometrica*, 55, 251–76.

Englund, P. (1990), 'Financial Deregulation in Sweden', *European Economic Review*, 34.

Ermini, L. (1994), 'A Discrete-Time Consumption–CAP Model under Durability of Goods, Habit Formation and Temporal Aggregation', University of Hawaii, Manoa, mimeo.

Evans, M. K. (1969), *Macroeconomic Activity*, New York, Harper and Row.

Farrell, M. J. (1959), 'The New Theories of the Consumption Function', *Economic Journal*, **69**, 678–96.

Feldstein, M. S. (1977), 'Social Security and Private Savings: International Evidence in an Extended Life-cycle Model', in M. S. Feldstein and R. P. Inman (eds), *The Economics of Public Services* Davidson, J. E. H., Hendry, D. F., Macmillan, 174–205.

—— (1980), 'International Differences in Social Security and Saving', *Journal of Public Economics*, **82**(5), 905–26.

—— (1994), 'College Scholarship Rules and Private Saving', *American Economic Review*, **84**.

Ferber, R. (1965), 'The Reliability of Consumer Surveys of Financial Holdings: Time Deposits', *Journal of the American Statistical Association*, 148–63.

—— (1966), 'The Reliability of Consumer Surveys of Financial Holdings: Demand Deposits', *Journal of the American Statistical Association*, 91–103.

Flavin, M. A. (1981), 'The Adjustment of Consumption to Changing Expectations about Future Income', *Journal of Political Economy*, **89**(5), 974–1009.

Flemming, J. S. (1973), 'The Consumption Function when Capital Markets are Imperfect', *Oxford Economic Papers*, **25**, 160–72.

Friedman, M. (1957), *A Theory of the Consumption Function*, Princeton, NJ, Princeton University Press.

Friedman, M. (1963), 'Windfalls, the "Horizon" and Related Concepts', in C. Christ *et al.* (eds), *Measurement in Economics: Studies in Mathematical Economics and Econometrics in Memory of Yeluda Grunfeld*, Stanford, CA, Stanford University Press, 1–28.

Fuhrer, J. C. (1993), 'What Role Does Consumer Sentiment Play in the US Macroeconomy?', *New England Economic Review*, January/February.

Fuller, W. A. (1976), *Introduction to Statistical Time Series*, New York, Wiley.

Gali, J. (1990), 'Finite Horizons, Life-cycle Savings and Time-series Evidence on Consumption', *Journal of Monetary Economics*, **26**, 433–52.

—— (1991), 'Budget Constraints and Time-series Evidence on Consumption', *American Economic Review*, **81**, 1238–53.

Gilbert, C. (1986), 'Professor Hendry's Econometric Methodology', *Oxford Bulletin of Economics and Statistics*, **48**(3), 283–307.

Granger, C. W. J. (1986), 'Developments in the Study of Cointegrated Economic Variables', *Oxford Bulletin of Economics and Statistics*, **48**, 213–28.

Grossman, S. J., and Shiller, R. (1981), 'The Determinants of the Variability of Stock Market Prices', *American Economic Review, Papers and Proceedings*, May, 222–7.

Haig, R. M. (1921), 'The Concept of Income: Economic and Legal Aspects', reprinted in R. A. Musgrove and C. S. Shoup (eds.), *Readings in the Economics of Taxation*, Homewood, IL, Irwin, 54–76.

Hall, R. E. (1978), 'Stochastic Implications of the Life Cycle–Permanent Income Hypothesis: Theory and Evidence', *Journal of Political Economy*, **96**, 971–87.

—— Mishkin, F. S. (1982), 'The Sensitivity of Consumption to Transitory Income: Estimates from Panel Data on Households', *Econometrica*, **50**, 461–81.

Hansen, L. P., and Sargent, T. J. (1981), 'A Note on Wiener–Kolmogorov Forecasting Formulas for Rational Expectations Models', *Economics Letters*, **8**, 253–60.

—— Singleton, K. J. (1982), 'Generalized Instrumental Variables Estimation of Non-linear Rational Expectations Models', *Econometrica*, **50**, 1269–86.

—— —— (1983), 'Stochastic Consumption, Risk Aversion, and the Temporal Behavior of Asset Returns', *Journal of Political Economy*, **91**(2), 249–65.

Harvey, A. C. (1989), *Forecasting, Structural Time Series Models and the Kalman Filter*, Cambridge, Cambridge University Press.

—— Jaeger (1993), 'Detrending, Stylized Facts and the Business Cycle', *Journal of Applied Econometrics*, **8**, 231–48.

—— Scott, A. (1994), 'Seasonality in Dynamic Regression Models', *Journal of Econometrics*.

Hatsopolous, G. N. P., Krugman, R., and Poterba, J. M. (1989), *Overconsumption: The Challenge to US Policy*, Washington, DC, American Business Conference.

Hayashi, F. (1982), 'The Permanent Income Hypothesis: Estimation and Testing by Instrumental Variables', *Journal of Political Economy*, **90**, 895–916.

—— (1986), 'Why is Japan's Saving Rate so Apparently High?', *NBER Macroeconomic Annual*.

—— (1987), 'Tests for Liquidity Constraints: A Critical Survey and Some New Observations', in T. F. Bewley (ed.), *Advances in Econometrics: Fifth World Congress*, 1–120.

Heaton, J. (1993), 'The Interactions Between Time-Nonseparable Preferences and Time Aggregation', *Econometrica*, **61**, 353–85.

Hendry, D. F. (1988), 'The Encompassing Implications of Feedback versus Feedforward Mechanisms in Econometrics', *Oxford Economic Papers*, **40**, 132–49.

—— (1994), 'HUS Revisited', *Oxford Review of Economic Policy*, **10**(2), 86–106.

—— von Ungern-Sternberg, T. (1981), 'Liquidity and Inflation Effects on Consumer's Expenditure', in A. S. Deaton (ed.), *Essays in Theory and Measurement of Consumers' Behaviour*, 237–60, Cambridge, Cambridge University Press.

Hicks, J. R. (1946), *Value and Capital*, Oxford, Clarendon Press.

Horioka, C. (1990), 'Why is Japan's Saving Rate so High? A Literature Survey', *Journal of Japanese and International Economies*, **4**, 49–92.

Hubbard, R. G., Skinner, J., and Zeldes, S. P. (1994), 'The Importance of Precautionary Motives in Explaining Individual and Aggregate Saving', Carnegie-Rochester Conference on Public Policy.

Jappelli, T., and Pagano, M. (1989), 'Consumption and Capital Market Imperfections: An International Comparison', *American Economic Review*, 79, 1088–105

Johnson, P. (1983) 'Life-cycle Consumption under Rational Expectations: Some Australian Evidence', *Economic Record*, 59(167), 345–50.

Jones, S. R. J., and Stock, J. H. (1987), 'Demand Disturbances and Aggregate Fluctuations: The Implication of Near-Rationality', *Economic Journal*, 97, 49–64.

Kennedy, N. and Anderson, P. (1994), 'Household Saving and Real House Prices: An International Perspective', Basle, Bank for International Settlements.

Kimball, M. (1990), 'Precautionary Saving in the Small and in the Large', *Econometrica*, 58, 53–73.

Koskela, E., Loikkanen, H. A., and Viren, M. (1992), 'House Prices, Household Savings and Financial Market Liberalisation in Finland', *European Economic Review*, 36(2/3), 549–58.

Kotlikoff, L. J. (1988), 'Intergenerational Transfers and Savings', *Journal of Economic Perspectives*, 2, 41–58.

—— Summers, L. H. (1981), 'The Role of Intergenerational Transfers in Aggregate Capital Formation', *Journal of Political Economy*, 89, 706–32.

Lattimore, R. (1993), 'Consumption and Saving in Australia', Oxford, D. Phil. thesis.

—— (1994), 'Australian Consumption and Saving', *Oxford Review of Economic Policy*, 10(3), 54–70.

Lehmussaari, O. P. (1990), 'Deregulation and Consumption Saving Dynamics in the Nordic Countries', *IMF Staff Papers*, 37(1).

Lucas, R. E. (1976), 'Econometric Policy Evaluation: A Critique', *Carnegie–Rochester Conference Series Supplement to the Journal of Monetary Economics*, 1, 19–46.

—— (1978), 'Asset Prices in an Exchange Economy', *Econometrica*, 46, 1429–45.

Mankiw, N. G., and Shapiro, M. (1985), 'Trends, Random Walks and Tests of the Permanent Income Hypothesis', *Journal of Monetary Economics*, 16, 165–74.

—— —— (1986), 'Do We Reject Too Often? Small Sample Properties of Tests of Rational Expectations Models', *Economics Letters*, 20(2), 139–45.

Miles, D. (1993), 'House Price Shocks and Consumption with Forward Looking Households', Birkbeck College, London, mimeo.

Modigliani, F. (1975), 'The Life-cycle Hypothesis of Saving Twenty Years Later', in M. J. Parkin and A. R. Nobay (eds), *Contemporary Issues in Economics: Proceedings of the AUTE Conference, 1973*, Manchester, Manchester University Press, 2–36.

—— (1988) 'The Role of Intergenerational Transfers and Life-cycle Saving in the Accumulation of Wealth', *Journal of Economic Perspectives*, 2(2), 15–40.

—— (1990), 'Recent Declines in the Savings Rate: A Life-cycle Perspective', Frisch Lecture, Sixth World Congress of the Econometric Society, Barcelona (August), mimeo.

—— Brumberg, R. (1954), 'Utility Analysis and the Consumption Function: An Interpretation of the Cross-section Data', *Post-Keynesian Economics*, New Brunswick, NJ, Rutgers University Press.

—— (1979), 'Utility Analysis and the Consumption Function: an Attempt at Integration', in A. Abel (ed.), *The Collected Papers of Franco Modigliani*, Vol. 2, Cambridge, MA, MIT Press, 128–97.

Muellbauer, J. (1983), 'Surprises in the Consumption Function', *Economic Journal*, 93 (Suppl.), 34–50.

—— (1988), 'Habits, Rationality and Myopia in the Life-cycle Consumption Function', *Annales d'Economie et de Statistique*, 9, 47–70.

—— Bover, O. (1986), 'Liquidity Constraints and Aggregation in the Consumption Function under Uncertainty', Discussion Paper 12, Oxford, Institute of Economics and Statistics.

—— Lattimore, R. (1994), 'The Consumption Function: A Theoretical and Empirical Overview', in M. H. Pesaran and M. R. Wickens (eds), *Handbook of Applied Econometrics*, Oxford, Blackwells.

—— Murphy, A. (1990), 'Is the US Balance of Payments Sustainable?', *Economic Policy*, 347–82.

—— —— (1993a), 'Income Expectations, Wealth and Demography in the Aggregate US Consumption Function', unpublished paper presented to H. M. Treasury Academic Panel, Nuffield College, University of Oxford.

—— —— (1993b), 'Income Expectations, Wealth and Demography in the Aggregate UK Consumption Function', unpublished paper presented to H. M. Treasury Academic Panel, Nuffield College, University of Oxford.

—— —— (1994), 'Explaining Regional Consumption in the UK', Nuffield College, University of Oxford, mimeo.

Murata, K. (1994), 'The Consumption Function in Japan', Oxford, M. Phil thesis.

Muth, J. F. (1960), 'Optimal Properties of Exponentially Weighted Forecasts', *Journal of the American Statistical Association*, 55, 299–306.

Patterson, K. D. (1984), 'Net Liquid Assets and Net Illiquid Assets in the UK Consumption Function: Some Evidence for the UK', *Economics Letters*, 14(4), 389–95.

—— Pesaran, B. (1992), 'The Intertemporal Elasticity of Substitution in the US and the UK', *Review of Economics and Statistics*, 74, 573–84.

Perron, P. (1989), 'The Great Crash, the Oil Price Shock and the Unit Root Hypothesis', *Econometrica*, 57(6), 1361–401.

Pesaran, M. H., and Evans, R. A. (1984), 'Inflation, Capital Gains and UK Personal Savings: 1951–1981', *Economic Journal*, 94, 237–57.

Phillips, P. C. B. (1986), 'Understanding Spurious Regressions in Econometrics', *Journal of Econometrics*, 33, 311–40.

—— (1987), 'Time Series Regression with a Unit Root', *Econometrica*, 55, 277–301.

—— Durlauf, S. N. (1986), 'Multiple Time Series Regression with Integrated Regressors', *Review of Economic Studies*, 53, 473–95.

Pischke, J.-S. (1991), 'Individual Income, Incomplete Information and Aggregate Consumption', *Industrial Relations*

Section Working Paper no. 289, Princeton, NJ, Princeton University, mimeo.

Poterba, J. (1991), 'Dividends, Capital Gains, and the Corporate Veil: Evidence from OECD Nations', in B. D. Bernheim and J. B. Shoven (eds), *National Saving and Economic Performance*, Chicago, IL, Chicago University Press for NBER.

Sefton, J. A., and Weale, M. R. (1994), *Reconcilation of National Income and Expenditure: Balanced Estimates of National Income for the UK, 1920–1990*, Cambridge, Cambridge University Press.

Shefrin, H. M., and Thaler, R. H. (1988), 'The Behavioural Life-cycle Hypothesis', *Economic Inquiry*, 26(4), 609–43.

Simon, H. A. (1978), 'Rationality as Process and as Product of Thought', *American Economic Review*, 68(May), 1–16.

Simons, H. (1938), *Personal Income Taxation*, Chicago, IL, Chicago University Press.

Singleton, K. (1990), 'Specification and Estimation of Intertemporal Asset Pricing Models', in B. M. Friedman and F. H. Hahn (eds.), *Handbook of Monetary Economics*, Vol. I, Amsterdam, North-Holland, 583–626.

Skinner, J. (1988), 'Risky Income, Life-cycle Consumption, and Precautionary Saving', *Journal of Monetary Economics*, 22, 237–55.

Slesnick, D. T. (1992), 'Aggregate Consumption and Saving in the Post-war US', *Review of Economics and Statistics*, 74, 585–97.

Spiro, A. (1962), 'Wealth and the Consumption Function', *Journal of Political Economy*, 70, 339–54.

Stock, J. H. (1990), 'Unit Roots in Real GNP: Do We Know and Do We Care? A Comment', *Carnegie-Rochester Conference Series on Public Policy*, 32, 63–82.

—— (1993), 'Unit Root and Trend Breaks in Econometrics', in M. H. Peseran and M. Wickens (eds.), *The Handbook of Applied Econometrics*, Vol. IV.

Stone, J. R. N., Champerowne, D. G., and Meade, J. E. (1942), 'The Precision of National Income Estimates', *Review of Economic Studies*, 9, 111–35.

Stone, R. (1964), 'Private Saving in Britain: Past, Present and Future', *Manchester School of Economic and Social Studies*, 32, 79–112.

Thaler, R. H. (1990), 'Anomalies: Saving, Fungibility, and Mental Accounts', *Journal of Economic Perspectives*, 4(1), 193–205.

Tobin, J. (1972), 'Wealth, Liquidity and the Propensity to Consume', in B. Strumpel, J. N. Morgan, and E. Zahn (eds.), *Human Behaviour in Economic Affairs (Essays in Honor of George S. Katona)*, Amsterdam, Elsevier.

—— (1980), *Asset Accumulation and Economic Activity*, Oxford, Basil Blackwell.

Tversky, A., and Kahneman, D. (1974), 'Judgment Under Uncertainty: Heuristics and Biases', *Science*, 185, 1124–31.

Weale, M. R. (1990), 'Wealth Constraints and Consumer Behaviour', *Economic Modelling*, 7, 165–75.

Wickens, M. R., and Molana, H. (1984), 'Stochastic Lifecycle Theory with Varying Interest Rates and Prices', *Economic Journal*, 94(Suppl.), 133–47.

Wilcox, D. W. (1991), 'The Construction of the US Consumption Data: Some Facts and their Implications for Empirical Work', Board of Governors of the Federal Reserve System, mimeo.

Zellner, A., Huang, D. S., and Chau, L. C. (1965), 'Further Analysis of the Short-run Consumption Function with Emphasis on the Role of Liquid Assets', *Econometrica*, 33, 571–81.

PART IV

INFLATION AND UNEMPLOYMENT

8

Inflation policy

Keble College, Oxford

I. Introduction

Economists have not always provided helpful advice to policy-makers regarding inflation. Despite the fact that, in most countries, inflation is a particularly unpopular economic phenomenon, economists have often been hard pressed to find reasons why inflation matters at all! This is, perhaps, especially true of economists in the free-market tradition, such as monetarists (in both their original and New Classical forms). Traditional economic theory suggests that economic costs are likely to arise when *unanticipated* changes in the rate of inflation occur, but that stable and anticipated inflation should have few economic effects. Put another way, the 'real' variables in the economy—such as employment, investment, consumption, and output—should be invariant to 'nominal' variables—such as inflation or the money supply.

The theoretical foundations of this view that, at least in the long run, the real equilibrium of an economy is unaffected by nominal variables, are examined critically by Hahn (1990). However, this paper focuses, in the main, on the ways in which actual economies deviate from their theoretical counterparts. It is argued that few economies are, in fact, neutral to the rate of inflation, and that inflation is likely to have important real effects even when it is fully anticipated.

The possible causes of inflation have been the subject of considerable research. For example, Gilbert (1990) examines the evidence linking commodity prices to inflation, and Nickell (this volume) considers the influence of the labour market (and the supply side in general) on inflationary pressures. Patrick Minford (1990) reviews the operation of monetary policy in the

UK over the last decade and suggests that monetary policy should now concentrate on monetary base control. Minford also suggests that the UK should retain control over its domestic monetary policy and exchange rate, and argues against moves towards European monetary union.

This paper will, as a result, pay less attention to the causes of inflation, and will concentrate on the appropriate policy regarding inflation in general. In particular, are any inflation trade-offs likely to exist in the short run or long run; how high should the reduction of inflation be on the policy-makers' agenda (which will require an analysis of the economic costs of inflation); what problems are encountered in reducing inflation; and do institutional arrangements make some countries more inflation-prone than others, and make the reduction of inflation more difficult to achieve?

The plan of this paper is as follows. In section II the long-standing question of inflation trade-offs is discussed. The economic costs of inflation are discussed in some detail in section III, and the influence of institutional factors on the control of inflationary pressure are considered in section IV. Conclusions are presented in section V.

II. Inflation and equilibrium

Much attention has been paid over the years to the possible Phillips Curve trade-off between inflation and unemployment (Phillips, 1958). The idea that policy-makers could achieve a lower level of unemployment if they were prepared to accept higher inflation was questioned by Phelps (1967) and Friedman (1968), who noted that employees should be concerned with the level of their *real* wages rather than *nominal* wages. Employees also have an incentive to form expectations

First published in *Oxford Review of Economic Policy*, vol. 6, no. 4 (1990). This version has been updated and revised to incorporate recent developments.

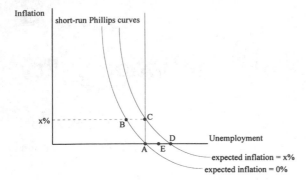

Figure 8.1. The Expectations—Augmented Phillips Curve

of future inflation rates when negotiating over nominal wage increases. They argued that a trade-off would exist only for as long as expectations of future inflation were not fulfilled. This is the essence of the expectations-augmented Phillips Curve, which suggests that policy-makers' attempts to reduce unemployment by accepting a higher inflation rate will result in failure—either through creating a potentially explosive wage-price spiral, or because expectations will eventually respond and offset any initial reduction in unemployment.

The conventional expectations-augmented Phillips Curve story is presented in Figure 8.1. Starting from a position of zero inflation (point A) the government increases demand, which increases inflation to x per cent, and reduces unemployment, as the economy moves from A to B. However, each short-run Phillips Curve incorporates a particular expectation of inflation. As workers realize that the inflation rate has risen to x per cent, they will eventually revise their expectations, realize that their real wages have fallen, and increase their nominal wage demands. The economy will, therefore, move from point B to C. The trade-off will only exist in the short run: unemployment will return to its original level, and inflation will have risen.

New Classical macroeconomists went further: if individuals form rational expectations, no *short-run* trade-off between inflation and unemployment will exist either. The New Classical approach is to suggest that the government policy to increase demand will be anticipated by workers, who will immediately increase their nominal wage demands by x per cent, so shifting the economy straight from A to point C.

There have been various reactions to this critique. First, it is important to consider what microeconomic assumptions have been made in the above analysis. Implicitly, the dual assumptions of market clearing and the uniqueness of equilibrium have been taken on board.

A second criticism that can be levelled against the

simple expectations-augmented Phillips Curve is that it makes strong assumptions about the *symmetry* of responses to changes in policy. Consider the effects of a demand expansion that is followed by an equal reduction in demand. In the expansionary phase, the economy would move from A to C via B, as described above. However, consider the contractionary phase of the cycle. At point C workers expect inflation to be x per cent. If the government wants to reduce inflation then, in the absence of rational expectations, increased levels of unemployment will be experienced as actual inflation falls. As the economy moves from C to D real wages rise and unemployment increases. In the final phase of the cycle, expectations adjust to the zero inflation and the economy returns to its initial position at A. The unemployment experienced at point D disappears as real wages fall back and those made redundant are welcomed back into the work force.

An obvious objection to this analysis is the assumed symmetry of response. Few would object to increases in their expected real wages in the expansionary phase of the cycle. If inflation turns out to be higher than expected, it is quite plausible that workers will aim to restore their real wages by demanding higher nominal wages. Up to this point the Phelps/Friedman analysis seems reasonable enough.

However, what will happen during the disinflationary phase of the cycle? How will firms react as workers demand high nominal wage increases which cannot be passed on in the form of increased prices, at least without losing sales? Will workers be prepared to accept the reductions in real wages required for the economy to move from D to C? Numerous theories have been suggested in recent years that incorporate the possibility of *hysteresis* effects, whereby the equilibrium position of the economy depends on the past history. Such *path dependence* constitutes an important class of reasons why many equilibria may exist in an economy; economic policy will be one factor determining which equilibrium is approached.

In the context of the example presented above, suppose that unemployment does not return to its original level, but instead falls only as far as point E. There are various reasons why this may occur. The unemployed may have become *outsiders* in the wage negotiation process, and the remaining *insiders* may be more concerned about maintaining their real wages rather than reducing unemployment (see Lindbeck and Snower, 1988, for a survey of models). The disinflationary stage of the cycle may have resulted in bankruptcy of firms, so physical capital may have been lost (as discussed in the next section). Similarly, if unemployment persists for some time, human capital, in the form of job-specific skills, may be lost. None of these effects is easily, or costlessly, reversible. To the extent that they are relevant, the expectations-augmented Phillips Curve, with its unique long-run equilibrium, may be a seriously misleading concept.

Of course, few of the above arguments would impress the New Classical economist. If individuals formed rational expectations then inflation could be immediately and costlessly reduced (the economy would move from C straight back to point A in Figure 8.1). There is something in this argument. Clearly one of the key problems policy-makers face is the need to convince firms and workers that disinflationary policies will be followed, and that inflationary pressures will not be accommodated. If these policies were viewed as credible, then expectations should change, and the cost of reducing inflation (in the form of higher unemployment) would be reduced.

However, it is important to disentangle the threads of the New Classical argument. Few would disagree that if all individuals formed rational expectations based on an economic model incorporating a unique natural rate of unemployment then disinflationary policy may well be costless to implement. Such 'new classical rational expectations' are quite capable of being self-fulfilling. However, individuals can form rational expectations based upon quite different economic models, incorporating, for example, imperfect competition, uncertainty over the strategies of other firms or workers, asymmetric information and so on. Whether or not these beliefs about the operation of the economy are correct, they also can be self-fulfilling (as demonstrated by the example of 'sunspot equilibria', Cass and Shell, 1983). In this case individuals will not necessarily expect disinflationary policy to have no effects on the real economy. In other words, rational expectations in themselves do not imply that the real economy will be invariant to inflation policy in either the short run or long run—such claims rely upon the additional, and far more questionable, assumption of a unique market clearing equilibrium.

In summary, I would suggest that macroeconomic policy towards inflation should not be concerned with choosing a position on a short-run Phillips Curve, but should consider ways in which policy may itself influence the equilibrium of the economy. As Summers (1988) suggested, Keynesian economics should dispense with the Phillips Curve, and instead embrace the idea that equilibria are fragile and dependent on history. Indeed, such hysteresis effects may suggest a firmly anti-inflationary stance, given the costs of reducing inflationary expectations once a wage-price spiral begins. On the other hand, some economists have used the problem of reducing inflationary expectations to argue that inflation policy should be directed towards

stabilizing rather than reducing inflation when prices are rising at moderate rates. However, as Hahn (1990) notes, the costs of reducing inflation ought to be compared with the costs of continuing stable inflation. This issue will be considered in detail in the next section.

Policy options regarding macroeconomic policy will, however, remain. Nickell (in this volume) demonstrates the sort of trade-offs policy-makers might encounter. In terms of unemployment, there is no simple Phillips Curve trade-off between unemployment and inflation in the Nickell model, but rather a choice between lower unemployment and increasing rates of inflation. In this sense the original Friedman critique of the Phillips Curve—that any trade-off between the rate of inflation and the level of unemployment ceases to exist once inflation becomes anticipated—has been completely taken on board. It is also clear that few would contemplate accepting ever accelerating inflation in return for a reduction in unemployment. Such a policy is ultimately unsustainable, would have to be reversed, and the process of reversal might itself impose significant economic costs, unless inflationary expectations could be costlessly changed.

However, Nickell also suggests a third component to the policy choice—the balance of trade. The ability to run a balance of trade deficit can partially offset domestic inflationary pressure, at least in the short run. Whenever domestic demand becomes excessive, a trade deficit will tend to occur, as demand cannot be satisfied by domestic production. However, demand that is directed towards imported goods also reduces domestic inflationary pressure. In a closed economy the excess domestic demand would all be absorbed domestically, resulting in higher levels of inflation.

An important development since the mid-1980s has been the emergence of large, often persistent trade imbalances—for example, the Japanese surpluses and the US deficit—and the increasing ease with which such imbalances have been financed. As a result, governments may now have far more discretion in choosing a particular level of the trade deficit/surplus as part of their overall macroeconomic policy. It will not always be optimal to aim for balanced trade (see the discussion in Allsopp et al., this volume).

Of course, if trade imbalances quickly resulted in exchange rate changes then such a trade-off might be short lived. According to purchasing power parity theory, a trade deficit should cause the exchange rate to fall, which will tend to increase the cost of imported goods, and hence inflationary pressure. However, such a theory is hardly relevant to a world in which flows across the foreign exchanges associated with financial transactions swamp those resulting from trade in goods and services. As a result, a trade deficit does not necessarily precipitate a rapid depreciation of the currency, especially if financial inflows can be encouraged by increasing interest rates or the yields on other assets. For as long as the exchange rate does not fall—which may be several years—a trade deficit will tend to absorb inflationary pressure.

Nickell suggests that the nature of the three-way trade-off between unemployment, the trade deficit, and changes in the rate of inflation will be determined by a variety of supply-side factors. By using monetary and fiscal policy, government can choose a position on this trade-off, but only by supply-side policies can the underlying trade-off be shifted. However, as Nickell notes, moving the supply-side constraint facing an economy is no quick or easy task. In the short run, therefore, policy towards inflation will depend, to a great extent, on the costs of inflation, to which we now turn.

III. Costs of inflation

In the mythical perfectly competitive economy with full inflation indexation, there remains one way in which inflation can have real effects—via the famous 'shoe leather' effect. Because inflation is a tax on money balances (which do not yield interest), individuals should economize on holding money, and should instead make more frequent trips to the bank. Whilst such effects can be important in cases of hyperinflation, they are likely to be insignificant in most other cases, especially as financial systems develop alternative means of payment. For too long economists have pointed to shoe leather costs in a desperate search for real effects of inflation. As a contribution towards redressing the balance, I will say no more about them.

One of the most important ways in which inflation can have real effects is via the tax system. In principle it is possible to construct a tax system that is entirely inflation neutral. This would require full indexation of allowances, thresholds, etc. In practice, most tax systems fail to satisfy these conditions. Consequently, inflation will often have real effects. In the following sections a variety of these effects are examined.

1. The tax system

(a) Corporate taxation

Corporate tax systems are, in practice, rarely inflation-neutral, although it is possible, in principle, to devise such systems (see the Meade Committee Report, 1978).

In the UK, for example, the rate of inflation influences the effective rate of corporation tax in a number of ways. Indeed, the 1984 reforms of corporate taxation, which became fully operational in 1987, made the system considerably *more* sensitive to the rate of inflation. As a result, many companies face higher effective rates of corporation tax when inflation increases.

The system of corporate taxation in the UK (in common with many other countries) is based on profits (alternative tax bases include cash flow, turnover, the value of assets, etc.). An important issue, therefore, is whether the tax base is affected by the rate of inflation. If the tax base increases, say, with inflation, then so will corporation tax payments, and this may affect the incentive to invest in various ways.

The first important way in which the tax base can increase with inflation is through stock appreciation. The value of stocks will tend to increase in nominal terms as a result of inflation, and if such stock appreciation is included in taxable profits, tax liabilities will increase. However, profits resulting from stock appreciation do not constitute real profits, and are certainly not associated with increased cash flow. On the other hand, the increased tax payments certainly do represent a cash outflow for firms.

The potential problems of not giving tax relief for stock appreciation were demonstrated in the mid-1970s in the UK: as inflation increased rapidly, company profits were increased by stock appreciation, which exacerbated the liquidity problems firms faced as the economy went into recession. To combat this, temporary tax relief on stock appreciation was introduced at the end of 1974. The 1984 reforms of corporation tax removed stock relief and hence left the corporate sector vulnerable to inflation.

A second way in which the tax base can be influenced by inflation is through the treatment of capital allowances. How the tax system treats capital expenditure by firms will have important effects on the incentive to invest. The UK tax system has varied considerably over the years in this respect, but a key feature of the 1984 reforms was the removal of initial capital allowances, whereby a proportion (100 per cent in the case of plant and machinery) of the capital expenditure could be written-off against tax in the first year. This was replaced by a system of annual writing-down allowances based on the *historic cost* (rather than the *current cost*) of the asset. As a result, inflation will reduce the real value of these allowances, and the tax base will again increase with inflation.

A final factor, which will tend to offset those previously discussed, is the treatment of interest payments by firms. Many corporate tax systems allow interest payments on debts to be tax deductible. However, when inflation rises, nominal interest rates are likely to rise, although the real interest rate may remain unchanged. The rise in the nominal interest rate compensates the lender for the reduction in the real value of the loan caused by the inflation. Consequently, a proportion of the interest payments should, in times of inflation, be considered as repayment of principal, rather than interest. This factor will be considered in more detail in the next section, but the fact that tax relief is given on *nominal* rather than *real* interest payments means that inflation will tend to reduce the tax base in this way. This offsets, to some extent, the effects outlined above, but is itself an important source of non-neutrality. Firms' financing decisions may therefore be distorted by inflation, with high levels of gearing being encouraged as a means of reducing tax payments.

Bond, Devereux, and Freeman (1990) attempt to quantify these effects for the UK corporate sector. Using a sample of 750 individual firms they estimated that had inflation been constant at 10 per cent from 1990–93, corporate tax liabilities would have been as much as one-third higher in 1993 compared to a zero inflation scenario. They also found that the effect of 10 per cent inflation is to raise the cost of capital by nearly 2 percentage points over the level it would be at zero inflation. They conclude that, 'almost the whole of the corporate tax bias against investment at current inflation rates results from the inflation non-neutralities in the present tax system' (p. 27).

This section has concentrated on the corporate tax system currently operating in the UK. The obvious conclusion appears to be that, unless firms have very high levels of gearing and hence high nominal interest write-offs, the effective rate of corporation tax and the cost of capital will increase with the rate of inflation. However, this has not always been the case. As King and Wookey (1987) point out, the tax system operating in the UK in 1979 (with stock relief and 100 per cent first-year allowances) had the feature that many investments were less heavily taxed at higher rates of inflation. The important conclusion should be that, in practice, many tax systems are not neutral with respect to inflation, and these non-neutralities can have important effects on real investment decisions, financing decisions and even, at times, the solvency of firms. Inflation-neutrality should be one aim of a tax system, but, unless this is achieved inflation is likely to have real effects on the corporate sector which should not be ignored.

(b) Individuals

In most countries, surveys suggest that inflation appears to be just as unpopular with individuals as with

Table 8.1. Inflation and the Incentive to Save

	Zero inflation	15% inflation
Nominal interest rate	5%	20%
Real interest rate	5%	5%
£1000 savings		
Pre-tax interest	£50	£200
Post-tax interest:		
25% marginal rate	£37.50	£150
40% marginal rate	£30	£120
Change in real value of savings	£0	−£150
Overall post-tax real rate of return:		
25% marginal tax rate	3.75%	0%
40% marginal tax rate	3%	−3%

the corporate sector. Some of the effects of inflation identified in the previous section apply equally to individuals, although in a somewhat different form.

Consider first the effect of inflation on the consumption/savings decisions of individuals. As mentioned above, tax systems typically fail to make the distinction between real and nominal rates of interest. As a result, savings can often attract negative real post-tax rates of return as a result of inflation. Consider a simple example, set out in Table 8.1, in which we assume that the real interest rate remains constant at 5 per cent as inflation changes. When inflation is zero, the nominal interest rate is 5 per cent, which yields £50 for each £1,000 of savings. Most individuals then pay tax on this interest at their marginal rate of tax (in the UK at present this would be 25 per cent or 40 per cent). In the absence of inflation, there would be no reduction in the real value of savings, and hence the post-tax real rate of return on savings would be 3.75 per cent for basic rate tax payers and 3 per cent for those taxed at the higher rate.

Consider now the corresponding rates of return when inflation is 15 per cent. Since the tax system taxes nominal, rather than real, interest payments, pre-tax interest payments of £200 are reduced after tax to £150 or £120 for those taxed at the basic rate and higher rate respectively. However, the real value of each £1,000 of savings will have fallen by the rate of inflation. For those taxed at the basic rate, the real post-tax rate of return is precisely zero, whilst those who pay tax at the higher rate suffer a −3 per cent rate of return.

The solution would be, as before, to tax real interest payments rather than nominal interest payments. In this case the rates of return to savers would be invariant with respect to the inflation rate. However, at present the taxation of the income from savings increases with the rate of inflation, which may both

Table 8.2. The Effects of Inflation on Cash-Flow

	Zero inflation	15% inflation
Nominal interest rate	5%	20%
Real interest rate	5%	5%
Operating profit	£1,000	£1,150
Interest payments	£400	£1,600
Cash Flow	£600	−£450

discourage savings and have quite arbitrary distributional consequences.

There are many other ways in which personal taxation can vary with inflation. Whenever tax allowances and tax thresholds do not increase with inflation, there are likely to be income and/or substitution effects. These could be avoided by implementing full indexation of the tax system. Again, until this happens, inflation is likely to have real effects.

2. Financial markets

Whenever firms, or individuals, borrow a fixed nominal sum at a variable interest rate, inflation will have an effect upon cash flow. This is because nominal interests rates will tend to rise with inflation, even if real interest rates remain unchanged. As mentioned above, the rise in nominal interest rates compensates the lender for the erosion in the real value of the loan caused by inflation. However, the corollary of this is that borrowers are, in effect, required to repay the principal of the loan before the end of its term. This effect is sometimes known as 'front-end loading'. This can have a serious effect on cash-flow, as can be seen in the following example.

Consider a firm that borrows £8,000 at a variable rate of interest and assume that the real rate of interest remains unchanged at 5 per cent as inflation varies. The firm is assumed to make £1,000 operating profit when inflation is zero and £1,150 when inflation rises to 15 per cent so that the real profitability of the firm is invariant to inflation. Corporation tax and dividend payments are ignored in this example. As Table 8.2 shows, when inflation is zero, the firm produces a cash flow of £600. However, if inflation rises to 15 per cent, the nominal interest rate rises to 20 per cent, raising interest payments four-fold. This results in a negative cash flow of £450. The firm may have reserves from which to meet this negative cash flow, but in the absence of sufficient reserves will have to raise additional finance.

However, how will the firm's bankers and shareholders react? Of course, they should be concerned not only with the current profit and loss account of the firm, but also with the state of its balance sheet. The effect of 15 per cent inflation will be to reduce the *real* value of the outstanding loan from £8,000 to £6,800. Correctly measured, therefore, the firm is equally profitable whether inflation is 0 per cent or 15 per cent.

Clearly, if the firm had arranged for its loan to be index-linked no such problems would arise. The firm would simply pay a real rate of interest, and the principal of the loan would be increased in line with inflation. Under such circumstances, inflation would have no effects on either real cash flow or the real balance sheet of the firm. Similarly, if the firm's bankers increased their loan to the firm by £1,200, the firm would be able to pay the additional interest charges.

However, in practice, bankers are likely to hesitate before lending additional amounts to firms producing negative cash flow. Decisions to lend are often based upon rules of thumb that do not fully take into account the effects of inflation. For example, suppose that banks require a firm's interest cover (the ratio of operating profit to interest payments) to exceed some minimum figure before they will lend. In the example above, the interest cover when inflation is zero is 2.5, whereas inflation at 15 per cent reduces interest cover to 0.72. If additional funds cannot be raised, firms will experience financial distress, and be required to take other action. This might typically include sales of assets, layoffs, cuts in dividend payments, etc., all of which represent very real effects of inflation.

Of course, in times of trouble, firms could approach their shareholders to obtain additional equity finance. However, there is evidence to suggest that inflation also has an adverse effect on stock market values. For example, Modigliani and Cohn (1979) suggest that share prices might be depressed by inflation to the extent that investors do not use proper *current* cost accounting estimates of profits (as discussed above) and use a nominal, rather than a real, rate of return in discounting profits. Wadhwani (1986) presents empirical evidence for the UK which suggests that the effect of inflation on the stock market is two-fold: not only are valuations of firms reduced along the lines suggested by Modigliani and Cohn, but inflation also increases the default premium required on equity, to compensate for the increased probability of bankruptcy when inflation increases.

The above analysis of the effects of inflation on nominal interest rates, and hence on liquidity, clearly applies with equal force to the household sector. In many countries, loans are neither index-linked nor at fixed nominal rates (the US is an exception to the latter rule) with the result that inflation will often result in significant increases in nominal interest payments on mortgages and other loans, causing cash-flow problems. The knowledge that the real value of your outstanding debt is falling may be of little comfort if your house is being repossessed.

Particular problems may arise in both the corporate and household sectors when disinflationary policies are being pursued. As explained in the previous section, many economies encounter severe problems in reducing inflation, due to the problems of reducing inflationary expectations. Fiscal and monetary policies are often used to push the economy towards recession, in order to reduce inflation. This was certainly the experience of the UK in 1980 and 1990. However, during recessions the value of certain assets, such as houses and equity, tends to fall. Both individuals and firms are thus often caught by the pincer movement of higher outgoings and dwindling collateral. In these circumstances banks may well be loathe to lend additional funds, even if the real value of outstanding debts is falling, as the debt to equity ratio may increase to unacceptable levels.

3. Competitiveness

The theory of purchasing power parity (PPP) suggests that the nominal exchange rate should move to offset differential inflation rates between countries so that the real exchange rate remains unchanged. If this theory held at all points of time, inflation would have no effect on competitiveness, and hence on real decisions to export or import.

There are many reasons why PPP is unlikely to hold, at least in the short run (see, for example, Begg in this volume). However, the most important in the present context are likely to be inflexibility of prices and wages, and the impact of the capital account on the exchange rate.

First, consider the decision facing a UK exporter when domestic inflation rises relative to foreign inflation. Should he simply translate the new (higher) price of his product into foreign currencies at the prevailing exchange rates, or should he maintain his foreign prices and take a lower profit margin on exports? For PPP to hold, the exporter should follow the former course and raise the foreign currency prices of his product. This should reduce foreign demand for the product, and hence reduce the demand for sterling. The exchange rate should then fall sufficiently to offset the initial inflation differential such that PPP is re-established.

The problem with this account is that it makes most sense when markets are perfectly competitive and firms have little control over their pricing strategy. However,

few markets for manufactured goods operate in this way. Markets are often oligopolistic, and pricing policy may form but one part of a complex strategic game. There are often significant sunk costs involved in establishing overseas markets—advertising, establishing dealer networks, etc.—such that firms are not prepared to lose market share by simply passing any domestic inflationary effects (or, for that matter, exchange rate changes) immediately on to foreign prices. In the short term firms may prefer to maintain foreign prices and reduce profit margins on exports. Of course, over time the firm must respond somehow to the reduced profitability of exports, but until PPP is re-established inflation may have serious effects on real competitiveness.

The second factor which reduces the plausibility of the PPP account is the increased importance of capital account flows in determining the exchange rate, especially in the short term. As foreign exchange markets have been liberalized, current account imbalances have become easier to finance. As capital account flows have increased in volume, and especially when the flows have been in the form of short-term deposits, interest rate policy becomes the dominant factor determining short-run exchange rates. Consider again the problem facing a government attempting to reduce inflation. A disinflationary policy of restricting monetary growth will tend to raise interest rates. Higher interest rates will, if uncovered interest parity (UIP) holds, immediately increase the exchange rate, which will tend to exacerbate the loss in competitiveness caused by domestic inflation. PPP may eventually be restored, but until it is domestic firms will be less competitive.

The effect of differential inflation rates on competitiveness will be especially pronounced when currencies are not allowed to freely float. This is most obviously true in rigidly fixed exchange rate systems, but can also be important in adjustable peg systems, such as the European Exchange Rate Mechanism (ERM). Firms and employees can no longer be protected from the consequences of inflationary price and wage increases by a fall in the exchange rate. Such inflationary pressures will result in a loss of *real* competitiveness, unemployment, falling output, and, ultimately, may undermine the credibility of government efforts to defend the parity. All these phenomena were witnessed in numerous European Union countries during the effective breakdown of the ERM in late 1992.

Meade (1990) looks one stage further into the future and considers the impact of Economic and Monetary Union on the control of inflation. He argues, in particular, that assigning the responsibility for the control of inflation to a central monetary authority, whilst leaving budgetary policy largely in the hands of national fiscal authorities, could be a recipe for economic instability.

4. Inflation variability

Much has been written over the years regarding the possible relationship between *high* inflation and more *variable* inflation (early examples being Okun, 1971, and Lucas, 1973). Most of the empirical evidence has tended to support the view that high average levels of inflation tend to be associated with a high variance of inflation. Such a relationship does not, of course, imply causality. Government policy towards inflation will clearly play an important role. The observation that high rates of inflation have tended to be more variable over time does not necessarily imply that such rates could not be relatively stable, with appropriate policies. There is no reason to believe that a policy of stabilizing the inflation rate could not be successful, particularly if it was widely announced and anticipated. The observation that high inflation has tended to be more variable may simply reflect policies that have attempted to reduce inflation when it exceeded certain levels, rather than stabilize inflation.

Of course, variability *per se* does not necessarily matter, it is only if variable inflation is less *predictable* that economic costs might arise. When inflation is unanticipated there are likely to be various economic costs. Lucas (1972) argued that unanticipated inflation renders the price mechanism less efficient at co-ordinating economic decisions. In his model economic agents know more about their own prices than about prices in the rest of the economy. If inflation is unanticipated, each agent may wrongly believe that the rise in nominal demand (and hence price) for his good or service represents an increase in real demand. The price signal may be interpreted as an increase in relative prices, rather than as a general increase in the absolute price level. As a result, real economic decisions may be taken (for example to increase output or labour supply) which would not be taken in the absence of inflation, or if the inflation had been anticipated. If inflation is both volatile and unanticipated it becomes hard to extract the signal—relative prices—from the noise.

In his Nobel Lecture Milton Friedman (1977) used such ideas to suggest that long-run Phillips Curves might be *positively* sloped—in other words higher rates of inflation might result in higher levels of unemployment. As noted above, the original expectations-augmented Phillips Curve implied that *any* stable rate of inflation was compatible with the natural rate of unemployment. The case had been made for *stabilizing* inflation, but not for *reducing* inflation. This rather weak policy prescription was significantly strengthened by the suggestion that higher inflation would increase unemployment in the long run by reducing the efficiency of the price mechanism. This is likely to be true if high rates of inflation increase *uncertainty*, which itself will depend upon the variance of the unanticipated part of inflation. However, Friedman suggested that institutional arrangements should change over time, such as the indexation of nominal contracts, so that real economic decisions should in the 'long-long run' be neutral with respect to inflation. As argued in the previous section, if hysteresis effects exist, even such long-run inflation-neutrality is unlikely.

This section has considered a number of ways in which actual economies may not be invariant to the rate of inflation. Some of these effects may not be relevant to a particular economy, and doubtless many other inflation non-neutralities exist. However, the message remains that economists should not be too sanguine regarding inflation, even if it is relatively stable. Of course, given that some of the effects of inflation discussed above are arbitrary, unintended, and unjust—such as the effects via taxation—these should be addressed directly (for example by reforming the tax system). However, in the meantime the perfectly inflation-neutral economy is probably as much of a fiction as the perfectly competitive economy.

IV. Inflation, co-ordination, and the 'going rate'

Some countries simply appear to be more inflation-prone than others. As Nickell (in this volume) points out, in the post-war era, aside from the odd year, the inflation rate in Britain has been higher than the average of the OECD countries. In most OECD countries a wide variety of economic policies have been followed, and the political balance has frequently changed. Yet the UK has consistently experienced a higher rate of inflation. Why should this be?

Soskice (in this volume) looks in detail at the possible impact of *institutional factors* in explaining comparative economic performance, both in terms of inflation and such factors as unemployment and growth. In particular, Soskice re-examines and extends the work of Calmfors and Driffill (1988) which considered the importance of wage bargaining institutions in explaining the levels of unemployment experienced by different countries. The particular institutional feature that Calmfors and Driffill focused on was the degree of centralization of wage bargaining. They argued that decentralized systems *and* centralized systems can be successful at achieving wage restraint, but that intermediate systems are less successful. It is worthwhile considering, briefly, the reasons for such a classification.

A belief in decentralized decision making is, in many respects, a belief in the efficacy of the free market. Wage bargaining, it is argued, should take place at the *lowest* level in order that workers can be rewarded for productivity increases and firms are able to ensure that they remain competitive. If productivity in firm A rises, workers should be rewarded accordingly. However, if wages rise too fast, employment in firm A is likely to be reduced, as capital is substituted for labour. Similarly, if wage costs rise more slowly elsewhere, firm A will become less competitive, demand for its output will fall, and employment will again be reduced. However, such unemployment is likely to put downward pressure on wages in other firms. The labour market should clear as the unemployed workers are quickly absorbed into rival firms, who expand their output as customers move away from firm A.

The arguments in favour of decentralization stress the benefits of flexibility in relative wages, the motivating effect of productivity-linked pay awards, and the efficiency of the market in allocating labour between firms or industries.

Consider, now, the argument in favour of more centralized wage bargaining. The crucial flaw in the case for decentralization, it is argued, lies in linking wages to productivity at any level below the national level. Consider again firm A, where productivity has risen. Why should the wages of workers in firm A rise? The obvious alternative option would be for firm A to take advantage of its lower costs by decreasing prices and increasing output and employment. There is no necessary reason for wages to follow productivity at the individual firm level. Indeed, classical economics says as much: labour should be employed up to the point at which the marginal (revenue) productivity of labour equals the wage rate. This should not be taken to imply that the wage rate should rise to achieve this equality rather than the level of employment.

Of course, if productivity across the whole economy is rising then if wages do not rise there would, sooner or later, become excess demand for labour (depending upon the initial level of unemployment). Hence, wages should be linked to *average* labour productivity in the whole economy, but not to productivity at an individual plant, or firm, level. In other words, a 'going rate' should be established. In relatively 'corporatist' economies, such as Austria or Sweden, representatives of workers, employers, and government negotiate centrally to determine how far wages can rise without damaging competitiveness and employment. Advocates of centralized bargaining claim that such institutional arrangements result in lower levels of unemployment and the ability to control inflation more effectively.

In many ways this debate can be restated in terms of the potential benefits of co-ordination. Inflation is seldom generally beneficial or popular, and yet is often both difficult and costly to reduce. Simple game theory shows us that unco-ordinated decision making can often lead to Pareto inefficient results, and this may be especially true in the case of wage bargaining. Consider the following simple application of the Prisoner's Dilemma problem.

Suppose there are two groups of workers and two firms in an economy. Each group of workers bargains separately with their respective employer; no co-ordination occurs either between firms or groups of workers. To simplify matters, assume that the workers choose a wage and the employers determine the amount of employment (the so-called 'right-to-manage' model). Consider, now, the strategies and payoffs facing each group of workers.

Both groups of workers can choose between a 'high' strategy, where they submit a nominal wage demand at the prevailing rate of inflation plus the increase in productivity over the last year, or a 'low' strategy, where they submit a productivity-only nominal wage demand. Assume that productivity growth has been identical in each firm. If both groups of workers submit low (high) nominal wage demands, overall inflation will be low (high). The payoffs are presented in the matrix below and show that both groups of workers prefer low inflation to high inflation (the actual magnitudes are unimportant, only the relative pattern of payoffs matters). However, for each group the worst possible outcome would be if it submitted a low nominal wage demand and the other group submitted a high demand. This is because the group following a 'low' strategy will experience a reduction in their relative wages as well as their expost real wages, to the extent that the 'high' wage demand of the other group results in higher overall inflation.

Group 1

		High	Low
Group 2	High	−5 −5	7 −7
	Low	−7 7	5 5

Suppose group 1 sets its wage first in the wage round. It might consider its options as follows. If group 2 submits a high demand, group 1 would obtain a higher payoff by also submitting a high demand (as −5 is better than −7). If, on the other hand, group 2 submits a low demand, group 1 would again obtain a higher payoff by submitting a high demand (as 7 is better than 5). The situation is entirely symmetric, and so group 2 will face similar payoffs. The outcome of this simple non-co-operative game is known as a Nash equilibrium, and results in the famous Prisoner's Dilemma: both groups would prefer to move to the low inflation position, and yet cannot. In other words, the outcome is Pareto inefficient—both groups could be made better off.

The clear problem here is the lack of co-ordination between the groups of workers. If they could only come to an agreement whereby they both submitted a low wage demand, everyone would be better off. Inflation could be reduced with far less economic cost (in terms of the transitional unemployment). These benefits are, perhaps, more likely to be achieved in countries with stronger corporatist institutions where bargaining takes place on a more co-ordinated basis and all parties are able to agree on a 'going rate'.

An important issue is whether co-ordinated bargaining can cope with the need to motivate workers and attract labour to sectors where there are skill shortages. Whilst productivity increases are not necessarily due to greater effort on the part of workers (output per worker often rises as a result of technological advances or increases in the capital–labour ratio), there may, none the less, be a role for productivity, or skills, contingent wage increases at a decentralized level. However, there is no reason why such local adjustments cannot be made against the background of a national going rate.

Recently in the UK there have been signs that the trades unions favour a more corporatist approach to wage bargaining. However, the employers' association the, Confederation of British Industry, has consistently opposed such suggestions. In 1990 the Director-General of the CBI wrote,

Flexibility in setting pay levels reinforces the awareness among employees that their well-being depends on the performance of the enterprise that employs them. This should be encouraged not stifled . . . CBI members have no desire for such a return to the corporate state mentality of the 1970s. (*Financial Times*, February 1990)

The potential problem with such productivity related pay is that it can quickly result in increased wage pressure. Individuals care about wage differentials. If sectors experiencing high productivity growth (perhaps as a result of technical progress) increase the wages they pay, the effect is likely to be propagated to other sectors of the economy in the form of higher wage claims. Indeed, why should workers in service industries or the public sector accept falling relative wages? The result of such a system of wage bargaining can be increased wage pressure (perhaps resulting in a wage-price spiral) and considerable industrial unrest. Equally worrying (although outside the scope of this paper) may be the tendency for high productivity sectors to become internationally uncompetitive.

As Soskice shows, many countries with more effectively co-ordinated wage bargaining appear better able to both achieve and maintain low inflation, without the need for periods of large-scale unemployment to reduce inflationary pressures. Powerful employer organizations often play an important role in maintaining wage restraint in these countries; arrangements seldom resemble UK style corporatism of the 1970s.

V. Conclusions

In this paper a selection of issues regarding economic policy towards inflation have been examined. Despite the various theoretical models which suggest that inflation does not affect the real economy, this paper has examined a variety of ways in which inflation, and policy towards controlling inflation, can have important real—often detrimental—effects on an economy. The main conclusions are as follows.

First, there are unlikely to be any simple long-run trade-offs between unemployment and the level of inflation. However, this should not necessarily lead us to accept the idea that inflation will only be stable at some unique 'natural' rate of unemployment of an economy. Other trade-offs may well exist, for example between the rate of inflation and the balance of trade, but these will only be sustainable to the extent that external deficits or surpluses can be financed.

Second, inflation is easy to start, but difficult to reduce. One of the most significant costs of pursuing an inflationary policy is likely to be the cost of reducing inflationary expectations if the policy is reversed. The process of reducing inflationary expectations is normally associated with increased unemployment and reduced output. There are likely to be important asymmetries in the operation of labour and product markets. For example, workers who are laid off during a disinflationary period may become outsiders to the wage-bargaining process, or may lose job-specific skills. Firms may scrap capacity or go bankrupt. Such hysteresis effects suggest that economic policy towards inflation will be important in determining the long-run equilibrium to which an economy converges.

Third, the above two points together suggest that although there may be beneficial short-run effects of inflation (in terms of higher output or employment) these are likely to be more than outweighed by the economic costs involved in reducing inflationary expectations, if the policy is reversed. This suggests that economic policies which risk igniting inflation should be strictly avoided. However, if inflation starts, should it be stabilized or reduced?

The case for stabilization is strongest if there are few economic costs of steady inflation. However, there are significant ways in which most economies are not inflation-neutral. For example, in the UK at present, stable inflation tends to increase the rate of corporate taxation, encourage high levels of gearing in firms, create cash-flow problems for companies and individuals alike, discourage saving, and reduce competitiveness. These are real effects of inflation. Of course, there may also be significant costs involved in reducing inflationary expectations, depending upon the ways individuals form expectations and the credibility with which government disinflationary policy is viewed. These costs need to be compared before deciding which policy to pursue, and how quickly to pursue it.

However, the case for stabilizing inflation is only really an option for countries with floating exchange rates. For example, when countries join the ERM they effectively agree to reduce inflation to the levels experienced in other ERM countries. This increases the credibility of the government's disinflationary stance (provided re-alignments do not occur) and yet increases the likelihood of competitiveness problems if inflation is not quickly reduced (as the extent to which the nominal exchange rate can fall is strictly limited).

Fourth, the problems encountered in achieving, and maintaining, low inflation may be exacerbated, rather than reduced, by a move towards more decentralized bargaining. Workers care about their relativities. They may happily accept a lower nominal wage rise if all other workers do likewise, but may not be prepared to reduce wage claims unilaterally. Viewed this way, the reduction of inflation, or inflationary pressure, can be interpreted as a co-ordination problem. This problem

is addressed in some countries by employer organizations, unions, and, perhaps, the monetary authorities establishing a going rate. Such institutional influences on wage bargaining may well go some way to explain why some countries manage to achieve lower, more stable inflation, with less economic cost—in terms of lost growth and unemployment.

References

Bond, S., Devereux, M., and Freeman, H. (1990), 'Inflation Non-Neutralities in the UK Corporation Tax', *Fiscal Studies*, 21–8.

Calmfors, L., and Driffill, J. (1988), 'Bargaining Structure, Corporatism and Macroeconomic Performance', *Economic Policy*, 7, 13–61.

Cass, D., and Shell, K. (1983), 'Do Sunspots Matter?', *Journal of Political Economy*, 91, 193–227.

Friedman, M. (1968), 'The Role of Monetary Policy', *American Economic Review*, 1–17.

—— (1977), 'Inflation and Unemployment', *Journal of Political Economy*, 85, 451–72.

Gilbert, C. (1990), 'Primary Commodity Prices and Inflation', *Oxford Review of Economic Policy*, 6(4), 77–99.

Hahn, F. (1990), 'On Inflation', *Oxford Review of Economic Policy*, 6(4), 15–25.

King, J., and Wookey, C. (1987), 'Inflation: The Achilles' Heel of Corporation Tax', IFS Report Series, No. 26.

Lindbeck, A., and Snower, D. J. (1988), *The Insider-Outsider Theory of Employment and Unemployment*, London, MIT Press.

Lucas, R. E. (1972), 'Expectations and the Neutrality of Money', *Journal of Economic Theory*, 4, 103–24.

—— (1973), 'Some International Evidence on Output-Inflation Trade-offs', *American Economic Review*, 63, 326–34.

Meade, J. (1990), 'The EMU and the Control of Inflation', *Oxford Review of Economic Policy*, 6(4), 100–7.

Meade Committee (1978), *The Structure and Reform of Direct Taxation*, London, Allen and Unwin.

Minford, P. (1990), 'Inflation and Monetary Policy', *Oxford Review of Economic Policy*, 6(4), 62–76.

Modigliani, F., and Cohn, R. (1979), 'Inflation, Rational Valuation and the Market', *Financial Analysts Journal*, March/April, 24–44.

Okun, A. (1971), 'The Mirage of Stable Inflation', *Brooking Papers on Economic Activity*, 2, 485–95.

Phelps, E. S. (1967), 'Phillips Curves, Expectations of Inflation, and Optimal Unemployment over Time', *Economica*.

Phillips, A. W. (1958), 'The Relation between Unemployment and the Rate of Change of Money Wage Rates in the United Kingdom, 1861–1957', *Economica*, 283–99.

Summers, L. H. (1988), 'Should Keynesian Economics Dispense with the Phillips Curve?', in R. Cross (ed.), *Unemployment, Hysteresis and the Natural Rate Hypothesis*, Oxford, Blackwell.

Wadhwani, S. B. (1986), 'Inflation, Bankruptcy, Default Premia and the Stock Market', *Economic Journal*, 96, 120–38.

9

Inflation and the UK labour market

STEPHEN NICKELL

Institute of Economics and Statistics, University of Oxford

I. Introduction

Inflation is endemic in Britain. Aside from the odd year, inflation in Britain has been higher than the average of the OECD countries for the last forty years. This is as true for the anti-inflationary 1980s as for the inflationary 1970s and the 'libertarian' 1960s. In the light of this fact, we look at the problem of inflation both in general terms and in the British context. Furthermore, we shall consider inflation from a particular point of view, namely that of the labour market, although it would perhaps be more apt to describe the viewpoint as the supply-side.

In much previous work on inflation, it has been thought convenient to distinguish between demand-pull inflation and cost-push inflation. This is, in a sense, an unfortunate distinction because it gives the impression that there are two competing theories of inflation underlying these two expressions. This is not the case. So, although our viewpoint is from the supply-side, we do not present a model of inflation which is, in any sense, inconsistent with demand-side theories such as 'monetarism', for example. All we do is demonstrate where the supply-side, in general, and the labour market, in particular, fit into the story.

At the outset, therefore, it is worth emphasizing that inflation is a demand-side phenomenon in the sense that, by appropriate demand-side (monetary and fiscal) policies, inflation can *always* be controlled. Of course, there may be serious technical problems in exercising such control, relating to such matters as the operation of financial markets and the like. Nevertheless, in principle, the statement is correct. If so, where does the labour market enter the story? Surely, as

Peter Wiles argued so passionately nearly twenty years ago in the *Economic Journal* (Wiles, 1973), it is obvious that if union leaders press strongly for higher pay rises, higher inflation will result. Despite the fact that it is apparently obvious, it is fundamentally wrong. With an appropriately tough monetary and fiscal stance, this rise in inflation can be strangled, if not at birth, at a very tender age. What is the story? Clearly higher pay demands generate inflationary pressure. But when this meets the tough policy stance, then one of two things can happen. Either union leaders will note the tough policy stance and modify their demands at the outset or, and this is more likely, unemployment will rise rapidly and, very soon, union members, fearing for their jobs, will ensure that their leaders change their position. The potential inflationary pressure leads, not to higher inflation, but to higher unemployment.

Suppose, however, that the inflationary pressure comes from a source which cannot be modified. A sharp rise in imported commodity prices, for example. How do things work in this case? A rise in imported commodity prices makes the country poorer. This is an inescapable fact. Since wages make up by far the largest part of National Income, it is more or less inevitable that they must decline in real terms. In the light of this, there are three possible outcomes following the rise in commodity prices. First, wage demands are modified to offset the commodity price rise, overall cost pressures remain unchanged and inflation is contained. This tends to happen if the labour market is highly competitive or, interestingly enough, if wage bargaining is highly centralized, as in Scandinavia. In this latter case, all the parties to the centralized wage bargain can see what is happening from their economy-wide perspective, and wages tend to be modified accordingly. The second possibility is that wage demands are not immediately modified but the government takes a tough policy stance. Unemployment rises rapidly and this

First published in *Oxford Review of Economic Policy*, vol. 6, no. 4 (1990). This version has been updated and revised to incorporate recent developments.

forces the appropriate adjustment. The third possibility is that wage demands are not modified and the government takes an accommodating policy stance. Inflation then starts to rise. The tough stance and consequent unemployment then typically follows at a later date in order to prevent the inflation getting out of control.

These scenarios reveal the fundamental picture. Inflationary pressure may arise from the supply-side and may thus quite sensibly be described as 'cost-push'. But whether or not such pressure actually turns into rising inflation depends on the policy response on the demand-side. In the simple framework sketched above, the supply side of the economy determines the position of the trade-off, in this case between rising inflation and unemployment. The demand side, which includes the macroeconomic policy stance of the government, then determines the precise mix of inflation increase (or decrease) and unemployment which comes about. Notice that the trade-off we have here is not one between particular *levels* of inflation and unemployment, but is one between *rates of change* of inflation and *levels* of unemployment. While the former kind of trade-off was thought to exist in the early 1960s, following the work of Phillips (1958), this rather rapidly proved not to be the case.

Finally, it is worth remarking that while our stories have referred solely to inflationary pressure arising from the supply side, this is not always the case. Clearly inflationary pressure and rising inflation can appear directly from the demand side simply by having a more expansionary policy stance. Thus we might describe demand-pull inflation as a situation where inflation rises via a demand expansion in the face of a fixed supply-side trade-off. Cost-push inflation, on the other hand, arises when the trade-off between the rate of change of inflation and unemployment 'worsens' and the resulting inflationary pressure is turned into rising inflation by an accommodating policy stance on the demand side. There is only one theory here. The rate of change of inflation is always determined by the demand-side stance relative to the position of the supply-side trade-off.

In what follows, we shall focus on the supply-side trade-off. In an open economy such as Britain's, this trade-off is rather more complicated than that described above, since it contains a third element in addition to inflation and unemployment, namely the trade deficit. So we shall first set out a simple model to demonstrate how the trade-off arises and then present some numbers which illustrate how this trade-off has moved in Britain over the last thirty years. We can then see how the position on the demand side has deter-mined the actual outcomes in terms of inflation, unemployment, and the trade deficit. We shall, also, briefly look at the current position.

II. A model of inflation

1. The demand side

In this section, we construct a simple model of the economy starting with the demand side. The idea here is to demonstrate that, in a world where prices are sticky, the demand side can be thought of as determining the level of real demand in the economy and the level of international competitiveness. Furthermore, we suppose that real demand is always satisfied and is, therefore, the same as real output (GNP).

In log-linear form, the demand side consists of the following equations:

$$\text{Money demand: } m - p = y_d. \tag{1}$$

$$\text{Goods demand: } y_d = \sigma_1 x - \sigma_2 r + \sigma_3 c. \tag{2}$$

$$\text{Competitiveness (definition): } c = e + p^* - p. \tag{3}$$

$$\text{Uncovered interest parity: } i = i^* + \Delta e^e. \tag{4}$$

$$\text{Expected depreciation: } \Delta e^e = \Delta p^e - \Delta p^{*e} - \delta(c - c^{LR}). \tag{5}$$

m = real money stock; p = price level (GDP deflator); y_d = real demand = real GDP; x = exogenous real demand factors including fiscal policy stance, world economic activity, autonomous expenditure (e.g. autonomous shifts in consumption); r = domestic real interest rate; c = competitiveness; e = exchange rate measured as the domestic currency value of one unit of foreign currency, so a rise in e represents a depreciation of the domestic currency; p^* = world price of output in foreign currency; i = domestic nominal interest rate; i^* = foreign nominal interest rate; Δp^e = expected domestic inflation; Δp^{*e} = expected foreign inflation; c^{LR} = expected long-run equilibrium competitiveness; and Δ represents the time difference.

This is more or less the simplest open economy model which can be constructed, and any number of complications can be incorporated without changing anything which is important for our purposes. Equation (1) is a standard quantity theory demand for money function, and equation (2) is a simple IS curve. Equation (3) defines competitiveness, and (4) is an arbitrage condition which holds if investors are risk neutral and there are no impediments in the currency market. That is, the return from investing in domestic securities is the same as that from investing in foreign securities correcting for expected movements in the exchange rate. Finally, equation (5) says that expected currency depreciation reflects the inflation differential modified by a tendency for the exchange rate to help competitiveness towards its long-run expected level.

These equations solve out to yield

$$y_d = m - p \tag{6}$$

$$c = c_1 (m - p) - c_2 x + c_3 r^* + c_4 c^{LR} \tag{7}$$

where $r^* = i^* - \Delta p^{*e}$, the foreign real interest rate, and $c_1 = 1/\Omega$; $c_2 = \sigma_1/\Omega$; $c_3 = \sigma_2/\Omega$; $c_4 = \sigma_2 \delta/\Omega$; $\Omega = \sigma_2 \delta + \sigma_3$. If prices are sticky in the short run, then (6) and (7) illustrate how exogenous demand-side shifts will influence real demand (output) and competitiveness. The key variables here are exogenous demand factors, x, and monetary policy, m. Different combinations of these will produce different combinations of real demand and competitiveness. In particular, competitiveness depends crucially on the tightness of monetary policy relative to exogenous demand factors. If money is tight relative to exogenous demand, then competitiveness will be low, essentially because interest rates and hence the exchange rate are high. So, for example, if monetary policy is reasonably expansionary but tight relative to exogenous demand factors (which are, therefore, very expansionary), then we have high real demand and low competitiveness (a high exchange rate). This kind of policy was pursued by the US in the period 1983–4. If, on the other hand, monetary policy is very expansionary and hence loose relative to reasonably expansionary exogenous demand factors, then real demand is high and competitiveness is high. This reflects the UK situation in the period after the stock market crash. The opposite of this, with very tight money and contractionary demand factors, yields low demand and low competitiveness, as in the UK in 1980–1.

We have now set the demand-side scene and we can turn to the supply side. This will tell us the consequences of demand side shifts in real demand and competitiveness on our key variables, namely unemployment, the trade deficit, and inflation.

2. The supply side

The supply side of our model has the following simple log-linear form:

$$\text{Production function: } y_d - y^{PO} = -u \tag{8}$$

$$\text{Price setting: } p - w = \beta_0 + \beta_1 [y_d - y^{PO}] + \beta_{12} c - \beta_2 \Delta^2 p \tag{9}$$

$$\text{Wage bargaining: } w - p = \gamma_0 - \gamma_1 u + \gamma_{12} c - \gamma_2 \Delta^2 p + z_w \tag{10}$$

$$\text{Trade deficit: } c = \delta_0 + \delta_1 [y_d - y^{PO}] - \delta_2 td - z_c \tag{11}$$

y^{PO} = potential output; w = wage; $\Delta^2 p$ = the rate of change of inflation; u = unemployment rate; z_w = exogenous factors raising wages; td = trade deficit as a proportion of potential output; z_c = exogenous factors tending to improve the trade deficit.

Again we have a very simple framework which omits long-run growth factors and any unemployment dynamics or hysteresis effects. (A more detailed analysis may be found in Carlin and Soskice, 1990, for example.) The production function (equation 8) simply relates unemployment to deviations of demand (output) from its (exogenously given) potential level, y^{PO}. The price-setting model (equation 9) expresses the mark-up of (GDP) prices on wages as a function of three variables. Prices are increasing in demand, for obvious reasons. The price mark-up is also increasing in competitiveness because it seems likely that, as firms become more competitive in world markets, demand becomes less elastic and higher mark-ups can be sustained (i.e. world sales become more profitable). The role of changes in inflation ($\Delta^2 p$) is very important. As we have already noted, prices are sticky in the short run. If inflation is rising, firms find that costs are rising more rapidly than expected and, to the extent that prices are sticky and tend to lag behind costs, this tends to squeeze profits and reduces the mark-up of prices on wages. The period 1974–5 in Britain provides a classic example of this process in action.

The wage bargaining equation (10), expresses the wage outcome as a mark-up on (GDP) prices although the actual bargaining procedure concerns nominal wages. The real wage outcome is decreasing in the level of unemployment which is, of course, an inverse measure of labour market tightness. The role of competitiveness here is very important. A rise in competitiveness generates upward pressure on wages via real wage resistance. As competitiveness increases, imports become more expensive and the cost of living rises (note that the cost of living is a weighted average of GDP prices and import prices). In so far as workers resist the consequent fall in living standards, this will exert upward pressure on wages relative to GDP prices. The role of increases in inflation ($\Delta^2 p$) is similar here to that in price setting. If inflation is rising, prices typically turn out higher than was expected when the wage bargain was struck and, as a consequence, wages turn out lower relative to GDP prices. Rising inflation tends to squeeze real (product) wages, *ceteris paribus*.[1]

Finally, we have a whole group of variables represented by z_w which reflect autonomous sources of wage pressure. These can be grouped conveniently under three headings. First, union effects. These capture any changes in the level of pressure exerted on wages by trade union militancy. Second, real wage resistance effects. As we have already noted with competitiveness, any factors which tend to produce autonomous falls in living standards will lead to upward pressure on wages. Such factors will include taxes on labour (e.g. income tax) and goods (e.g. VAT), and adverse changes in the international terms of trade (e.g. a rise in the world relative price of commodities). Finally, we have reductions in the effective supply of labour at given unemployment. Such reductions can arise via increases in the generosity of the unemployment benefit system or increases in the degree of occupational or regional mismatch between the unemployed and available vacancies. Both of these will tend to reduce the effective availability of labour.

The last equation (11) determines the trade deficit, although it has been written with competitiveness on the left-hand side for future expositional convenience. This simple equation says that the trade deficit is increasing in demand and decreasing in competitiveness. The final term, z_c, captures any exogenous factors which tend to improve the trade balance, the most notable of which, in the British context, is North Sea oil.

By manipulating these equations, we can now determine the impact of demand and competitiveness, which appear from the demand side, on unemployment, the trade deficit, and changes in inflation. More specifically we have

$$u = -\,[y_d - y^{PO}] \text{ (from (8))} \qquad (12)$$

$$td = 1/\delta_2\,[\delta_0 + \delta_1\,\{y_d - y^{PO}\} - c] - z_c \text{ (from (11))} \qquad (13)$$

$$\Delta^2 p = 1/\alpha_2\,[\alpha_0 + \alpha_1\,\{y_d - y^{PO}\} + \alpha_{12}c + z_w]$$
$$\text{(from (8), (9), (10))} \qquad (14)$$

where $\alpha_0 = \beta_0 + \gamma_0$; $\alpha_1 = \beta_1 + \gamma_1$; $\alpha_{12} = \beta_{12} + \gamma_{12}$; $\alpha_2 = \beta_2 + \gamma_2$

---

[1] One point is worth clarifying here. We have just noted that rising inflation tends to squeeze real wages. But when we were discussing pricing, we found that rising inflation tends to squeeze profits and reduce the mark-up of prices on wages. But this is tantamount to raising real wages! How can rising inflation do both things at once? The answer is simple. Real wages and changes in inflation are both determined within the system along with many other things. Inflation cannot change unless something else changes as well. An exact analogy is provided in a simple supply and demand system. On the supply curve, quantity is increasing in price and on the demand curve it is decreasing. This ensures that price cannot change unless something else shifts.

These equations tell us first that demand directly determines unemployment, so when demand is low, unemployment is high and vice versa. Second, they reveal that the trade deficit is increasing in demand and decreasing in competitiveness. Third, we see that inflation rises with both demand and competitiveness. The latter effect is worth some comment. When competitiveness is high, imported goods are more expensive and this enables firms which produce goods in competition with imports to raise their prices and also exerts upward pressure on wages as workers attempt to compensate for increases in the cost of living.

We can now see the implications of various demand-side scenarios. If demand and competitiveness are both high, we have low unemployment and rising inflation with the trade deficit being small or even negative. If demand is high but competitiveness is low, we have low unemployment, a large trade deficit and more or less stable inflation. These two possibilities immediately reveal a trade-off between inflation and the trade deficit in the sense that, with low unemployment, one can have more of one and less of the other with different demand-side scenarios.

3. The fundamental supply constraint

In order to examine such trade-offs in more detail we need to do some manipulation on our three equations (12), (13), (14). Let us simply eliminate demand, $[y_d - y^{PO}]$ and competitiveness, c. This yields the fundamental trade-off equation

$$[\alpha_1 + \delta_1\alpha_{12}]u + \alpha_2\Delta^2 p + \alpha_{12}\delta_2 td = \alpha_0 + z_w - \alpha_{12}z_c \quad (15)$$

How should we interpret this rather complex looking expression? What it does is to tell us those combinations of unemployment, u, rates of change of inflation, $\Delta^2 p$, and trade deficit, td, which are available. Shifts on the demand side can move the economy around on this constraint but they cannot get the economy off it. However, the constraint can and does move. Any rise in autonomous wage pressure, z_w, will worsen the trade-off in the sense that at least one of u, $\Delta^2 p$, and td must be higher. On the other hand, an increase in any exogenous factor which improves the trade balance, z_c, will improve the trade-off.

The existence of this constraint has very important implications. For example, for some time in Britain we had a trade deficit and rising inflation ($td > 0, \Delta^2 p > 0$). The constraint tells us that however subtle are the demand-side manipulations of interest rates, exchange rates, taxes, government expenditures, savings incentives, joining the ERM, etc., etc., we cannot stabilize inflation and reduce the trade deficit without a substantial increase in unemployment, unless either z_w goes down or z_c goes up. Since it is hard to shift these in the short run, the constraint tells us we simply have to lump it. If we want stable inflation and a lower trade deficit, we must have higher unemployment. If we want to bring inflation down ($\Delta^2 p < 0$) we must have even higher unemployment.

It is clear from this discussion that the role of the wage pressure and trade balance factors, z_w, z_c is absolutely crucial since these are the only things which can move the constraint around. When we come on to our discussion of the British economy we shall consider them in some detail. In the meantime, however, their effect can usefully be summarized by the notion of the equilibrium level of unemployment, $\hat{u}$ (sometimes known rather misleadingly as the 'natural rate', although an object which is less embedded in nature is hard to imagine). This equilibrium level may be defined as that unemployment rate which is consistent with constant inflation ($\Delta^2 p = 0$) and balanced trade ($td = 0$). From (15) we thus have

$$[\alpha_1 + \delta_1 \alpha_{12}]\hat{u} = \alpha_0 + z_w - \alpha_{12} z_c \quad (16)$$

and we see that $\hat{u}$ summarizes the impact of z_w and z_c on the fundamental trade-off. Indeed (15) can now be rewritten as

$$[\alpha_1 + \delta_1 \alpha_{12}]u + \alpha_2 \Delta^2 p + \alpha_{12} \delta_2 td = [\alpha_1 + \delta_1 \alpha_{12}]\hat{u} \quad (17)$$

So as $\hat{u}$ rises, the fundamental trade-off gets worse and the economy is more beset by problems. We now have a framework within which to analyse the British economy and in the next section we shall attach some actual numbers to our equations.

III. Britain's inflation problem

In order to apply this framework to the British economy we must have some parameter values. The ones we present here are derived partly from equations presented in Layard and Nickell (1986) and partly from new estimates. Full details may be found in chapter 9 of Jackman, Layard, and Nickell (1991). It is important to recognize that the numbers which are presented are subject to considerable error. They are based on averages of the past behaviour of the economy and on models which are clearly a gross simplification of reality. Nevertheless, they provide a useful flavour of what is actually going on. The equation which corresponds to (17) above has the form

$$0.091 \log u + 0.05\, u + 1.07\, \Delta^2 p + 1.25\, td$$
$$= 0.091 \log \hat{u} + 0.054\, \hat{u} - 1.27\, \Delta u \quad (17')$$

where u is the unemployment rate (OECD standardized measure, which is currently slightly above the UK published rate), $\Delta^2 p$ is the rate of change of annual inflation, td is the trade deficit as a proportion of potential GDP, and Δu is the annual change in the unemployment rate.

This equation differs significantly from its theoretical counterpart in two respects. First, it is not linear in the unemployment rate. This is simply an inevitable consequence of our attempts to fit the data. Second, the change in the unemployment rate also appears in such a way that if unemployment is rising (falling), the fundamental trade-off improves (worsens). This is known as a hysteresis effect. Why does it happen? The main reason is that when unemployment is actually going up, pressure on wages is significantly reduced whatever the level of unemployment. This is hardly surprising, for when workers actually see job losses taking place all around them, this leads naturally to their modifying wage demands. So when unemployment is actually going up, the trade-off temporarily improves.

What are the trade-offs implicit in $(17')$? At constant unemployment, a rise in the trade deficit of 1 per cent of potential GDP is worth just over one percentage point per annum off the rise in the inflation rate. In this sense, therefore, the trade deficit can be thought of as suppressed inflation. Alternatively, at constant inflation, a one-point rise in the trade deficit as a percentage of potential GDP is worth around three-quarters of a percentage point off unemployment (from a baseline level of 6 per cent, which is around the current rate). Finally, at constant trade deficit, a one-point rise in unemployment from the 6 per cent baseline will reduce the rise in inflation by around 1.3 percentage points per annum.

We can also use equation $(17')$ to produce estimates of the equilibrium rate of unemployment and so, in Table 9.1, we present some results for the past 40 years. During the first two periods we consider, the economy was rather close to equilibrium on average with a low level of unemployment, but by the early 1970s the equilibrium rate of unemployment was starting to creep up. During this period, this produced higher unemployment and increasing inflation which was offset by a significant trade surplus. After the first oil shock, in 1974, equilibrium unemployment moved up very sharply and with actual unemployment lagging significantly behind, there was a rapid increase in inflation and a sharp movement into deficit on the trade front. This was a key period for the British economy, for the fundamental

Table 9.1. Estimates of Equilibrium Unemployment in Britain Based on Equation (17′)

	1956–9	1960–8	1969–73	1974–80	1981–7	1988–90	1991–5
$u(\%)$	2.24	2.62	3.39	5.23	11.11	7.27	9.30
$\Delta u(\%)$	−0.06	0.035	0.43	0.30	0.76	−0.90	−0.14
$\Delta^2 p(\%)$	0.58	−0.11	1.00	1.51	−1.45	1.03	−1.12
$td(\%)$	−0.57	−0.22	−0.81	1.06	−1.39	1.44	0.90
$\hat{u}(\%)$	2.2	2.5	3.6	7.3	8.7	8.7	8.9

Note: The values of Δu, td are lagged two years and that of $\Delta^2 p$ is lagged one year to take account of the time it takes for these factors to feed through the model into unemployment. Since the reaction of unemployment to general economic conditions appears to be more rapid in the recent past, we use current values for 1991–5. The 1995 data use forecasts published in Goldman-Sachs UK *Economics Analyst*.

Table 9.2. The Contribution of Various Factors to Equilibrium Unemployment in Britain (Percentage Points)

	1956–9	1960–8	1969–73	1974–80	1980–7
$\hat{u}(\%)$	2.2	2.5	3.6	7.3	8.7
Rise in $\hat{u}$ from 1956 to 1959	—	0.3	1.4	5.1	6.3
Contributing Factors					
North Sea oil	—	0	0	−0.28	−2.86
Mismatch	—	0.09	0.44	0.99	2.53
Terms-of-trade effects	—	−0.41	−0.50	0.99	2.26
Benefit system	—	0.34	0.93	0.64	1.12
Unions	—	0.34	0.60	1.42	1.50
Taxation	—	0.09	0.09	0.12	−0.20
Unmeasured	—	−0.15	−0.16	1.22	1.95

Note: The contributions of the various factors to the rise in $\hat{u}$ assume there is a two-year lag before they impact on unemployment because of the time taken for them to feed through the system. Furthermore, because of the non-linearity in the unemployment effect in equation (17′), any specific change in a contributory factor has a bigger effect on equilibrium unemployment if the baseline level of unemployment is higher. This accounts for the larger numbers in the right-hand columns.

trade-off worsened dramatically and the attempt to avoid a very sharp increase in unemployment had severe consequences for inflation and the trade deficit. In the early 1980s, there was then a very rapid reversal of this process. Unemployment was allowed to rise to unprecedented heights and, as a consequence, inflation fell rapidly and the economy moved into a large surplus on the current account. In the late 1980s the situation moved rapidly in the other direction. Unemployment came down to well below the equilibrium rate, inflation rose, and we had a huge current account deficit. Since that time we have had a further significant recession followed by part of a recovery. Unemployment went above the equilibrium rate, inflation came down substantially, and the current account deficit improved, although it has not moved into surplus.

It is clear from this analysis that the problems of the economy since the first oil shock arise essentially from the large rise in the equilibrium rate of unemployment.

This has made the policy choices very much tougher in the sense that, while it is still possible to control inflation by appropriate demand-side policies, the costs of doing so on the unemployment or trade front are politically very much harder to bear. In the light of these remarks, we must attempt to shed some light on why equilibrium unemployment has risen, particularly the dramatic change in the 1970s. So, in Table 9.2, we set out some rough estimates of the contribution of various factors. Recall that these reflect autonomous shifts either in wage pressure variables or in factors influencing the trade balance.

The actual numbers in Table 9.2 refer to the contribution of each factor to the change in equilibrium unemployment since the late 1950s.

Considering each factor in turn, the first reflects the beneficial impact of North Sea oil on the trade balance. This has produced a dramatic improvement in the trade-off in the most recent period considered. Mismatch refers to a measure of the labour market mismatch between the skills of the unemployed and those required by employers. By adding to pressure on wages at given levels of unemployment, this has made a significant contribution in recent years. Terms-of-trade effects are another important factor arising essentially from the behaviour of commodity prices. Their impact comes about via real wage resistance and in earlier years it was favourable, with real commodity prices declining secularly from the Korean War to the first oil shock. Their dramatic rise in 1974 and again in 1979–80 generated significant upward pressure on wages and this was enough to offset the favourable effects of North Sea oil. The increasing generosity of the unemployment benefit system has had a relatively small impact but that of increased union militancy has been fairly substantial. There appears to have been no reduction in this latter effect in the 1980s, which is quite surprising given the pressure on unions over this period. There are two points worth noting here. First, the lags in the system mean that the impact of unions on equilibrium unemployment in 1980–7 is caused by their activities at least two years earlier (i.e. 1978–85) because of lags in the system. Given that the defeat of the miners in 1985 was the key symbol of the reduction in union power, the effects of this had yet to show through to any great extent. Second, it has frequently been noted that while unions have been much more co-operative in achieving productivity improvements in the 1980s, the anti-union legislation appears to have had little effect on wage-bargaining activities, once account is taken of the general labour market situation (see Matthews and Minford, 1987). So, for example, now that the labour market is buoyant, we see the traditional union sectors again leading the way in wage demands, as they did in the early 1970s.

The contribution of changes in taxation, via real wage resistance, has had a relatively minor impact but there remains a considerable residual contribution arising from factors which we are unable to measure and about which we can only speculate. In this regard, there are several points worth bearing in mind. First, the breakdown in Table 9.2 is very rough and ready. The number of factors which influence real wage resistance and the effective supply of labour are simply enormous and most of them are more or less impossible to measure. For example, obtaining accurate measures of labour market mismatch is very difficult, particularly since the consequence of various efficiency drives in the Government Statistical Service meant that detailed figures on the occupational breakdown of unemployment and vacancies were no longer produced after the early 1980s. There is some evidence from the CBI sur-

veys that skill mismatch is considerably worse now than it was in the late 1970s and much worse than in the early 1970s,[2] but accurate quantitative information is very hard to come by.

The second point concerns the inadequacy of the measures used. Thus the benefit effect, for example, is based on the benefit replacement ratio, that is the ratio of unemployment benefits to post-tax earnings. But this is not, in fact, the key feature of the benefit system for our purpose. Far more important is the pressure exerted on the unemployed to take up jobs, which is difficult to measure, although there is some evidence that it was reduced with the separation of Benefit Offices and Job Centres in the 1970s and significantly increased again with tougher rules in the late 1980s. For example, international comparisons reveal that benefit systems which are structured to avoid the build up of long-term unemployment, are very effective in reducing equilibrium unemployment. This does not mean that the system has to be one in which the unemployed are simply starved back to work after a certain period. Thus in Sweden, a country not noted for any absence of generosity in its welfare system, the benefit replacement ratio is far higher than in Britain but after a certain time (around 14 months) unemployed individuals are offered either reasonable jobs at the going rate or retraining, which they more or less have to take up. Consequently, there are hardly any long-term unemployed in Sweden, which has favourable consequences both for the individuals concerned and for the workings of the macroeconomy.

Finally, with regard to Table 9.2, it is worth speculating on the current position. Since the mid-1980s, the favourable impact of North Sea oil has been much reduced but this has been offset by the improvements in the terms-of-trade effects via the fall in real commodity prices, and the increase in the severity of the benefit system which we have already mentioned. However, the weakness of the training system ensures that we have persistent skill mismatch problems and, overall, we should not be surprised if the equilibrium unemployment rate remains around 8 per cent. So we are still facing the same unpleasant trade-off which has been with us for some fifteen years now. Furthermore,

---

[2] Thus, the CBI Industrial Trends Surveys report the following numbers for manufacturing:

	Percentage of firms reporting shortages of skilled labour (1)	Percentage of firms reporting shortages of unskilled labour (2)	(1)/(2)
1969–73	24.4	9.0	2.71
1974–80	19.3	5.1	3.78
1988–9	22.4	4.0	5.60

unless we can organize the labour market so as to reduce the upward pressure on wages which remains with us despite apparent labour market slack (as measured by unemployment),[3] then this same unpleasant trade-off will be with us for a good few years yet.

IV. Summary and conclusions

In the preceding analysis we have demonstrated the existence of a fundamental supply-side constraint in the economy which takes the form of a three-way trade-off between unemployment, the trade deficit, and increases in inflation. The mix of monetary policy and exogenous real demand factors then determines which combination of these three outcomes actually occurs. Demand factors cannot, however, shift the underlying trade-off in any simple way.[4]

In the British context, we have demonstrated that the fundamental constraint has shifted adversely, particularly over the last two decades. As a consequence policy-makers have been confronted with ever more difficult choices and this has resulted in a persistent problem of high unemployment and relatively high inflation. We have enumerated some of the factors

underlying this adverse shift, although we have by no means obtained a complete picture.

Finally, it is worth correcting an impression which may have been gained by the reader, namely that given the existence of the fundamental supply-side constraint, the government's task is very simple. In the short run, it chooses monetary and fiscal policies to select the desired point on the constraint, and in the long run, it undertakes appropriate supply-side policies to move the constraint in a favourable direction. Unfortunately the former is very difficult because we do not know precisely where the constraint is, and given the other shocks on the demand side, the actual consequence of any particular monetary and fiscal policy mix cannot be predicted with any great accuracy. The latter is even harder because most of the relevant factors are outside the direct control of government.

References

Carlin, W., and Soskice, D. (1990), *Macroeconomics and the Wage Bargain*, Oxford, Oxford University Press.

Jackman, R., Layard, R., and Nickell, S. (1991), *Unemployment*, Oxford University Press, Oxford.

Layard, R., and Nickell, S. (1986), 'Unemployment in Britain', *Economica* (special issue on unemployment), 53, August, S121–70.

Matthews, K., and Minford, P. (1987), 'Mrs Thatcher's Economic Policies 1979–87', *Economic Policy*, 5, October, 57–102.

Phillips, A. W. (1958), 'The Relation between Unemployment and the Rate of Change of Money Wage Rates in the United Kingdom 1861–1957', *Economica*, 25, 238–99.

Wiles, P. (1973), 'Cost Inflation and the State of Economic Theory', *Economic Journal*, 83, 377–98.

---

[3] It is worth emphasizing that the upward pressure on wages which we currently observe, despite 6 per cent unemployment, is not simply a matter of militant unions and feeble managers. In large parts of the labour market firms are forced to pay substantial wage increases simply to retain and recruit workers, which suggests that mismatch problems of various kinds must be severe.

[4] In the longer term, demand factors may influence the position of the fundamental constraint by their impact on capital accumulation and hence on the available level of capacity. Any effects arising from this source are, however, likely to be short-lived.

10

Wage determination: the changing role of institutions in advanced industrialized countries

DAVID SOSKICE

Wissenschaftszentrum, Berlin[1]

I. Introduction

What influence do institutions have on wage-setting in the advanced industrialized countries in the 1990s? How and why has this changed over the past two decades? What have been the consequences for inflation and unemployment? Are there any lessons which can be drawn for the UK as it enters the exchange rate mechanism of the European Monetary System?

Until recently, and since the early 1980s, discussion of institutional influences on wage restraint—which had been a major academic industry in the 1970s and early 1980s—has been close to non-existent. This reflected what was widely believed, that decentralization and deregulation of labour markets had made the subject redundant. Indeed, this still represents, and for good reason, a major view of the role of institutions in the current wage-setting process.

The possibility of institutional wage restraint, however, is now back on the agenda. That this is the case owes much to the work of Calmfors and Driffill. In a remarkable paper (Calmfors and Driffill, 1988) they made a persuasive argument that, if wage-bargaining institutions in different countries could be placed on a scale ranging from decentralized to centralized, both *decentralized* and *centralized* would work well (to reduce the level of unemployment) but systems in

between would work badly: they showed, in other words, a hump-shaped relation between the degree of centralization and the level of unemployment, with low unemployment being associated with both decentralized systems (e.g. in their classification, Japan and Switzerland) and centralized systems (e.g. Austria and Sweden), while other countries with 'middling' systems had high unemployment (e.g. UK, Germany, France, Italy).

Thus there are at present two main views as to the role of institutional influences on wage restraint. The first is the view that decentralization of decision-making through the last decade has eliminated institutions external to the market from the wage-setting process. The second is the Calmfors–Driffill argument that both decentralized and centralized systems, but not those in between, can deliver wage restraint; and that centralized systems continue to exist.[2]

In this paper it is argued that neither the 'decentralization' hypothesis, nor that of Calmfors and Driffill (C–D), is correct, but that both contain important elements which need to be embodied in a more satisfactory account. A simple version of C–D is set out in the next section, followed by a critique in section III. Section IV explains the decentralization hypothesis and its limitations. And in section V an alternative approach is developed, building on the previous two sections. The concluding section VI asks what lessons can be drawn for the UK.

First published in *Oxford Review of Economic Policy*, vol. 6, no. 4 (1990).

[1] I should like to thank for their helpful comments on the paper and for discussions on these issues Peter Burgess, Andrew Glyn, Andrew Graham, Michael Knetter, Peter Lange, Richard Layard, Marino Regini, and Philippe Schmitter. Also the invaluable monthly *International* (now *European*) *Report* of Incomes Data Services.

[2] Another argument not considered here relates institutions to other dimensions of performance, including labour market participation and income distribution, see Rowthorn and Glyn (1990).

II. The Calmfors–Driffill model

This section sets out a simple version of the C–D model for those unfamiliar with the theoretical argument. In the simplest case the union is in a monopoly position to set the money wage for its members, leaving it to the employer to choose the level of employment. (Although this case is unrealistic, the basic argument is the same as in bargaining models.) The union may be operating either (*a*) at company level (decentralized wage-setting), or (*b*) at industry level (intermediate wage-setting), or (*c*) at national level (centralized wage-setting).

The argument can be summarized as follows. The union is concerned both about the real wage in terms of consumer goods and the level of employment of its members. Consider first the union setting a wage at company or industry level. Raising the money wage has two effects: first, it raises the real consumption wage; but, second, it reduces employment, since it pushes up the company or the industry price level relative to the general price level, and this lowers product demand in the sector. Assume the industry and the company are each small relative to the economy as a whole, so that an increase in the money wage at company or industry level has no perceptible effect on the general price level: in that case, the first effect of raising the money wage—namely to increase the real consumption wage—is the same at both company and industry level. But as regards the second effect—the reduction in employment—a given increase in the money wage is likely to lead to a larger cut in employment in the company than in the industry: this reflects the higher product elasticity of demand for the company as compared to the industry. Thus more moderate behaviour can be expected from union wage setting at company level than at industry level.

But at national level another factor is introduced. When the union sets the money wage for all workers in the economy, an increase in the money wage leads directly to an equiproportionate increase in the price level. Hence the first effect does not operate: there is no increase in the consumption real wage. As regards the second effect: relative prices cannot change, but the real money supply will fall and hence reduce employment. The monopoly union chooses the real money supply consistent with full employment. So again, though for different reasons, a monopoly union setting the wage at a centralized level will behave moderately.

The next step in the argument is to relate wage-setting behaviour more explicitly to equilibrium unemployment. Consider first why the union at industry or company level will demand a higher expected real wage as aggregate demand in the economy increases.

At a particular level of aggregate demand, the monopoly union sets some money wage relative to the expected price level; at higher aggregate demand, the same expected real wage will be associated with higher employment than before: so given a trade-off between employment and the real wage, the union will now set a higher expected real wage. Thus there will be an increasing function relating the expected real wage set by the monopoly union to the level of aggregate demand and hence of employment. This function will be lower in the company union case as compared to the industry union case, since at any aggregate demand level the expected real wage chosen by the company union will be below that chosen by the industry union.

Still confining attention to company and industry wage-setting, Figure 10.1 shows these two expected real wage functions. Equilibrium employment (and hence unemployment) is given by the intersection of the relevant function with the real wage implied by price-setting behaviour. A simple assumption of cost-plus pricing (e.g. constant product market demand elasticity and constant marginal labour productivity) implies a horizontal price-determined real wage schedule—or 'feasible real wage' in Layard–Nickell terminology—in this diagram (Nickell, 1990; Carlin and Soskice, 1990). (Note that prices are set by companies, whether the wage is set at company or industry level, so that the elasticity of product demand used in price-setting does not depend on the level of wage-setting.) In Figure 10.1 the equilibrium levels of employment are E_{ind} in the industry wage-setting case and the higher level E_{com} in the company case. Now consider centralized bargaining, where the union chooses the money wage for the economy as a whole. The union is assumed to know the model of the economy, so that it knows the equilibrium real wage which will result from price-setting; the nominal money supply is fixed and known to the union. By setting a money wage, the union implicitly chooses the price level, hence the real money supply, real aggregate demand, and employment. In terms of the diagram, the union chooses a position on the horizontal real wage schedule implied by price-setting. In Figure 10.1, equilibrium employment in the centralized case is shown as higher than in the intermediate case, as it is in the decentralized wage-setting example. It can be argued that equilibrium employment with centralized wage-setting is higher than in the decentralized case (as shown in Figure 10.1): that is because the centralized union can in principle choose any point to the left of supply of labour schedule, E^s, and so might be expected to choose 'full employment'.

This point (though it is not central to the C–D argument) can be clarified by superimposing labour demand curves and union indifference curves on to

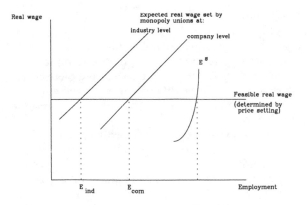

Figure 10.1. Wage Functions and Equilibrium Employment

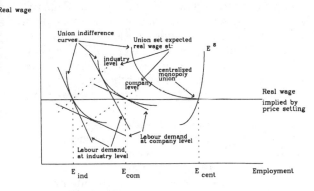

Figure 10.2. The Impact of Bargaining on Wages and Employment

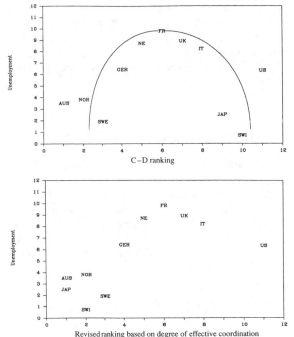

Figure 10.3. The Calmfors–Driffill Model

Note: Unemployment is averaged over 1985 and 1989. Italian unemployment refers to North and Central Italy (see text). *Source*: OECD, *Economic Outlook*.

Figure 10.1, cf. Figure 10.2. In fact this may be of use in understanding the whole C–D model. The union has the same indifference curves at whatever level it is setting wages: these are the preferences of the representative worker. The curves are flat as they approach the labour supply schedule; and it is assumed that to the left of E^s the trade-off between the real wage and employment becomes steeper, at any given real wage, as employment declines. The labour demand curve of the individual company or industry is assumed to be linear; its slope reflects inversely the product elasticity of demand, higher for the company (thus a flatter slope), lower for the industry; and its position depends on the level of aggregate demand. In equilibrium the monopoly union must choose the expected real wage so as to be maximizing against the relevant labour demand schedule, and the expected real wage must be equal to the real wage implied by price-setting.

In Figure 10.2, the expected real wage set by the monopoly union with industry-wide bargaining is traced out, for different levels of aggregate demand, by tangency points of the union indifference curves with the industry labour demand functions. The equilibrium

employment rate, E_{ind}, is the level of employment which equates this expected real wage to the real wage implied by price-setting; i.e. where the higher upwards-sloping dashed line cuts the horizontal real wage line. Similarly, the equilibrium rate of employment with company-level bargaining, E_{com}, is given by the intersection of the expected real wage schedule at the company level with the horizontal real wage line. Finally, for the centralized union, the only constraint is the real wage implied by price-setting: so it chooses equilibrium employment at E_{cent} where the highest union indifference curve just touches the horizontal real wage line.

These results can now be translated into the relationship between the degree of centralization in the wage-setting process and the equilibrium unemployment rate. In Figure 10.3, the C–D hypothesis is illustrated with eleven countries, showing the relation between the average of unemployment rates in 1985 and 1989 and their degree of wage-setting centralization as measured by Calmfors and Driffill. This is six fewer countries than C–D, and a different time period: the countries chosen are those where I am reasonably clear about the functioning of wage-setting and labour market-related institutions; and the years are chosen to represent unemployment in the second half of the 1980s. As can be seen from the top diagram in Figure 10.3, these data choices in no way undermine the dramatic hump-shaped relation between unemployment and bargaining structure.

III. Reassessing Calmfors–Driffill

The critique of the C–D argument is divided into four parts. The first, and by far the largest, is that key countries are wrongly classified. Second, that there are major problems with unco-ordinated company-level bargaining. Third, that the pushfulness or otherwise of unions at local level is neglected. And finally that the effect of wage restraint on unemployment is different in open and closed economies.

1. The formal level of bargaining versus the effectiveness of economy-wide co-ordination

When C–D use 'the degree of centralization' as their institutional variable, they are concerned to measure the extent to which wages in different sectors of the economy are set in a co-ordinated fashion. As far as their theoretical argument goes, their 'centralized' wage-setting case is simply the case where wage setting is co-ordinated across the economy. Their theory therefore implies nothing about *where wages are technically set*, only that wherever they are set the process is co-ordinated. The C–D theory relates to co-ordination, but the empirical measure they use—the degree of centralization of bargaining institutions—relates to the actual location of bargaining.

This measure is unlikely to be wrong when bargaining institutions are in fact centralized: that is likely to be associated with co-ordinated wage-setting across the economy. This is, indeed, the case of the most centralized bargaining systems under discussion here, namely Austria, Norway, and Sweden.[3]

Japan and Switzerland

The problem which arises is that less centralized systems (at least on a formal level) may in fact be highly co-ordinated. Critically, the two countries which do most empirical work in C–D as examples of well-performing decentralized economies fall into this category: Japan and Switzerland. In both countries, most formal wage negotiations are concluded at the company level (though by no means all). But there are effective mechanisms, different in the two countries, which ensure a high degree of co-ordination across the economy.

The common element in these two systems of co-ordination is the role of powerfully co-ordinated employer organizations and networks. Switzerland has one of the most powerful sets of employer organizations among advanced industrialized countries. In Japan, the power of the employers' organization is reinforced by strong links and understandings between the large companies and within and between the Zaibatsu groups. In Switzerland the channel of transmission of co-ordination is via the employer-organization-dominated arbitration system, through which company wage disputes are settled (in key industries), as well as via industry-wide bargaining in others. In Japan co-ordination takes place through informal wage cartels in the main industries in effect working out company settlements.[4] The Swiss system gives companies somewhat more latitude than the Japanese one does.

The Japanese system of co-ordinated wage bargaining focuses on the so-called Spring Offensive or Shunto. This is the annual bargaining round, traditionally in April, in which most company and industry settlements are announced. The most important settlements in the private sector are company settlements, negotiated formally between enterprise management and the enterprise union. *De facto*, however, the key decisions are made by a small number of the largest companies grouped on an industry or multi-industry basis, after prolonged discussions between large companies across industries, and between business and the government. Although exports are a relatively small share of GDP, these discussions are based on the requirements of maintaining as far as possible cost competitiveness. They therefore involve working out with government likely exchange rate movements. The institutional framework for these interchanges benefits from the membership of many of the large companies in *zaibatsu* groupings, from the employer associations (though they are less important than in say Germany or Switzerland), and from a tradition of easy informal intercourse across companies and between the business sector and government.

Of particular importance are the group of metal-

---

[3] For descriptions of these systems up to the early 1980s, see Flanagan *et al.* (1983); a good picture of developments in the 1980s of Sweden and Norway is in Elvander (1990).

[4] See Dore (1987), Shirai (1984), and Shimada (1983) on Japan, and Aubert (1989) on Switzerland.

using exporting industries, iron and steel, electrical appliances, ship-building, and automobiles. The close co-operation between the largest companies in these sectors leads to the large companies in one of these industries, traditionally the five iron and steel majors, starting the Spring Offensive by making identical offers in terms of percentage increases. These are nearly always accepted without question by the enterprise unions, and are immediately followed by similar offers from companies in the other industries. This rate of settlements then generalizes across the economy. In the words of one of the leading Japanese industrial relations experts:

Although such group bargaining remains informal, it has practically the same effect as more systematic procedures of industry-wide or multi-industry negotiation since the industries concerned are closely related and exert a decisive influence in wage settlements in other branches such as chemicals, metal engineering, private and national railways, telecommunication and postal services, and national and local civil services. (Shirai, 1984)

The role of unions is secondary in this process. In fact, the national unions, of which the enterprise unions are usually members, are more actively involved than the enterprise unions themselves. Most important are the federations of metal-working industry unions, who negotiate with the groups of large companies at industry level. Combined in the International Metalworkers Federation–Japan Council, they have moved from a militant position in the 1950s (when Spring Offensive meant offensive) to accepting fully the centrality of maintaining international competitiveness for their members' long-term employment.

Two additional points should be made. First, the public sector follows the private in the wage-setting process, as a result of widespread use of arbitration designed to maintain parity between the two sectors. Generally, the government has resisted the temptation to intervene in this procedure, (though not always). Second, the non-unionized sector also follows the union sector. The non-unionized sector of small companies, etc. does not operate like a free market, even if there is more flexibility than in larger companies. Moreover, the close links between small and large companies enables another mechanism to be of relevance.

This mechanism is that large companies can operate their own type of wage restraint on small companies by setting as monopsonists the prices at which they are prepared to buy from them. This enables large companies to extend the same discipline which they can impose on their own wages to the less organized sectors of the economy.

If we put 'degree of effective co-ordination' on the horizontal axis instead of 'centralization' and simply switch Japan and Switzerland, while keeping all other values as before, the C–D result collapses in the sample of countries selected. This is illustrated by the bottom diagram in Figure 10.3.

The role of powerful employer organizations

Little attention has been paid in the economics literature to the role of employer organizations. (This contrasts with the strong interest they have had for political scientists and some industrial relations experts, notably William Brown and more recently Richard Layard.)[5] Most discussion by economists on wage restraint and incomes policy has focused on unions, their ability to reach agreements with governments, and their ability to impose those agreements on their members. Calmfors and Driffill are more sophisticated, and certainly take account of employer organizations: but they remain preoccupied by unions. Moreover, they suggest that strong unions breed strong employer organizations and vice versa (Calmfors and Driffill, 1988, p. 37, n. 8); perhaps this implicitly explains their concentration on unions.

Powerful employer organizations can impose wage restraint to greater or lesser degree on companies, because in systems in which employer organizations and networks are important, companies depend on them in a variety of ways. One set of ways relates to the influence which business associations and employer organizations have in the disbursement to companies of public money and resources. This ranges from areas such as technological innovation to export marketing to vocational training. Thus companies need to keep well in with their associations. This is bolstered by the advisory services which employer organizations provide. Also by the need to maintain good relations with other companies.

There is indeed an institutional equilibrium, in economies in which business associations and employer organizations are important, whereby governments use them to carry out policies on innovation and training, for example, because they are in a position to get companies to behave appropriately; and these associations can continue to remain powerful because they control access to such government resources.

Generally, where employer organizations or networks are powerful they play an important role in

---

[5] Layard (1990) develops a strength of employer co-ordination variable which has considerable explanatory power in explaining comparative unemployment rates. The most valuable book on comparative employer organizations is Sisson (1987); also useful is Windmuller and Gladstone (1984).

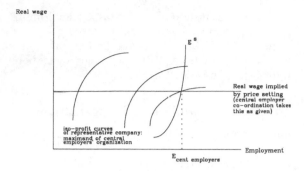

Figure 10.4. The Effect of Employers' Organizations

collective bargaining. Their sanctioning ability is usually informal. It may take the form of quite indirect warnings of a tacit sort related to other areas of activity: generally letting it be known that such and such a company is not a good citizen. Or it may involve more explicit actions, such as financial sanctions, as the Swedish SAF has power to impose. Or it may not involve sanctions but support when a company is strike-bound: such as the strike insurance funds run by German employer organizations.

(It may be objected that the theory of employer-imposed wage co-ordination is different from that of the monopoly union. We argue in subsection 4 that the closed economy is not the right framework to examine these questions: but if we stick for the time being with the closed economy Layard–Nickell model, it is not difficult to see in Figure 10.4 that—if (*a*) co-ordinated employers do not interfere with the price-setting process, so that they are constrained by the price-determined real wage schedule; if (*b*) the real wage/employment combination must be on or to the left of the E^s schedule; and if (*c*) profits are increasing in employment and decreasing in the real wage, so that iso–profit curves have a positive slope and increase in value as they move rightwards—organized employers will choose the same point as the monopoly union. For (*c*) to hold in the relevant region requires that the marginal productivity of labour curve cuts the price-determined real wage schedule to the right of the labour supply schedule.)

Germany and the Netherlands

Both Germany and the Netherlands are wrongly classified in the C–D paper. The basic mistake which is made is the same as that made in respect of Japan and Switzerland. That is, the incorrect assumption by Calmfors and Driffill that the level of co-ordination of

the system is the same as the formal level at which bargaining takes place. The industry is the main formal level of bargaining in Germany and the Netherlands: in Germany, industries at regional level; in the Netherlands, large companies as well as industries. But in neither Germany nor the Netherlands is it true that bargainers believe that they have no effect on bargains in other industries, and that no co-ordination takes place across industries. Co-ordination in both countries is across industries, at least tacitly. With powerful employer organizations in both countries, and with *de facto* co-ordination across unions in Germany and to a lesser extent in the Netherlands, the degree of effective co-ordination is high; not as high as in Japan, Austria, or Switzerland, but close to Sweden and Norway.

We look at the German system of wage determination, to see how it is co-ordinated. Because there are many misunderstandings about the German system of wage determination, and because of the importance of the German economy, the system will be discussed in some detail. In particular, it is worth saying that *there is no formal system of joint discussions at national level* between the social partners, and between them and the government or the Bundesbank; such a system existed between 1967 and 1977, with formal Concerted Action meetings between both sides of industry, the government, and the Bundesbank. It broke down for reasons unconnected with wage determination, the unions withdrawing from the system as a symbolic protest at the employers' challenge in the courts to the legislation in 1976 by the Social Democrat government extending co-determination. Moreover, most observers never treated it as much more than ritual; the unions would have been unlikely to have withdrawn had it been. However, despite the lack of importance of Concerted Action meetings when they existed, and their subsequent demise, the German system works informally to provide a high degree of economy-wide co-ordination.

The most important formal locus of wage bargaining in Germany is the industry at regional level, and secondly the company and/or plant level. German unions are organized on an industry basis, so that with minor exceptions they do not compete for members. The largest and most important union is the engineering union, I.G. Metall. Union coverage is not overall great, around 35 per cent of dependent employees. But nearly all large and most medium-sized companies in manufacturing are heavily unionized; and industry agreements can be legally extended to cover all companies in an industry by agreement between the signatories to the collective bargain and the regional government. Moreover, employer organizations are very powerful: organized on an industry basis, 80 per cent of employers are members. These industry-based organizations are reinforced at local and regional level by geographically organized chambers of commerce and industry, membership of which is legally required. And both employers and unions have powerful peak organizations, the BDA representing employer organizations and the DGB industry unions.

Wage determination is on an annual basis with settlements tending to take place in the first quarter of the year. The tone of the wage round is usually set by a key leading settlement, which is often a regional engineering agreement, such as the one reached recently in Baden-Wurttemberg. Before that settlement—which other industry settlements will broadly follow—a great deal of informal discussion will have taken place. The discussions which take place involve the DGB and the BDA, as well as other industry unions and employer organizations; and there are certainly extensive contacts particularly on the employers' side with the government and with the Bundesbank; but the discussions are primarily within the I.G. Metall and within the engineering employers' organization and between the two. This can mislead uninformed commentators into thinking that co-ordination takes place within but not between industries. To understand why this is not the case it is simplest to look at the decision problem which confronts I.G. Metall.

I.G. Metall's main objective is to increase, or at least minimize the reduction in, employment in the engineering industries, which include automobiles and steel. This reflects its concern to maintain union membership. It therefore needs to be sure that money wage developments in the engineering sector do not erode the competitiveness of engineering in the world economy. There are three elements involved here, given the evolution of world prices and costs: first, German unit labour costs in DMs; second the DM cost of non-labour input costs into engineering; and third, the nominal effective exchange rate.

Much time is taken up in discussing with the engineering employers organization the development of world engineering prices and export prospects, as well as prospective growth in labour productivity in engineering in Germany. Assuming the exchange rate remains constant, and given the growth of nominal non-labour unit costs, the permissible growth rate of nominal wages is determined.

The exchange rate, however, cannot be taken as exogenous in these calculations because of the perception by I.G. Metall and by the engineering employers of the Bundesbank's strategy. This strategy is not, of course, cut and dried, but the rough belief of the social partners is that increases in the rate of inflation will be met by exchange rate appreciation. Why is this relevant to wage bargaining in engineering? The key settlements set the broad percentage increase which, give or take 1 or 2 per cent, most sectors of the economy follow (for reasons which will be explained below). Thus, with a fixed exchange rate, price inflation is substantially determined by the nominal increase in engineering wages less economy-wide productivity growth, or at least it is seen by participants in that way. To avoid the risk of a deterioration of international competitiveness as a result of the Bundesbank responding to higher inflation by tightening monetary policy and thus bringing about an appreciation, engineering wage settlements have to be in line with existing inflation plus economy-wide labour productivity growth.

This point is reinforced by the need to hold down inflation of non-labour inputs into engineering. This inflation rate will likewise be (indirectly) determined by the key wage settlements on the above argument. Thus too high a rate of nominal wage growth in engineering will also lead to an increase in non-labour costs by this process, which will worsen competitiveness.

An important element of this process is that settlements in other industries and supplementary settlements in companies are sufficiently in line with the initial key settlements. Were that not the case—if for instance the effect of even important settlements on the 'going rate' was individually small as in the UK—the sanctioning effects of the Bundesbank's strategy would be nothing like so powerful. To understand why these other settlements usually stay roughly in line, we look first at other regional-industry settlements and then at company negotiations.

There are temptations for unions at industry level to settle above the going rate. Settlements later in the wages round have less effect on the inflation rate; and settlements in the public sector and the non-trading sectors generally have little to fear from the Bundesbank threat to raise the exchange rate: indeed, they gain from it since it raises real wages without affecting their employment position. Such temptations can cause great problems. For either it raises the going rate—especially if it is a key settlement at the start of the round—or it puts strains on unions and employer organizations in the next round. Thus pressure would be put on any sector which did attempt to settle at a significantly higher than normal rate. These pressures could come from either side at national level. The ability of the centralized employers' organization, the BDA, and/or the union confederation, the DGB, to exert pressure is substantial, since many contacts between industry organizations and central or regional government are via the two peak associations. More generally, especially for the smaller industry unions and employer organizations, there is much reliance on the DGB and the BDA respectively for provision of services. Industry organizations would not wish to fall out with them.

There is somewhat more flexibility about bargaining at company level. Almost all basic rate agreements are arrived at at regional-industry level (Volkswagen is an exception, but its basic rate agreement is very similar to the engineering agreements for automobile sectors). Company agreements are supplemental, and take place as a result of negotiations between senior management and the senior members of the company or plant works council. (In a technical legal sense, works councils are not allowed to negotiate over wages, so these agreements are formally unilateral management decisions without legal force.) These company agreements are carefully watched by the social partners at industry and regional and local level. Both industry employer organization and industry union are keenly aware that a generous settlement in one company can lead to pressures for increases in other companies; and aware also that companies using high wages as a recruitment strategy can be destabilizing for local labour markets, especially if incentives are set up for skilled workers to move.

On the other hand, both employer organization and union want, for different reasons, to allow some flexibility to the company. Employer organizations believe in giving companies as much freedom as possible, so long as it does not have damaging externalities, to develop internal incentive structures. Unions want to help works councils, which in large and medium-sized companies are generally close to the union (although technically independant); the negotiation by the works council of a supplementary increase strengthens its members' position among employees.

These supplementary agreements are, however, carefully controlled. Both employer organization and union have a range of possible sanctions they can use against management and works council respectively. In the German system, employer organizations and the business associations to which they are closely related provide a far greater set of infrastructural services than their counterparts in the UK: for instance in export marketing, research and development co-operation, and vocational training. They function as conduits to government at regional and national level. They run extensive strike insurance funds. They provide assessors, with the unions, for the Labour Courts, which in turn have extensive powers *inter alia* over questions of individual and collective dismissals. In an industrial system in which long-term but frequently non-contractual relations exist between companies

for product development and other purposes, they can function as mediators and facilitators. And with the banks, they embody a collective memory of strong and weak points of companies, including their 'misdemeanours'. Thus if need be employer organizations are in a position to restrain companies. Equally, unions supply a wide range of services to works councils, and can hold them back in company-level wage negotiations.

There are problems with this system, which will be referred to in discussing the trends to decentralization in the next section. But overall it provides a clear enough incentive structure.

The union and employers organization involved in the key settlement or settlements at the start of the wages round believe that subsequent regional-industry settlements will more or less follow the initial ones, and that company supplementary agreements will not significantly distort this pattern. Given the exchange rate and world price developments, this key settlement has a major impact on the inflation rate. They believe that the Bundesbank has an inflation objective and will tighten monetary policy with a likely exchange rate appreciation if actual inflation looks to exceed the target.

Given these beliefs, two things are critical: first, that the union or unions involved in the key settlements are employment-oriented; and that they are concerned to get or keep as low a real exchange rate as possible. (In any case they must not be concerned with maximizing short-term real wage growth: the incentive structure set up by the Bundesbank would then work in exactly the wrong direction, for it would translate high nominal wage growth into high real wage growth as a result of exchange rate appreciation. This is why it is problematic for the system if public sector unions with high employment security and limited concern for international competitiveness become too important in the wage-bargaining process.) Second, it is critical that employers can sufficiently co-ordinate their activities to ensure that other sectors and profitable companies are held roughly in line.

France, Italy, and the UK

In this section we make some further amendments to C–D's classifications. These concern primarily France, Italy, and the UK. C–D's ordering, in increasing degree of centralization, is Italy, the UK, and France.

At the start of the 1990s, there is no doubt that the UK is the least co-ordinated of the three economies. Outside the public sector, collective bargaining is largely conducted at plant and company level. More-over, this is not merely the formal location: it does not conceal a hidden system of higher-level employer or union co-ordination of wage bargaining. In the C–D classification, the UK now counts as a decentralized wage-bargaining system.

In both the other two countries, there is some degree of overall co-ordination. Thus there is a qualitative difference between the UK on the one hand and France and Italy on the other. France and Italy are both countries which attempt co-ordination which is designed to have *economy-wide* effects. They do not fit C–D's classification of wage bargaining on an industry-by-industry basis, with no co-ordination across industries. The very different methods of co-ordination in France and Italy are not as effective as those of Sweden or Austria, say, at low levels of unemployment: actually the Italian system is superior to the French in this respect, and the male unemployment rate in Northern and Central Italy has remained at around 5 per cent through the 1980s (the male unemployment rate being more relevant for the effectiveness of co-ordination since powerfully organized workplaces are substantially male). But both systems have the ability to think in 'rational expectations' terms about co-ordinated reductions in wage inflation; paradoxically, the way the French system of co-ordination works makes it more able to engineer quick reductions in wage inflation than the Italian, at least at a high enough rate of unemployment, while the Italian system can hold down expected real wages more effectively at lower rates of unemployment.

The French system of co-ordination is unlike any other country in our sample. It has operated in the 1980s as a result of government determination to hold down the pace of wage inflation, together with the weight of the public sector in the 1980s in the wage-setting process and with the collapse of union opposition. The nationalized industry sector has dominated the economy and the wage-setting process, since the French government nationalized some of the most important companies in the economy in 1981–2. Thus the co-ordination has been government run, or at least strongly government influenced. It works because of the great weight which the public sector has had in the crucial areas of French industry in the 1980s. It operated initially and formally with increasingly reluctant union support, while the Communist Party was part of the governmental coalition until 1984. From then, as unemployment rose, it was maintained informally and tacitly and tightened despite union opposition. This was made possible for two reasons: first, in general the insider problem is less severe in France than

elsewhere because of a more hierarchical structure of work organization and promotion which limits the possibility for collective action within companies. Second, and specific to the mid-1980s on: the French union movement has been collapsing, mainly because the largest union, the communist CGT, has been pulled down by the collapse of the Communist Party; and because traditionally anti-union French employers have had no incentive to build up the other non-communist unions in its place.

Italy also has a capacity for overall co-ordination, but of a quite different sort to France. Before explaining this system, it is necessary to make one correction to the Italian unemployment figures. In my view these are the most misleading of all unemployment figures, and have become more so as the 1980s have progressed. The bulk of effective union strength in Italy (outside public services) lies in unionized male workforces in manufacturing industry—as it does in most industrialized countries. The great bulk of this workforce is in Northern and Central Italy. The unemployment rate of male workers in these areas has remained quite remarkably low: around 3.5 per cent in 1977 and 5 per cent in 1987. Moreover, the increase in unemployment in the second half of the 1980s, which is sometimes pointed at to illustrate the shortcomings of the Italian economy, is completely confined to the South: the overall unemployment rate in Italy rose from 10.0 per cent in 1985 to 12.2 per cent in 1988; but the unemployment rates of both men and women fell in the North and Centre, the male rate being 5.2 per cent in 1985 and 5.0 per cent in the first half of 1988, and the combined rate 8.4 per cent and 8.1 per cent. What Italy witnessed was a massive increase in unemployment in the South over the period from 13.6 per cent in 1984 to 20.4 per cent in the first half of 1988 (pp. 197 and 81 of the Italian Ministry of Labour's 1988 report, *Labour and Employment Policies in Italy*).

The system of co-ordination in Italy is far more a phenomenon of the North and Centre than of the South. Thus it needs to be judged against performance in the North and Centre. On the employers' side there are three important types of network, with links between them. First, the major companies (Fiat, Pirelli, Olivetti, etc.) are embedded in a complex web of interconnecting ownership, with close links to the banking system (indeed, the modern Italian banking system developed at the end of the last century was modelled on the German). These large companies play a major policy-making role in the main employers' organization, Confindustria. Second, the leadership of the largest Italian state holding company, IRI, has made a conscious attempt in the 1980s to develop co-ordinated collective bargaining strategies; and IRI in turn maintains close informal connections with the leading private companies and Confindustria. Finally, the regional employer organizations, especially in Lombardia, the Veneto, and Emilia-Romagna, play a role in informal supervision of wage-setting agreements in smaller companies.

In addition to employer networks, the unions have become more centralized in the 1980s. This has not been without problems as attempts at informal wage restraint have led to the development of unofficial unions in the public sector. But in general the official unions have been at least privately supportive of wage restraint. Co-operation between the government and employers, with initial union support, led in 1983 and 1984 to national agreements responsible for a significant reduction in wage inflation. Since then inflation has been contained. Apart from these national agreements, wage bargaining is based on triennial agreements, technically at industry level; but they are seen in a national framework. They are supplemented by company agreements. The system does not work in a synchronous way: hence the cost of reducing inflation is high for the participating institutions. None the less, the system has been at least moderately effective in keeping inflation down.

To summarize this section: the UK has been the least co-ordinated of these three countries in the 1980s. The French and Italian systems both have the capacity for overall co-ordination: neither are as effectively co-ordinated as Germany or the Netherlands, or the low-unemployment countries. But the nature of the French system has permitted low inflation (though this has required highish unemployment). The Italian system has enabled inflation to be substantially reduced without great unemployment costs in the North and Centre of the country; but its ability to reduce inflation requires a higher cost than the French system.

2. The weaknesses of decentralized bargaining

This subsection re-examines C–D's claim that unco-ordinated company-level wage-setting may work effectively. The theory employed by C–D to show the merits of decentralized wage-setting is on its own assumptions of course correct. What is argued here is that these assumptions seldom match the reality of unco-ordinated company-level wage-setting.

Two main issues are raised here: the first relates directly to the wage-setting process; and the second to the relationship between collective bargaining, training, and decentralized wage-setting institutions.

(a) Profits and relative wages

The following pattern of wage-setting in a decentralized system is commonplace: a profitable company concedes a high wage increase to its employees; this wage increase is then used by employees in other less-profitable companies in the same industry to get a higher wage increase than they would have got had their increase simply had reference to their own company's profits; the net result is a higher average wage in the industry than in a decentralized system in which profits were the only factor taken into account in each company. Beckerman and Jenkinson (1990) produce evidence for this effect in the UK. (This might be called a perversely co-ordinated system.)

(It is worth a parenthetical pause to ask exactly what is happening in this process. The process is asymmetrical. That is, a symmetrical increase in the dispersion of profits (*ex ante*), with the average unchanged, implies an increased average wage. This is because the ability of management in the profitable company to demand wage restraint on account of the implied *increase* in relative wages compared to wages elsewhere in the industry is less than the inability of management in the less profitable companies to prevent wage increases to compensate for the *decrease* in relative wages compared to wages in the profitable company. Thus the psychological basis for the asymmetry in the bargaining positions is that workers are more upset over a loss of relative position than they are pleased by a relative improvement.)

In this respect industry bargaining will produce better results. The industry bargain will relate to the average level of profitability in the industry, and give the same wage increases across the board. Of course, things are not that simple, since with industry bargaining there is likely to be supplementary bargaining at company level; but the type of problem which the UK motor industry faces as a result of company bargaining is—at the very least—far less acute in, say, Germany.

This does not mean that industry bargaining is overall better, since the product-elasticity of demand point will generally favour company bargaining. But there is no longer any reason, without empirical knowledge of the relative importance of these contrary effects, to favour decentralized over the intermediate case.

On the other hand, the superiority of effective co-ordination of wage-setting across the economy is reinforced in relation to both company- and industry-wide bargaining. Indeed, this is a standard argument for co-ordination across the economy, though C–D do not make it. Just as co-ordination within an industry can reduce the knock-on effect of high wage increases from profitable companies to less profitable ones, so co-ordination across the economy reduces the effect of leading industries or leading companies determining the going rate.

(b) Inadequate training and mismatch

The second reason for doubting the efficacy of decentralized wage-setting systems is less direct, though of no less importance. Again, C–D do not take it into account. Companies who find it difficult to secure an adequate supply of skilled workers are generally in a weakened bargaining position in relation to their own skilled insiders. Decentralized systems aggravate this situation in two ways.

The first is standard and well known: that the provision of training is a public good which companies will be reluctant to provide unless there are some external pressures to do so. These pressures are usually associated with strong employer organizations capable both of helping companies to provide the facilities for training manual and low-level white-collar workers and of imposing sanctions if they do not do so adequately. Thus decentralized systems, if we go beyond pure wage-setting institutions, usually fail to provide sufficient high-quality skilled labour, and so further reduce the degree of restraint possible in such systems. Since the C–D paper is frequently used to show the case for decentralization in relation to economic performance generally (whatever the intentions of the authors), it seems justified here to refer to decentralized institutions beyond pure wage-setting institutions.

Even if attention is confined to decentralized wage-setting institutions, i.e. unco-ordinated company wage-setting, such wage-setting carries a direct problem for training. This is that if companies can respond to their skill shortages by offering higher wages to skilled workers outside, this makes poaching of skilled workers from other companies easier and reduces the incentive for companies to train their own skilled workers. Some form of wage co-ordination, and co-operation between companies is necessary to avoid this problem.

There are two common forms of such co-ordination. One is the Japanese with (roughly) tenure-based wage scales. Hence the individual worker has a disincentive to move, since in a new company he/she starts at zero tenure. Note *critically* that this system only works so long as all companies co-operate in playing by the same rules. Co-operation in this way is not an optimal strategy for a company seeking skilled labour. Thus there

are implicit sanctions or understandings in the Japanese system.

A second way is the Northern European collective bargaining system, where, at least as far as manual workers are concerned, companies are constrained by the collective agreement as to what they can offer to bring in workers from outside. Those agreements are either legally binding or buttressed by employer organizations which would be concerned by companies using wage incentives to poach skilled workers.

To summarize this subsection, the argument for the effectiveness of decentralized systems put forward by C–D, namely the greater product market elasticity facing companies than industries, is one among a number of arguments; and as suggested here there are important arguments going in the opposite direction. Thus if the empirical results cast doubt on the efficacy of decentralized arrangements, it may be that the arguments presented here outweigh the Calmfors–Driffill point.

3. Local pushfulness

C–D take no account in their analysis of the degree to which workforces at plant level are capable of and choose to push their claims to higher wages. One element in this is simply the power of unions at plant or more generally local level. This determines their capability to secure a higher real wage. Thus there is a great difference between the UK and the US situation at plant level: the UK still has rather strong plant level unions in key wage-setting sectors (despite the last decade); the US does not (because of the last decade). Although both have decentralized bargaining systems in the second half of the 1980s, the expected real wage bargaining schedule in the UK is likely to be higher than that in the US. Or, to compare highly co-ordinated bargaining systems, such as Sweden and Germany with Japan and Switzerland: the former have strongly organized local unions (at least *de facto*) at plant level in the key wage-setting sectors; while in Japan and Switzerland they are weak. We would expect, and it is the case, that it is significantly harder for Swedish and German national unions and employer organizations to impose wage moderation on their lower-level affiliates than it is for the Japanese and the Swiss.

But a second factor needs to be taken into account to explain fully the behaviour of local unions in the wage-setting process. This second factor is the desire of local unions, leaving aside any pressures which national unions and employer organizations may be able to impose on the company wage-setting process, to make maximum use of their bargaining power. This depends on the incentive structures which face workers. There is a significant difference between the position of a German worker and a British worker: the German worker has considerable employment security and prospects of further training within the company and can take a long-term view of wage bargaining, treating the forgoing of short-run real wage gains as an investment in the company; the British worker, with less employment security and virtually no prospects of further training, will wish to maximize short-run real wage gains.

These two factors, capability to push for real wage gains at local level as summarized in plant-level union power in key wage-setting areas, and desire to do so reflecting the incentive structure the company gives the worker, are developed into an index of 'local push-fulness' in section V.

4. Open and closed economies

The C–D argument is set in the framework of a closed economy. This implies there is a unique equilibrium rate of unemployment corresponding to a given bargaining structure, as shown in section II. A fall in unemployment below the equilibrium rate leads to nominal wage growth above the preexisting rate of price inflation, as unions exploit their increased bargaining power, and then increasing price inflation as businesses seek to restore the feasible real wage. Thus maintaining unemployment below the equilibrium rate results in continuously rising inflation.

The open economy operates differently, at least with quasi-fixed exchange rates. So long as the economy is in current account equilibrium, at least over the medium run, it can choose its level of unemployment without fear that accelerating inflation will result. This can be seen as follows. Think of the real wage as the consumption real wage, a ratio of the nominal wage to a weighted average of the GDP deflator and the import price level. Assume world prices are constant, there is a fixed exchange rate and hence constant import prices. Then businesses no longer have the ability to set the consumption real wage, even if they can still set the product real wage for domestic goods. This is because a rise in the nominal wage increases the real consumption wage, even if the GDP deflator rises equiproportionally with the nominal wage. Thus in the open economy with fixed exchange rates, if the real consumption wage is below the expected bargainable real consumption wage, unions will bid up nominal wages until the real wage has risen to its bargainable level. The consequence is that the international cost competitiveness of the

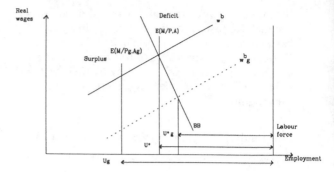

Figure 10.5. The Open-Economy Model of Wages and Employment

economy declines, since domestic wages and prices have risen relative to world prices.

Thus any rate of unemployment is consistent with stable inflation, so long as the exchange rate is not put under pressure. In the medium run this means that unemployment rates which implied significant current deficits (without long-term inflow) are not sustainable in equilibrium. So the minimum equilibrium unemployment rate, in very loose terms, is that associated with current balance. In Figure 10.5 the downwards sloping line, BB, represents current account equilibrium; it slopes downwards because a higher level of employment implies a higher level of GDP and hence imports, and this requires a lower real consumption wage and thus increased cost competitiveness to maintain a balance. The minimum equilibrium unemployment rate, U^*, is then given by the intersection of the bargained real wage schedule, w^b, with the BB schedule.

In Figure 10.5, actual unemployment is determined by aggregate demand. $E(M/P, A)$ is the level of employment associated with a particular level of the real money supply, M/P, and exogenous expenditures, A. If the union wage-bargaining schedule is w^b, the level of actual unemployment implied by $(M/P, A)$ will be the minimum equilibrium rate. But it is clear from the diagram that the benefits associated with institutionalized wage-bargaining are not captured by the actual unemployment rate, since that depends on the level of aggregate demand in the economy.

Imagine that the German wage-bargaining schedule is $w^b g$ (implying more effective wage restraint than w^b), but that the German government and the Bundesbank choose a more deflating level of public expenditure and the real money supply, Ag and M/Pg. The actual German unemployment will be Ug. But Ug is greater than U^*, which was achieved with w^b. So the effectiveness of co-ordinated wage-bargaining in Germany (or the Netherlands) is not picked up in actual unemployment. The minimum equilibrium unemployment rate, $U*g$, on the other hand, does measure the greater effectiveness of the German bargaining system.

To reiterate the point, C–D make two mistakes in their discussion of Germany and the Netherlands: first, they are classified as industry-level bargaining systems; secondly, they are evaluated by their actual unemployment performance. The two mistakes together play a significant part in the empirical success of their paper: industry-level bargaining being associated with high unemployment. But if Germany and the Netherlands are reclassified as having reasonably effective economy-wide co-ordination of wages; and if the minimum equilibrium unemployment rates consistent with external balance are seen as the test of institutional performance; then Germany and the Netherlands no longer provide empirical support to the C–D thesis.

IV. The decentralization hypothesis: a reassessment

The second major position on the role of institutional restraint in the process of wage determination in the 1980s is a sceptical one, and until recently very widely held. Underlying processes of economic change, on this view, have interacted with market-oriented government policies to bring about radically more decentralized behaviour than in the preceding decades since the Second World War; and this decentralization has so weakened the role of unions that it has spelt the end of economy-wide co-ordination of wage bargaining.

There is no serious doubt about the importance of decentralizing forces during the last decade, and only slightly less about their durability. We sketch briefly the reasons which have brought them about, and then discuss how they should modify our views of the role of extra-market institutions in the wage determination system.

Three underlying forces are largely exogenous, have become more clearly apparent as the 1980s have proceeded, and will probably be in operation for a considerable time:

(*a*)   The technological revolution based on the microprocessor has made it possible for most industries to produce customized, differentiated products, involving rapid product modification and innovation. This has led to a move away from the production of standardized goods, produced under conditions of mass production, with a relatively standardized labour force, where there was a natural presumption of collective bargaining. More idiosyncratic production has required more idiosyncratic labour, and hence more individualized reward packages.

(*b*)   Long-term shifts in demand in advanced industrialized countries have been away from manufactured goods towards services. This has led to the increased employment of women, in smaller workplaces and to a lesser extent in industrial centres—all trends associated with a decline in unionization.

(*c*)   The increased openness of economies has reduced the possibility of companies to pass on cost increases in the form of higher prices, and more generally limited the bargaining power of unions. In addition to these trends, there are two sets of factors which are in principle reversible:

(*d*)   The set of factors associated with the world recession which started in the mid-1970s, and has included a slow-down in economic growth and an increase in unemployment. These factors have had a major impact on union bargaining power and organizing ability.

(*e*)   A shift in the governments of major countries from centre-left or centre in the 1970s to centre-right in the 1980s. These include the US, the UK, Germany, and (to a lesser extent) Italy. This has combined with a pervasive ideological shift towards free markets and against trade unions. In policy terms, unions have been weakened by three types of strategic policy: (i) reduction in public sector employment; (ii) deregulation of financial markets, putting pressure on companies to improve financial performance; (iii) deregulation of labour markets, ranging from anti-union legislation to removal of constraints on hiring and firing.

Whether changes are durable or not, it is clear that they have brought about a major weakening of trade unions over the last decade, and some weakening of employer organizations. Three points, however, need to be made to clarify this picture.

First, the ability of unions to protect the weaker elements of the labour force has significantly decreased in almost all countries. These weaker elements include women and part-time workers; they also include unskilled and semi-skilled workers. But this part of the labour force is not important from the viewpoint of economy-wide wage co-ordination in the 1980s, because their bargaining power is limited.

Second, and much more important with respect to wage co-ordination, is the bargaining power of skilled workers. The bargaining power of skilled workers within companies has increased as a result of the importance of company-specific skills associated with teamwork, rapid changes in work organization, and so on. This is the insider phenomenon. And it has coincided with a reduced power of external unions.

Third, a related phenomenon is the increased power of companies relative to employer organizations. This reflects the need companies have to work out differentiated solutions for the more individualized problems they find themselves facing in a world of faster technological change and customization.

These last two points have without question made it significantly harder for countries to have systems which effectively co-ordinate wage determination, than it was in the 1960s. There is another difference which has increased the difficulty of wage co-ordination in the 1980s. That is the decline in productivity growth between the 1960s and the 1980s. With high productivity growth even weak systems of wage co-ordination were able to have some effect in restraining real wage growth: it is easier to get agreement to restrain real wage growth by 1 per cent from 4 per cent to 3 per cent, than from 1 per cent to 0 per cent.

There are certainly at the start of the 1990s many strains within the Swedish system of wage co-ordination, with an economy operating at low levels of unemployment. The same is true in relative terms of the Japanese system. Even at higher unemployment, the German employer organizations are under considerable pressure to allow more decentralization. It is fair to say that all economy-wide wage co-ordination systems have introduced some increased flexibility through the 1980s.

The mistake of the decentralization hypothesis is the belief that economy-wide co-ordination no longer exists. The mistake is in one sense understandable since co-ordination is no longer accompanied by the fanfares of 'social contracts', 'concerted action', and the like, as it was in the 1960s and 1970s; unions are everywhere visibly much weaker than they were; and the intellectual climate has been one of deregulation and free markets.

What has in fact happened is more interesting. In economies in which co-ordination was well-established in the 1960s and 1970s it still exists (and for reasons connected with changing industrial relations structures has even developed in the case of France and Italy). But unions now play a less important role and employer organizations and networks a more important one. In part, indeed, this accounts for the lack of fanfare: for employer organizations are more reticent than unions.

But in those countries in which co-ordination was already fragile in the 1960s and 1970s, in particular the US and the UK, co-ordination has now become too difficult—short of significant structural changes. This has been reinforced by the differential pattern of deregulation. Where employer organization was strong, for instance in Germany, employer organizations both had a major role in policy advice on the labour market, and were loath to see deregulation which eroded their own position as well as that of unions. Similarly, a powerful industrially oriented banking community in Germany, also with a position in the policy process, led to a cautious approach on financial deregulation. By contrast, weak employer organizations in the US and the UK had a very limited role in policy-making on labour market deregulation, and less to fear from it. And US and UK banks believed they would benefit from financial deregulation.

Thus the important pressures towards decentralization have not ended economy-wide wage co-ordination except in initially weakly co-ordinated countries such as the UK and the US. It is certainly true that where economy-wide wage co-ordination has been maintained it has become less effective. But, and this is the important point for the UK as it enters more closely into European monetary union, our main competitors in the system may have become less effective than they used to be at co-ordination: they have become more effective relative to the UK.

V. Institutions, equilibrium unemployment, and inflation

The first part of this section puts the critique of C–D's explanation of unemployment into a more systematic form. It assigns numerical values to the degrees of economy-wide co-ordination and local pushfulness in the eleven countries in the second half of the 1980s. Unemployment rates are regressed on these two variables in an elementary exercise designed to show that the argument is consistent with the statistics. The second part of the section discusses what further institutional factors are needed to explain inflation success.

1. Unemployment

The explanation of unemployment in this section uses only institutional variables relevant to wage determination.

We focus here on two variables. The first is the extent and strength of economy-wide co-ordination. The approach differs from C–D in that it measures the level at which co-ordination actually occurs, as opposed to the formal location of bargaining. It differs, in addition, in that it asks to what extent co-ordination is

Table 10.1. A Classification of Economy-Wide Co-ordination

Country	Economy-wide Co-ordination in mid- to late 1980s
United States	Zero employer and union co-ordination (0).
United Kingdom	Zero employer and union co-ordination (0).
France	Tacit government co-ordination via public services and large nationalized industry sector (1.5).
Italy	Informal employer co-ordination via big employers, especially Fiat, IRI, and some regional employer associations; some help from union confederations, CGIL and CISL (2).
Netherlands	Strong employer organizations and informal co-ordination between giant companies; occasional differences between giants and industry organizations; medium union co-ordination (3).
Germany	Strong employer organizations, with considerable co-ordination across industries; medium-strong union co-ordination (3.5).
Sweden	Powerful centralized employers organization; generally strong co-ordination across industries, with some divergence of interests; centralized union confederations with some internal conflicts (4).
Norway	As Sweden, with government playing additional co-ordinating role (4).
Switzerland	Very powerful employer organizations, playing tacit co-ordinating role; unions weak and pliant (4).
Austria	Very powerful union, with centralized co-ordinating role; medium-strong employer organizations (5).
Japan	Very powerful tacit employer co-ordination across large companies, in more or less centralized way, with backing from industry employer organizations; weak and pliant unions (5).

economy-wide, without drawing a distinction between atomistic company bargaining and industry-wide co-ordination. And, in line with explaining across-country unemployment rates averaged between 1985 and 1989, the degree of co-ordination is assessed in the mid- to late 1980s.

The approach differs more generally from most other attempts to measure co-ordination in two main ways. First, the importance of formal or informal co-operation among employers is emphasized. Second, the measure is not a weighted average of employer and union co-ordination. The reason for this can be seen from an example: Japan has strong employer co-ordination and weak union co-ordination; it is wrong to conclude that co-ordination is on average medium: instead it takes place through the stronger partner. Thus, roughly, the degree of co-ordination is taken to be the degree of co-ordination of the stronger partner (see Table 10.1).

In addition to this variable designed to measure the extent of economy-wide co-ordination (EWC), there is a second institutional factor to be taken into account. This is the extent to which plant-level or local-level activity is potentially disruptive. This variable will be called local pushfulness (LOC). What this seeks to measure is the answer to the following hypothetical question: 'In the absence of external pressures by employer organizations and national unions towards restraint, how restrained would wage settlements be?' This requires imagining unemployment to be at the same level across each of the countries in the sample: it has been taken to be a hypothetical 5 per cent.

There are two major factors involved in the answer. The first is whether or not work-forces have the capac-

ity to take autonomous collective action at plant—or local—level in those sectors of importance in the wage-setting process: this still includes larger companies in manufacturing, and also the public services. Where work-forces are relatively strongly unionized at plant-level, they clearly have this capacity: countries in this category in the sample are UK, Netherlands, Germany, Sweden, Norway, and Austria. Weakly unionized or non-unionized work-forces do not generally have a capacity for collective action in support of wage claims at plant level; this is the case for Japan, Switzerland, and the US. France is (in a way) an exception to this: the reasons will be examined below. Italy in the late 1980s is somewhere in between.

Whether the capacity for autonomous action is used to maximize current real wages depends on a second factor. That is the nature of the implicit underlying incentive structure which employed workers face in different countries. The two main incentive structures for workers in medium to large establishments are as follows. First, the Northern European and Japanese model: reasonable employment security, continuing training in broad skills, and effective systems of grievance resolution, representation, and participation are offered in exchange for adequate performance and co-operation. In this system the incentive for group action to maximize current real wages is tempered by the concern for the future development of the company. The second incentive system is the UK model: instead of employment security and continuing training in broad skills, adequate performance is exchanged for wages over a shorter-term horizon, typically with stricter monitoring and controls. The US situation is similar to that of the UK; in both countries, of course,

Table 10.2. Constructing an Index of Local Pushfulness

Worker incentive structure	Plant-level union strength	
	Strong	Weak
Long-term consensual	2 Germany Netherlands Norway Sweden Austria	1 Japan Switzerland
Long-term hierarchic		3 France
Short-term	4 United Kingdom	1 United States

there are some companies which use Japanese-type incentives: the proportion in the US is larger than in the UK.

In addition to these two systems, a third type of incentive structure is the French one. By contrast to the short-term systems of the UK and the US, workers have considerable employment security. By contrast to the long-term consensual systems, they have relatively narrow tasks, narrow company specific skills, limited representation, and often ineffective grievance procedures; management decision-making tends towards the hierarchical rather than the consensual, and there is a higher degree of monitoring and control than in the Northern European systems. Because of the lack of representation and consensus, workers have relatively low trust in management. Therefore, despite long-term attachment to a company, workers do not see current wage restraint as an investment as easily as their counterparts in long-term consensual systems.

These two factors can now be used to classify countries according to their degree of local pushfulness. In Table 10.2, union strength or weakness is measured along the top, and the three incentive structures are measured down the side.

Values are assigned from the highest (4) to the lowest (1). In the top right-hand box are countries in which unions are weak and in which incentive structures for workers are long term: both lead to wage moderation, hence the value of 1 assigned to Japan and Switzerland. In the top left-hand box, plant unions have the capacity for collective action, but (even in the absence of outside pressure) would not maximize current real wages because of the long-term incentive structure: so these Northern European countries are given a slightly higher value, namely 2.

Now look at the bottom left-hand box. The United Kingdom, uniquely, is in the worst of both worlds:

plant-level unions remain strong in the key wage-setting sectors of the economy; but at the same time the incentive structure is short term: the UK is given the highest value for local pushfulness, of 4. By contrast in the bottom right-hand box, the US is given a value of 1: with few exceptions, workers at plant level in the US have no capacity for collective action over wages in the late 1980s. The contrast between the US and the UK reflects the major difference in the changes in the power of unions in the 1980s: the strongholds of American unionism suffered dramatically from the combined assaults of foreign competition, deregulation of transportation, and the sunbelt strategy of anti-union employers. In Britain, labour legislation and deregulation has weakened unions at national level more than local. There have been few attempts by employers to get rid of plant unions; and though the power of shop stewards is weaker than in previous decades, local unionism is still well-entrenched in the key wage-setting areas of manufacturing. If we take a hypothetical level of unemployment of 5 to 6 per cent, local pushfulness in the UK is high.

The French system is peculiar. The capacity for independent collective action by an individual plant or by a group of workers in a plant is very limited, and would be so even if unemployment were at a hypothetical 5 per cent. But the lack of effective representation of employees lends itself to a build-up of unresolved grievances (a situation which has somewhat improved through the 1980s as a result of the Auroux laws of 1982). The French problem, if unemployment were to fall substantially, is that of waves of strikes, rather than actions by individual plants. How serious a problem is this? France and the US are the two countries in which unionism was greatly weakened in the 1980s: but French strike waves tend not to be initiated by unions. Thus we have assigned a value of 3 to local pushfulness in France at hypothetical unemployment level of 5 per cent. Italy (Centre and North) is assigned a value of 2.5: Italian industrial relations are in a state of change. Transformed in the plants at the end of the 1960s from something like a French system to something like a British one, the period from the late 1970s to the present has seen a further major transition. This is to somewhere between the top two boxes, i.e. between the Japanese and German. But there are significant remnants of the past. In particular, local pushfulness remains extremely high in the public sector.

We now turn to relating these variables to unemployment across the eleven countries.

In the following equation, the dependent variable is unemployment taken as the average of the unemployment rate in 1985 and 1989. For reasons explained in III.1 Italian unemployment is for Central and Northern Italy. The independent variables are EWC (economy-wide co-ordination) and LOC (local pushfulness).

Equation 1: Dependent Variable: Average of 1985 and 1989 unemployment.

Constant	EWC	LOC
5.09	−0.91	1.50
R^2=0.66	t=2.18	t=1.81

Equation 1 explains the actual average unemployment rate in the late 1980s. But as explained in III.4, in an open economy the appropriate measure of unemployment is that consistent with current balance equilibrium. In concrete terms it explains why, when Germany and the Netherlands both have a reasonable degree of economy-wide co-ordination of wages, their unemployment rates are so high. The explanation is that the institutional variables are relevant to the minimum sustainable unemployment rate, i.e. that consistent with medium-run current account equilibrium, while the actual unemployment rate is determined by aggregate demand.

We have therefore made a simplisitic transformation, by subtracting the current balance of payments surplus as a percentage of GDP from the unemployment rate. In a very rough and ready way this gives a better idea of the sustainable unemployment rate.

It also has a second purpose, in that it acts as a measure of more general economic performance. It is used to measure this by C–D, as a 'misery' index of the unemployment rate *plus* the percentage current account deficit.

The results are shown in Equation 2:

Equation 2: Dependent Variable: Unemployment rate *plus* percentage current account deficit (unemployment/external balance 'misery' index).

Constant	EWC	LOC
5.51	−1.64	1.92
R^2=0.78	t=3.31	t=1.95

This correction, consistent with the open economy version of the Layard–Nickell model produces a significantly sharper result.[6]

The purpose of this sub-section has not, obviously, been to nail down an econometrically persuasive ar-

gument. What it has done has been to summarize the critique of C–D, first by showing very briefly how numerical values might be constructed for economy-wide co-ordination and local pushfulness. And second, to show that this parsimonious approach is at least consistent with the cross-country behaviour of unemployment in the second half of the 1980s.

2. Inflation

Clearly the degree of economy-wide co-ordination relative to local pushfulness will go a long way to explaining the ability of a system to choose a low rate of nominal wage inflation. But there are other factors which need to be taken into account. This can be seen by posing the following question: How easily can a system produce a sustained cut in nominal wage inflation?

There are two types of condition which determine how easily such a reduction can be met. The first relates to fears which workers and unions may have that relative or real wages may be cut. The second is the extent to which appropriate incentives are built into such economy-wide co-ordination as exists.

Fears about relative wages

These are going to be greater, the more asynchronous is the wage-setting process.

Fears about real wages

This is more complex. Cost-of-living adjustments reduce fears about declining real wages, but, depending on how the COLA is set up, may make nominal wage reductions harder. The most satisfactory system is where workers have enough trust in union involvement in a centralized wage bargaining system.

In respect of both of the above, the systems with a high degree of economy-wide co-ordination are superior to the more decentralized. Both the decentralized systems, the US and the UK, have asynchronous wage-setting; and there is a low trust element about the possibility of reduction in the nominal wage inflation rate without loss of real wages. Of the systems with some element of economy-wide co-ordination, Italy is probably the least well-placed, with some asynchronous wage-setting and some degree of indexation.

Incentives and economy-wide wage co-ordination

In the next section we will look at the reasons why incentives in the UK to reduce nominal wage inflation

[6] To check that there was some statistical basis to this transformation of the dependent variable, we got the following result when the percentage balance of payments surplus was included as an explanatory variable:

$$u = 5.23 + 0.34\,BP - 2.31\,EWC + 3.28\,LOC \qquad R^2 = 0.71$$
$$t=1.09 \quad t=2.46 \quad t=1.98$$

are weak despite membership of the exchange rate mechanism. Where there is economy-wide wage co-ordination, both employers and unions are concerned about maintaining a low real exchange rate; what matters for choosing low nominal wage inflation is then the nature of the exchange rate system. The Swedes have not had a fixed exchange rate constraint. Higher rates of inflation in Sweden reflect this absence rather than showing co-ordination does not work. Of course, moving to lower rates of inflation is costly for unions: but if the Swedes were to enter seriously into the ERM, one might predict a relatively quick fall in their inflation rate.

VI. Lessons for the United Kingdom?

In this concluding section, we ask whether there are any lessons for the UK from the preceding analysis. Before addressing this question, the main conclusions of the analysis as it applies to the UK are summarized.

(i)   Of the many countries examined in this article, most have mechanisms which introduce some element of economy-wide co-ordination into the wage-setting process. For the reasons set out in section IV, this co-ordination has become harder for most countries to achieve in the 1980s, than in previous decades. The leading actors in the 1980s are employer organizations or networks, with unions and government playing a supporting role; co-ordination is generally less formal and unaccompanied by highly publicized social contracts. Sometimes the co-ordination is explicitly centralized, as in Japan, and sometimes, as in Germany, it works via a structure of incentives which implicitly ties industry bargaining together with monetary policy. In the above countries, economy-wide co-ordination, while more difficult than in the 1970s, remains effective.

France and Italy also attempt economy-wide co-ordination of wages, although—apart from the early 1980s—not explicitly. For reasons set out in III.1, their ability to do so has increased during the last decade, while remaining limited compared to Germany and the more centralized countries. It is, however, worth stressing the Italian success which is masked by the aggregate unemployment figures: the strength of Italian unionism is in the two-thirds of the male work-force located in the North and Centre of Italy; the male unemployment rate for the North and Centre rose from 3.5 per cent in the mid-1970s to only 5 per cent in 1985, and has since slightly declined. This is in comparison

to the UK male unemployment rate which rose from 3 per cent to 13.6 per cent over the same period.

In contrast to these countries, in the UK and the US there has been no economy-wide co-ordination in the last decade. With weak employer organization and weakly co-ordinated national unions in the 1970s, the forces of decentralization of the 1980s would have made co-ordination extremely difficult, short of major structural institutional changes. Thus, while economy-wide co-ordination has become less effective in the more centralized economies in the 1980s compared to previous decades, it has become more effective relative to the UK and the US, where no such co-ordination has been possible.

(ii)   The second institutional factor of importance in wage-setting, operating to counter the beneficial effects of economy-wide co-ordination, is what we have called local pushfulness, the capacity and choice of local unions to use their bargaining power to maximize current real wages. Two variables determine the extent of local pushfulness: whether plant-level unions are strong or weak in the main wage-setting industries; and whether or not workers have a structure of incentives, including employment security, effective representation and participation, continuing retraining in broad skills, which gives them a long-term perspective in plant-level bargaining. The United Kingdom is unique in our sample of countries to have the worst combination: relatively strong plant-level unionism in industries such as engineering and automobiles; and a structure of incentives—limited employment security, participation, and retraining possibilities—which leads to a short-term perspective in wage-bargaining. The UK thus has the highest value of local pushfulness.

(iii)   The third conclusion is that none of the three factors conducive to the effectiveness of the ERM as an anti-inflation incentive system operate in the UK. First, wage bargaining is asynchronous, so that relative wage problems are thrown up by attempts to reduce nominal settlements. Second, while some important wage-setting companies are solely in the tradable goods sectors, the manufacturing sector is relatively small, and some of the largest companies are in services or spread between services and manufacturing. Finally, UK unions' interest in manufacturing centres more on short- to medium-run real wage increases than on long-term employment preservation. This means that a fixed exchange rate, if credible, has the *opposite* effect to that desired, as far as union bargaining behaviour is concerned. By contrast to the desire of the German engineering union for a low real exchange rate to preserve long-term competitiveness and hence employment—favouring low nominal wage growth—UK unions will prefer high nominal wage growth since a fixed exchange rate translates that into a higher real wage.

Is it possible to change this system? The desirability of change is hardly in question: a reduction in nominal wage growth from around the current 9–10 per cent down to 2 or 3 per cent would avert a sharp deflation in the near future, and more important a permanent increase in the rate of unemployment. The remaining paragraphs of this article discuss the feasibility of change. It is written in the spirit that change is possible, but difficult, and in the current climate unlikely. But as the implications for the UK labour market of closer involvement with Europe become clearer, it will be important to discuss the basic institutional structures of collective bargaining. The following are some of the elements which will need to be borne in mind.

Almost all discussions of institutionalized wage restraint in the UK have suffered from one or more of three limitations:

(*a*) 'Quick-fix' solutions. That is, little or no attention has been paid to the need to change the basic institutional structures of collective bargaining. Yet without such a change, any attempt at wage restraint will be short-lived, with the likelihood of a subsequent reaction eliminating some or all of the gains. In the UK system neither unions nor organized business have ever been able to rely on much more than moral suasion.

(*b*) Wage restraint in the UK has been seen as primarily an agreement between the government and the unions, with the government initiating policy, the unions taking the major responsibility for enforcement, and organized business playing at most a supporting role. The weight of evidence from our successful competitors abroad is that this ordering is wrong. Business, through formal or informal co-ordination, should take the primary responsibility: it is their profits and markets which are most directly in danger from wage increases (the collective action problem is addressed below). National unions need to support business, and to bargain improvements on non-wage issues such as training and employment security in exchange for that support; but for national unions in the UK to have primary responsibility for wage restraint—the opposite of what most of their members believe to be their main function—imposes intolerable institutional strains. The government is necessarily involved in wage determination as employer, and as macroeconomic policy-maker; but once it begins to lay down explicit pay guidelines it can easily get caught in politically damaging situations.

(*c*) Problems of wage restraint have been divorced in UK discussions from a range of other collective action infrastructural problems of business. The arguments for formally or informally co-ordinated business in other countries do not rest on economy-wide wage co-ordination. They concern vocational training, technology transfer, export marketing, product quality norms, co-operation between companies, as well as larger questions of restructuring industries, and so on. The case for improved co-ordination across businesses, for augmenting the power of employer organizations and business associations, and at local level chambers of commerce, is very strong in any long-term strategy of improving international competitiveness. The institutional debate on wage restraint should be part of a wider debate on this range of infrastructural issues. In particular, the question of wage determination should be linked to the adequate provision of training (see Soskice 1990*a*, *b*, and *c*).

What suggestions can be made? The fundamental problem is how to develop effective co-ordination between employers. This co-ordination does not require to be widespread. One of the features of the last decade has been that many large companies in the UK, while decentralizing wage bargaining to plant or department level, have kept actual control of wage determination at a central level. Moreover, there is something which looks quite like the 'going rate' of wage increases still in operation. Thus it may be that co-ordination among a relatively small group of the largest companies, say between thirty and fifty, would be sufficient. If they could agree to bargain at the same period of time each year, and could each agree to set roughly the same rate of wage inflation, there would be a mechanism in operation which could reduce the rate of nominal inflation.

The key question is how to make such co-ordination effective. It should be clear that simply explaining this to the largest companies or to the CBI would have virtually no effect, even if they agreed with the analysis. This is because of the standard collective-action dilemma. Individual large companies will certainly be happy if other companies control wage increases more effectively, but if they face damaging strikes when they do so themselves they will be reluctant to do so; so companies will be unlikely to co-operate. And the CBI will make little difference (even if it tried to) since it has no powers.

As we move closer to a Europe in which macro performance will depend on an effective supply-side, the British government will eventually be forced to intervene in both the wage-setting process and in vocational training if industry cannot reform itself. Similar situations (for example in Sweden in the 1930s) have galvanized large companies and their representatives into action when the threat of government intervention becomes explicit enough. A British government could intervene in large companies both with respect to wage-setting and training, for instance by imposing large financial sanctions on companies with wage increases above some norm, together with detailed monitoring of legally required training systems. Such solutions would be both distasteful to companies and less efficient than self-regulation solutions.

One element, then, in a strategy of a government which wanted to effect serious institutional change, might then be to threaten the type of intervention described in the last paragraph if business is unable to transform itself. The velvet glove containing the threat needs to give employer organizations resources (particularly in relation to training), large enough to begin to give them some power in relation to large companies. There is a strong case for giving the TUC some control over training expenditures to channel to individual unions: the electricians, for instance, who have made vocal their opposition to any sort of restraint, might take a different position if the TUC controlled resources which the EEPTU wished to gain access to. And perhaps there is a case for allowing the CBI to impose sanctions on a limited number of large companies if they refused to schedule their wage-bargaining into a synchronized system.

This approach answers three collective action problems. First, it would give some power to the TUC to bring national unions into line; and to develop with them bargaining strategies on training within companies, employment security, and so on in exchange for wage moderation. Second, it would give the CBI the ability to sanction large companies who refused to support a move towards economy-wide wage co-ordination. Third, the threat of government intervention if industry could not put its own house in order, would act as enough of an incentive on individual companies to get together to work out a more sensible wage determination system.

Such a strategy would need to be accompanied by many other changes. Of these the most important would be the gradual development of longer-term labour market incentives for workers which matched those in Northern Europe and which would ease the task of economy-wide co-ordination.

The type of institutional goals discussed here appear out of place in the UK. But as awareness grows of the way in which some of our most effective competitors organize their institutions, these goals will appear less unrealistic.

References

Aubert, G. (1989), 'Collective Agreements and Industrial Peace in Switzerland', *International Labour Review*, 3.

Beckerman, W., and Jenkinson, T. J. (1990), 'Wage Bargaining and Profitability: A Disaggregated Analysis', *Labour*, 4(3), 57–77.

Calmfors, L., and Driffill, J. (1988), 'Centralization of Wage Bargaining', *Economic Policy*, April.

Carlin, W., and Soskice, D. (1990), *Macroeconomics and the Wage Bargain*, Oxford, Oxford University Press.

Dore, R. (1987), *Taking Japan Seriously*, University of California Press.

Elvander, N. (1990), 'Incomes Policies in the Nordic Countries', *International Labour Review*, 1.

Flanagan, R., Soskice, D., and Ulman, L. (1983), *Unionism, Economic Stabilization and Incomes Policies: The European Experience*, Brookings.

Layard, R. (1990), 'Wage Bargaining and Incomes Policy: Possible Lessons for Eastern Europe', Centre for Economic Performance, LSE, DP no. 2.

Nickell, S. (1990), 'Unemployment: A Survey', *Economic Journal*, 100.

Rowthorn, R. E., and Glyn, A. (1990), 'The Diversity of Unemployment Experience', in A. Bhaduri and S. Marglin (eds.), *The End of the Golden Age*, Oxford, Oxford University Press.

Shimada, K. (1983), 'Wage Determination and Information Sharing in Japan' in *Collective Bargaining and Incomes Policies in a Stagflation Economy*, Proceedings of the 6th World Congress of the International Industrial Relations Association, 1983.

Shirai, T. (1984), 'Recent Trends in Collective Bargaining in Japan', *International Labour Review*, 3.

Sisson, K. (1987), *The Management of Collective Bargaining: An International Comparison*, Oxford, Blackwell.

Soskice, D. (1990a), 'Reinterpreting Corporatism and Explaining Unemployment: Co-ordinated and Unco-ordinated Market Economies', in R. Brunetta and C. Dell'Aringa (eds.), *Labour Relations and Economic Performance*, International Economic Association Conference Volume 95, Macmillan.

—— (1990b), 'Skill Mismatch, Training Systems and Equilibrium Unemployment: A Comparative Institutional Analysis', in K. Abraham and F. Padoa-Schioppa (eds.), *Mismatch and Equilibrium Unemployment*, CEPR Conference Volume.

—— (1990c), 'Providing the Infrastructure for International Competitiveness: A Comparative Institutional Perspective', presented at the International Economic Association Conference on *The New Europe*, Venice, 1990.

Windmuller, J., and Gladstone, A. (eds.) (1984), *Employers Associations and Industrial Relations*, Oxford, Oxford University Press.

11

Explanations of unemployment

ASSAR LINDBECK

Institute for International Economic Studies, University of Stockholm

DENNIS SNOWER

Birkbeck College, University of London

Economists commonly distinguish between two sorts of unemployment. Roughly speaking, the 'involuntarily unemployed' workers are unwillingly out of work (i.e. they feel that they would be better off with jobs). On the other hand, the 'voluntarily unemployed' workers prefer to be out of work (for any of a wide variety of reasons, e.g. because they are searching or training for a job or because they choose leisure over work).

This paper attempts a selective evaluation of what appear to be particularly noted or fruitful contributions to the literature on this subject, especially the most recent. The emphasis is primarily on involuntary unemployment.

Our ultimate concern lies with theoretical explanations of involuntary unemployment; empirical studies lie outside our range of purview. Strange as it may seem to the layman, economists have found it very difficult to pose coherent arguments which show why people who are willing and able to work at the prevailing wages in market economies cannot find jobs when they seek them. Thus, an evaluation of the theoretical explanations—even without consideration of empirical testing—turns out to be quite a challenge in its own right. Besides, some of the most intriguing explanations in the running are too recent in origin too have received significant empirical attention.

The paper is organised as follows. Sections I–III provide an overview of 'early' (pre-1970s) and recent contributions to the literature. The purpose is to isolate the salient strengths and weaknesses of rival explanations, thereby reducing the field to a few interesting contenders, which are to be considered in greater detail later. Section IV attempts to clarify what economists

First published in *Oxford Review of Economic Policy*, vol. 1, no. 2 (1985).

commonly mean by the term 'involuntary unemployment' and proposes several criteria which successful explanations of involuntary unemployment might satisfy. The few chosen contenders are evaluated in this light.

I. Early contributions

As noted, economists have had a difficult time explaining how involuntary unemployment comes about and why it may persist for substantial periods of time. They appear to have gone through all the various behaviours which doctors exhibit in the face of unresponsive patients: scepticism, diagnosis, refinement of the diagnosis, finding reasons for doubt, retracting the diagnosis, pronouncing the problem non-existent, formulating a new diagnosis, and so on.

Until the mid-thirties, the conventional economic wisdom was that unregulated market economies can overcome the problem of involuntary unemployment automatically, though temporary unemployment spells may occur. The standard story ran as follows. Changing economic conditions—arising out of, say, new tastes, technologies, patterns of world trade—mean shifting business fortunes, and the workers who happen to find themselves in the declining sectors may be put out of work. Naturally, they then seek work elsewhere and, if they do not find it reasonably promptly, they reduce their wage demands. This process continues until the unemployed have either found jobs or are no longer willing to work. Whatever the outcome, involuntary unemployment disappears, provided that wages are not regulated by government or strong unions.

The Great Depression of the thirties and the less severe (but nonetheless painful) recessions since then appear to have put the lie to this tale. Even if we accept that government and union controls may be responsible for some unemployment, it is far fetched to suppose that *swings* of the unemployment rate in major capitalist countries between, say, 2 and 10 per cent in the postwar period (or between 5 and 20 per cent in the interwar period) have depended on changes in such controls.

Thus, we are forced to reopen the question of why market economies are periodically unable to provide all the jobs which people look for? The answers to this question are of critical practical importance, since they suggest the economic policies needed to alleviate the problem.

1. The early Keynesian and 'classical' diagnoses

Keynes' (1936) diagnosis in the thirties attributed unemployment to a deficient demand for goods. Unemployment persists because the deficient goods demand arises from unemployment itself: firms employ too little labour because their customers buy too few goods. And the reason for the customers' deficient demand is their limited purchasing power which, in turn, is due to the circumstances that a proportion of them is unemployed. Thus, the vicious macroeconomic circle is closed.

One way to break it is through government demand-management policies. A rise in government spending induces firms to employ more labour, which raises purchasing power, and thereby leads to increased private-sector spending (which induces firms to employ even more labour, and so on). In general, a rise in the aggregate demand for goods (regardless of whether it arises from government expenditure increases, tax reductions, export increases, etc.) leads to a fall in unemployment and (if firms' production capacity thereby becomes exhausted) possibly also to inflation. This was the basic diagnosis-treatment combination which the Keynesians refined throughout the fifties and much of the sixties.

The Keynesians were opposed by economists who believed that unregulated markets have an inherent tendency to clear. These economists focused primarily on errors in wage-price expectations as the source of unemployment. The current-generation members of this group, commonly known as 'New Classical' macroeconomists, believe that the expectational errors which occur are predominantly unavoidable, arising from unpredictable economic events. Their predecessors assumed that expectations could be formulated in an *ad hoc* fashion and thus avoidable errors could occur. By way of contrast, let us call these people the '*Old Classical*' macroeconomists.

To begin with, the Old Classical macroeconomists indicated that the Keynesian diagnosis was not in accord with the then accepted body of microeconomic theory (e.g. Friedman (1968)) or with the actual movement of inflation and unemployment, particularly in the late sixties and early seventies (e.g. Flanagan, 1973, Gordon, 1972, Wachter, 1976). The microeconomic theory presupposed that economic agents buy and sell all that they wish to demand and supply at the prevailing wages and prices. In other words, markets were assumed to clear, so that full employment prevails. Moreover, the empirical evidence appeared to suggest that fluctuations in aggregate demand were not consistently driving inflation and unemployment in opposite directions.

Accordingly, the Old Classical macroeconomists proposed a different diagnosis, whereby unemployment was viewed as the outcome of either misguided government regulations or errors in people's price expectations. In this light, unemployment becomes either straightforwardly curable (by eliminating the harmful regulations) or voluntary *ex ante* (since the expectations are formulated freely, though the errors are regretted once they come to light). Of these two explanations, the one concerning expectational errors received the lion's share of attention. It was consistent with market clearing: regardless of what people's *expected* wages and prices may be, the *actual* wages and prices would move to bring demand and supply into equality in each market.

2. The reappraisal of Keynes

Throughout the seventies, these two broad stands of thinking about unemployment—the Keynesian and the Old Classical—were elaborated and reinterpreted in virtual disregard of one another.

The Keynesian camp attempted to develop microeconomic foundations for its macroeconomic unemployment theory (see for example, Barro and Grossman, 1976, Benassy, 1975, Dreze, 1975, Malinvaud, 1977, Muellbauer and Portes, 1978). This line of research has come to be called the 'Reappraisal of Keynes'. Here, economic agents pursue the same objectives under the same constraints as in the Old Classical world, except that they now take into account the possibility of being rationed. An economic agent is 'rationed' whenever his preferences, technology, and endowments lead him to demand or supply more than he is able to buy or sell at the prevailing wages and prices. For example, under involuntary unemployment, workers are rationed, so that the prevailing wages and prices induce them to supply more labour than they are able to sell.

In the Old Classical world, such rationing could not persist because wages and prices are assumed not only to be responsive to surpluses and shortages, but capable of eliminating them. The wage-price adjustment process continues until demand and supply in each market are identical. In the Reappraisal of Keynes world, involuntary unemployment persists because wages (and possibly also prices) are not viewed as responsive in this sense. Instead, they are assumed to be 'rigid' at levels which are incompatible with full employment. The 'rigidity' lasts at least long enough for agents to make their demand and supply decisions compatible with the rations they face. Thus, the reason why involuntary unemployment was seen to arise from deficient demand for goods is that wages failed to clear the labour market. Under rigid wages, firms may have no incentive to hire unemployed workers, since these workers would produce goods which the firms may be unable to sell.

This was a departure from Keynes' own position. Keynes did not make his explanation of unemployment entirely dependent on the assumption of wage-price rigidity. Rather, he argued that even if they were responsive to excess supplies and demands, involuntary unemployment would nevertheless refuse to disappear. In other words, once the vicious circle of unemployment and deficient goods demand was in place, it would take more than wage-price deflation to get rid of it. Thus, there was no need to look very closely at the how's and why's of wage-price rigidity.

The Reappraisal of Keynes was unable to provide a firm microeconomic justification for this view. One difficulty lay in the 'real balance effect': when wages and prices fall, the real values of monetary assets held by the private sector may rise, thereby stimulating private-sector spending and reducing unemployment. The real balance effect was debated as early as the 1930's, when Pigou offered it as a challenge to Keynes. Despite considerable (albeit sporadic) attention, the debate is still unresolved. Keynesian economists have argued that the monetary assets in question are not significantly large, that the effect of a rise in real balances on private sector spending is small, and that even if this effect were large, there are equally large counterveiling effects at work when the economy suffers from involuntary unemployment and deflation. Yet the empirical evidence is inconclusive.

It was only by assuming that wages (and sometimes also prices) were fixed that the Reappraisal of Keynes microfoundations could be constructed. Economic agents accept the prevailing wages and prices—together with the resulting rations—as given and formulate their demands and supplies on this basis. But not only do rations influence demands and supplies but—since one agent's purchases are another agent's sales—these demands and supplies also influence the rations. A major job in the Reappraisal of Keynes was to describe the circumstances under which all the rations are 'consistent' with each other, in the sense that a particular set of rations gives rise to a particular set of demands and supplies which, in turn, gives rise to the original rations.

For example, when firms' sales are rationed, they demand a limited amount of labour. The limited labour demand gives rise to job rationing. And when jobs are rationed, workers demand a limited amount of goods. This limited goods demand leads to a rationing of firms' sales. The big question is whether this sales rationing is the same as that with which we started out. If so, then the firms' sales rations are 'consistent' with the workers' job rations. The Keynesian 'multiplier' is a portrayal of how this happens. For example, when government expenditures rise, firms employ more labour and thereby raise workers' job rations. As a result, the workers buy more goods and the sales rations rise to become consistent with the previous job rations. But more sales call for more employment, and thus the job rations must rise to become consistent with the previous sales rations. At the end of this multiplier process, the two sets of rations are mutually consistent.

Wages (and possibly also prices) remain fixed long enough for ration-consistency to be achieved. The question that remains open is why. The Reappraisal of Keynes provided no answer. Early contributors suggested that since wages and prices could not be expected to clear their markets completely and instantaneously —and massive swings in unemployment as well as inventory and order-backlog statistics bear this out—it is reasonable to suppose that they are 'rigid'. But such rationalisation is unsatisfactory. It is one thing to agree that wages and prices do not clear their respective markets; it is quite another to suppose that they remain unaffected by the entire process in which rations become consistent.

Thus, there is a hole in the logic of the Reappraisal of Keynes theory: the assumption of wage-price rigidity was not rationalised in terms of agents' objectives and constraints. This is a serious drawback, since the Keynesian policy prescriptions are generally not independent of the prevailing wage and price levels. If government policies (such as changes in taxes and transfers) and their private sector repercussions do indeed affect wages and prices—and it would be a strange world in which they did not—then the Reappraisal of Keynes theory ceases to provide reliable guidelines for dealing with problems such as involuntary unemployment.

The only sort of economy which the Reappraisal of Keynes can cover in a logically consistent manner is one with a comprehensive system of wage controls (and perhaps also price controls). This may be a reasonable approximation of some centrally planned economies, but it is certainly a long way away from the market economies which had been the focus of Keynes' attention.

Nevertheless, since Keynes' time, most capitalist countries have evolved a variety of wage restrictions, as governments tried (effectively or ineffectively) to keep their citizens from falling beneath particular minimum standards of living. There have been minimum wage laws, public sector pay scales, and various other government regulations regarding labour remuneration. Could it be argued that these are responsible for involuntary unemployment of the Keynesian variety? Paradoxically enough, this would imply that the Keynesian explanation of involuntary unemployment it the *same* as that suggested by the pre-Keynesian economists.

The underlying argument is straightforward: if the legal minimum wage is greater than the wage required to clear the labour market, then labour supply will exceed labour demand. But at second glance, things become more complicated. If employers have monopsony power in an unregulated labour market, the imposition of a minimum wage may *raise* the level of employment, since it prevents the exercise of this monopsony power (e.g. Maurice, 1974, Stigler, 1946). When the minimum wage has incomplete coverage, its imposition merely drives labour from the covered into the uncovered sectors (Gramlich, 1976, Mincer, 1976, Welch, 1974). The empirical evidence is that the effect of the minimum wage on unemployment is extremely weak (e.g. Moore, 1971, Lovell, 1972, Ragan, 1977, and a survey by Brown, Gilroy and Kohen, 1982). The volatile statistics of wage inflation from year to year do not give the impression that there are rigid institutional forces driving wages steadily and inexorably upwards.

3. The 'Old' and 'New' Classical macroeconomics

Meanwhile, as the Reappraisal of Keynes developed in the seventies, the Old Classical camp refined its

analysis of how expectational errors cause swings in the unemployment rate. When the public's wage-price expectations are correct, unemployment is supposed to be at a unique rate, the 'natural rate of unemployment', which is voluntary *ex ante* and *ex post* (i.e. voluntary both when the expectations are formulated and when their accuracy has been assessed). All deviations of unemployment from its natural rate are allegedly related to errors in wage-price expectations. This relationship is known as the 'natural rate hypothesis'. A variety of microeconomic rationales were adduced for it.

In the '**misperceived real wage approach**' (related to the work of Friedman, 1968, 1976 and others), the real wage relevant to firms is the nominal wage they pay in terms of the product price they receive and the firms are assumed to have reasonably accurate perceptions of both elements. On the other hand, the real wage relevant to workers is the nominal wage they receive in terms of the expected consumer good prices they pay and they may misperceive the later element (i.e. the prices). The greater the expected price level relative to the actual one, the smaller the real wage perceived by the workers relative to that perceived by the firms. As a result, the lower is the level of employment. The difference between the full-information level of employment and the actual employment level (above) may be represented as voluntary unemployment.

The upshot of the argument is this. If the price level falls unexpectedly while the nominal wage clears the labour market, firms reduce their labour demand (since the real wage rises for any given amount of employment) and workers reduce their labour supply (since their expected real wage falls for any given amount of employment, due to a fall in the nominal rate). Thus, the greater the difference between the expected and actual price levels, the greater the rate of unemployment.

In the '**misperceived interest rate approach**' (inspired by Lucas and Rapping, 1970, Weiss, 1972, and others), workers misperceive real interest rates. The greater the expected rate of inflation relative to the actual inflation rate, the smaller the perceived real interest rate relative to the actual one. If workers respond to a fall in the real interest rate by reducing their current labour supply (since current work yields less purchasing power in the future), then employment falls. Here, the greater the difference between the expected and actual inflation rate, the greater the rate of unemployment.

In the '**capital gain approach**' of Lucas, 1975, economic agents know the current price level, but they can observe neither the money balances transferred by the 'young' generation to itself when it becomes 'old', nor the distribution of the population between 'young' and 'old'. A low current price level may be due to either a low money transfer or a large 'young' generation (which is assumed to produce all the goods). Since the 'young' people do not know which is the case, they suppose that each of them is possible. The smaller the current price relative to the expected future price, the greater the capital loss on the intergenerational money transfer anticipated by the 'young' people (since they ascribe a low current price, in part, to a large 'young' cohort). Consequently, employment falls. Once again, we find that the difference between expected and actual prices is positively related to the voluntary unemployment rate.

In the 'job search approach' (Alchian, 1970, Gronau, 1971, MacCall, 1970, Mortensen, 1970, Parsons, 1973, Siven, 1974, and many others), workers may misperceive the distribution of wages which they face. On the basis of their perceived wage distribution, they are viewed as calculating the marginal benefits and marginal costs of being unemployed while searching for a job. (Here the explanation of unemployment rests on the rather unrealistic assumption that job search can be performed only while unemployed.) The greater the workers' expected wages relative to the actual wages they face, the greater the marginal benefits of job search relative to the associated marginal costs. Consequently, the greater is the amount of job search and the greater is the level of unemployment.

Finally, the 'wage setting approach' of Phelps, 1970 goes far beyond the standard conceptual confines of the natural rate hypothesis. It explicitly recognises that when workers have imperfect information about wage offers, firms gain monopsony power in the labour market. Phelps assumes that the greater the unemployment rate and the lower a firm's vacancy rate, the lower the firm will set its wage relative to what it perceives the average market wage to be. Given that unemployment and vacancies are inversely related, there is a unique 'natural' rate of unemployment, at which each firm sets its wage equal to its perceived average market wage. Moreover, the lower the unemployment rate, the greater the difference between each firm's set wage and its perceived average market wage (and consequently the faster the average market wage rises). Note that, unlike the other approaches to the natural rate hypothesis, unemployment is not necessarily voluntary here, i.e. the analysis does not rely on the assumption that the labour market clears. This is also true of Phelps's seminal paper (1967), which provided the first formal statement of the natural rate hypothesis.

Mortensen, 1970 adopts a similar approach, except that he assumes firms to choose their profit-maximising wage-employment combinations from a sequence of short-run labour supply curves. Due to workers' imperfect information about wage offers, a 'high' wage generates an increasing supply of labour to the firm (at that wage) and a 'low' wage generates a shrinking supply. Since workers are not off their labour supply curves, whatever unemployment there is may be considered voluntary *ex ante* (i.e. voluntary, given workers' imperfect information).

The next path-breaking development was the 'rational expectations hypothesis' (see, for example, Muth, 1961 and Lucas, 1976). Economic agents are assumed to avoid predictable errors in expectations. What is 'predictable' depends on the particular economic model we have in mind. With reference to such a model, agents' expectations are 'rational' whenever they coincide with the predictions of the model itself, contingent on the information the agents are assumed to have.

In the so-called 'New Classical Macroeconomics', the rational expectations hypothesis is combined with the natural rate hypothesis, to yield dramatic implications for the nature of unemployment. Whatever the expectational errors are that generate unemployment (in accordance with the natural rate hypothesis), they must be unpredictable (in accordance with the rational expectations hypothesis). Consequently, all unemployment is not only voluntary *ex ante*, but also quite transient: it can last only as long as the unpredictable errors (and their repercussions) do. If the natural rate of unemployment does not change much through time, then the actual unemployment rate must move as randomly as the unpredictable errors which generate it (see, for example, Barro, 1976, Lucas, 1972, McCallum, 1980, Sargent and Wallace, 1975, Shiller, 1978).

This line of thinking ran into some severe difficulties. In empirical studies using any of the standard macroeconometric models of major market economies, the movement of the unemployment rate through time certainly does *not* appear to be predominantly random (e.g. Hall, 1975). Consequently, a number of economists (e.g. Lucas, 1973 and Sargent, 1976) have argued that the natural rate of unemployment moves cyclically and thereby explains the cyclical movement of the actual unemployment rate. Yet the suggested rationales for such activity (e.g. a prompt investment response to swings in wage-price expectations) are not convincing. Moreover, it is absurd to argue that the massive unemployment of the thirties, accompanied by the misery and sometimes even homelessness and starvation of the unemployed workers, was the result of their desire to choose more leisure or to engage in more job search on account of a change in their expectations about real wage rates or interest rates.

Recall that what is considered 'random' (viz. unpredictable) depends on the reference model and the appropriate reference model, in turn, depends on the economic information which economic agents possess. But how do we find out what economic agents know? It would be a logically and practically hopeless task to find out by asking them directly. Thus, it is difficult to see how the rational expectations hypothesis could be falsified. A stop-gap strategy would be to develop a theory of information acquisition and this theory could then be tested conjointly with the rational expectations hypothesis. To date, however, little has happened in this area.

In practice, this problem is commonly side-stepped by simply assuming that the public always knows the structure of the macroeconomic model used by the author of each successive article on this subject. Usually any disagreement between the public and the author is ascribed to the supposition that the public's economic statistics are one year (or one quarter) out of date. There is something rather curious about the view that households and firms in every country always change their perception of the world whenever a new (improved) rational-expectations macro model is constructed.

4. The stand-off

At this point in the development of Keynesian and New Classical macroeconomic thinking, the controversy over the causes of unemployment seemed to have stalled. The Keynesian economists had three major objections to the New Classical research program. First, they regarded the Neoclassical assumption of clearing markets as blatantly counterfactual. They found it impossible to believe that the unemployment of, say, the 1930s depression or the 1978–82 recession was largely voluntary.

Second, they had overwhelming doubts that errors in wage-price expectations could be large enough to account for the magnitude and length of witnessed fluctuations in unemployment. It seemed implausible that these errors should have such a powerful impact on workers' labour supply, that the unemployment rate could frequently double or halve in the course of a single year. It seemed even more implausible that these errors—occurring largely because macroeconomic data is not available to the people who need them—should be unavoidable for time periods long enough to explain the persistence of recessions. Yet if this objection is granted and thus (in accordance with the

natural rate hypothesis) unemployment fluctuations are attributed mainly to swings in the natural rate of unemployment, there is another difficulty. The natural rate of unemployment depends solely on the tastes, technologies and endowments of an economy and these are quite unlikely sources of cyclical macroeconomic activity.

Third, the New Classical economists provided no microeconomic reason for why the public does not acquire those pieces of information that it needs to avoid its expectational errors. (Why don't newspapers, government statistical bureaus, economic consulting agencies publish them?) Of course, there are costs of information acquisition and dissemination, but no attempt has been made thus far to show that these costs are sufficiently large to account for the costs of periodic recessions.

On the other side of the barricades, the New Classical macroeconomists pointed out that the Reappraisal of Keynes theory was incomplete in its scope, defective in its logic, and misleading in its policy implications.

As already noted, the quantities demanded and supplied in each market are formulated in accordance with agents' objectives and constraints, but prices are not explained in this way. Prices are neither set by individual agents, nor driven by market forces. A macroeconomic theory which covers quantity determination but not price determination—even though quantities demanded and supplied do, in general depend on prices—must be incomplete.

Furthermore, when markets fail to clear, agents face unexploited gains from trade: by changing the prices and quantities at which they trade, some agents could be made better off without any others being made worse off. For example, when there is involuntary unemployment, the marginal product of labour (which underlies the market-clearing labour demand) falls short of the marginal value of workers' time (which underlies the market-clearing labour supply). Thus, by employing additional workers, producing additional output, and selling this output to consumers, the agents could succeed in generating goods whose value—in terms of labour time—is greater than its cost. These are the sort of gains from trade which agents in market economies have obvious incentives to exploit. Yet in the Reappraisal of Keynes world, these gains are exploited only through quantity adjustments, not through price adjustments. This is a problem of logical consistency.

Finally, although the Reappraisal of Keynes theory implies that the way in which government policies affect unemployment depends (in part) on wage and price levels, the theory does not tell us how these wage and price levels respond to the government policies. In fact, wages and prices are assumed impervious to policy influence. Now, if this assumption turns out in fact to be false, then the theory provides no way of assessing what the effectiveness of government policies is.

Thus the Keynesians and New Classical macroeconomists had reached a stand-off. The most convincing arguments of the contestants seemed to be their criticisms of each other.

It is interesting to observe that the weaknesses of the two positions are in fact quite similar. The failure of agents to acquire information that would enable them to avoid expectational errors, and the failure to adjust prices when in the agent's interests to do so—these both are potential instances of unexploited gains from trade.

Another weakness (which has received somewhat less attention) concerns the assumption that agents accept the prevailing wages and prices as beyond their control. This assumption is standard both in the Reappraisal of Keynes and in the New Classical Macroeconomics, but it does not sit well in either. Price-taking is usually associated with perfect competition, which arises out of two conditions:

(a) there are many well-informed buyers and sellers in a market, each responsible for only a negligible proportion of the transactions in the market, and
(b) each agent (with given tastes, technologies and endowments) can buy or sell as much as he wishes to demand or supply under prevailing wages and prices.

These conditions imply that when a seller raises the price of his commodity above the prevailing market price, he loses all his buyers; and when the price is reduced, he gains more buyers then he can satisfy. (Similarly also for price-setting buyers.)

But when markets do not clear—as they do not in the Reappraisal of Keynes—the second condition does not hold. For example, when there is involuntary unemployment, workers are unable to sell all the labour they wish to supply. In that case, they are generally able to influence their employment rations (e.g. their places in job queues or their amount of part-time work) by adjusting their wages. Similarly, when there is deficient demand for goods, firms are unable to sell all the output they wish to supply, but they may well be able to influence their sales rations by adjusting their prices. Under these circumstances the incentive to be a price-taker disappears. However, the macroeconomic activity—and, in particular, the unemployment—which arises from price-setting behaviour (e.g. Hart, 1982 and Snower, 1983) is quite different from that described by the Reappraisal of Keynes.

In the same vein, when agents do not have perfect information about the wages and prices they face—as they do not in the New Classical Macroeconomies—the first condition does not hold. A firm that raises the price of its product above the prevailing market price does not lose all its customers, because some customers will not be aware that firm's product is overpriced. Similarly, a firm that reduces its wage payments beneath the prevailing wage does not lose its entire workforce, because some workers will not realise that they can get higher pay elsewhere. Once again, agents gain some latitude in setting their prices and price-taking behaviour disappears. Phelps, 1970a,b and Mortensen, 1970 are among the precious few who recognised the inconsistency of assuming that the agents have imperfect information *and* that they are price-takers. Apart from their work, little has been done to explore the macroeconomic implications of price setting within the New Classical framework of thought.

In this light, it would appear eminently sensible to investigate explanations of involuntary unemployment which recognise the existence of both imperfect information and price setting behaviour. Indeed, this has been the thrust of many recent contributions to the unemployment literature. But the joint assumptions of imperfect information and price setting do not go well with the market clearing foundations on which the New Classical Macroeconomics is built.

When price setters are imperfectly informed about their economic environment, there is no guarantee that the set prices will eliminate all surpluses and shortages. (For example, a monopolist who does not know precisely what demand curve he faces may offer to sell his output at a price that results in undesired inventories or delivery backlogs.) Furthermore, when the price setters have less information about their products than the price takers (on the other side of the market), then prices may be set so as to influence product quality. As shown below (under the heading of the 'efficiency wage hypothesis', Section III.4), this set-up may also be incompatible with market clearing. Consequently, it is not surprising that many recent contributions to the unemployment literature have moved away from the presumption that all markets clear or have rested on new concepts of what 'market clearing' means.

II. Recent contributions in the market-clearing tradition

From the mid-seventies to the present, both sides in the debate on the causes of unemployment began to address the criticisms they faced. By this time, loyalties to past schools of thought had faded, so that economists studying non-clearing markets no longer automatically considered themselves 'Keynesian' and those exploring clearing markets were no longer always called 'New Classical'. Nevertheless, the divergent lines of research remained.

On the market-clearing side, there were several significant developments, of which the following perhaps deserve special attention.

1. Implicit contracts

In their simplest guises, the 'implicit contract theories' (e.g. Azariadis, 1975, 1979, Azariadis and Stiglitz, 1983, Bailey, 1974, Gordon, 1974, Grossman and Hart, 1981, 1983, Hart, 1983) combine three perceptions of labour markets:

(a) workers are more risk averse than firms;
(b) workers have less access to capital markets than firms, and
(c) workers are immobile among firms.

The first two elements imply that efficient wage-employment contracts involve an insurance package by firms to workers. The third implies that both the firms and the workers must decide in advance whom to employ, at what wage, and under which circumstances.

One aim of this analytical set-up was to show that, under unpredictable, adverse business conditions, workers may be laid off in accordance with their previously agreed contracts. Here the underlying concept of 'market clearing' is different from that of the natural rate hypothesis.

In the latter, agents formulate their demands and supplies on the basis of their wage-price expectations and then actual wages and prices adjust so as to bring demand and supply into equality in each market. Thus, the market clears in the following *ex ante* sense: for the given expectations (which may or may not be mistaken), the offers to buy are matched by offers to sell. Yet *ex post* (when agents have discovered whether their expectations were mistaken), the market participants may conclude that it would have been better to have made different offers. But the past cannot be undone and, at every moment in time, demand and supply are equal.

By contrast, in the implicit contract theories, contingent trading agreements are made before the corresponding transactions take place and thus the market clears in a different *ex ante* sense: given the agents' expectations, the contingent agreements are efficient (in

that it is impossible to make one party better off without making another worse off). Yet *ex post* (when the agreements are put into effect on the basis of the contingencies which occur), it may turn out that offers to buy may not be matched by offers to sell. For example, labour demand may fall short of labour supply. In that event, the workers who turn out to be unemployed may wish to renegotiate their labour contracts, but do not have the opportunity to do so.

Whether such unemployment is called '*ex ante voluntary*' or '*ex post involuntary*' is a semantic issue. At any rate, the persistence of this unemployment now depends not merely on the survival of unavoidable expectational errors (as in the New Classical Macroeconomics), but also on workers' inability to move among firms and to renegotiate their labour contracts.

2. Imperfect competition

This line of research (e.g. Hart, 1982, Layard and Nickell, 1984, Snower, 1983) explores how imperfectly competitive behaviour may generate unemployment —quite independently of any other impediments to trade (such as costs of information acquisition or costs of price change). Sellers are assumed to have some market power, whereas buyers have none. Thus, there is a division of responsibility over price and quantity decisions. In the product markets, firms set prices and workers decide how much of the goods to buy; in the labour markets, workers (say, through their unions) set nominal wages and firms make the employment decisions. The firms face product demand curves and the workers face labour demand curves. Moreover, the demand curves which firms and workers perceive may or may not coincide with the ones they actually face, viz. they may be 'conjectures'. In this sense, Hahn's work (1977, 1978) on 'conjectural equilibria' fits into this framework as well.

At first glance, one may be tempted to consider the sellers to be 'rationed', in the sense that their decisions are constrained by the demand curves they face. However this is a quite different concept of rationing from the one used in the Reappraisal of Keynes. Workers who freely and unanimously set their wages— knowing full well what the resulting levels of employment will be—can scarcely be called involuntarily unemployed.

Nevertheless, unemployment may arise from another quarter. The above division of responsibility over price and quantity decisions may be inefficient (in that a benevolent dictator who determines wages and prices could make some agents better off without making others worse off). In particular, employment may be less when competition is imperfect than when it is perfect. The difference between the two unemployment levels may be considered voluntary unemployment.

3. Customer markets

Another line of research combines imperfectly competitive behaviour with the observation that, in some markets, customers compare the relative prices of similar products only when they perceive these prices to be unstable through time. This product search behaviour may induce the sellers to stabilise their prices, viz. not to adjust their prices completely and instantaneously to demand and supply shocks.

Okun, 1976, 1981 distinguishes between 'auction markets' (which always clear) and 'customer markets' (where product search may occur). In the latter, product search is costly to the customers and thus they are willing to pay a premium to sellers who keep their prices sufficiently stable to obviate the need for search. In practice, Okun suggests, this means that sellers may adjust their prices to permanent changes in cost, technologies or overall demand, but not to transient ones (and particularly not ones that affect sellers in the same market differently). In other words, these sellers may be induced to keep their long-run mark-up (of prices over costs) roughly constant through time.

In principle, these ideas could also be applied to 'customer labour markets', where workers find job search costly and are willing to work at a discount for employers who keep their wages sufficiently stable to make job search unnecessary. Yet, to date, there has been no methodical, detailed statement of this principle.

Moreover, the entire microeconomic rationale for customer markets remains incompletely explored. Given a customer's costs of product search, why should his search decision depend on the stability of a product price relative to the long-run costs of producing that product (and possibly also the long-run costs of producing demand), rather than on the marginal benefits of search (which is related to the probability of finding a cheaper product)? If the latter is the case, then the stability of price relative to long-run costs is entirely irrelevant; all that matters is whether a product price is above or below the 'acceptance price' (at which the marginal costs and benefits from search are equal). It is difficult to see in what respect this approach is an advance over the 'job search' and 'wage-setting' approaches to the natural rate hypothesis.

4. Inventories and delivery lags

Another reason why prices may be sluggish in response to demand and supply shocks is that it may be profitable to adjust inventory levels and delivery lags as well (see, for example, Blinder, 1980, 1981 and Carlton, 1978, 1979). Consider, for example, a profit-maximising firm deciding on inventory holdings of its finished goods. Its output is such that its marginal cost of production equals the shadow value of inventories; its sales equate the shadow value of inventories with its marginal revenue. (Naturally, the difference between output and sales represents changes in the inventory stock.) A temporary increase in demand (reflected in marginal revenue) has little effect on the shadow value of inventories (which depends primarily on long-run considerations) and thus sales rise while production remains sluggish. Consequently, inventories must decline. Similarly, a temporary increase in marginal costs reduces production relative to sales, causing inventories to fall as well. These inventory responses to demand and supply shocks mean that the associated price adjustments are more moderate than they otherwise would have been. An analogous story can be told regarding delivery lags.

Yet although inventories and delivery lags can explain price sluggishness, this does not in turn explain the existence of unemployment, but rather the reaction of unemployment to government policy. In the Reappraisal of Keynes models, temporary policy shocks have a smaller short-term impact on unemployment when inventories and delivery lags adjust than when they do not. In the New Classical models, unexpected policy shocks give rise to inventory movements and thus their effect on unemployment lasts beyond the time during which expectations are in error.

Yet to explain unemployment itself, the logic of inventory holdings and delivery lags must be applied to the labour market itself. When there are significant costs of hiring, firing, and training, the profit-maximising firm operates on several margins. In its hiring decision, it sets the marginal revenue product of inexperienced labour against the wage minus hiring and training costs. In its firing decision, it compares the marginal revenue product of experienced labour with the wage minus firing costs. And to find how much of its current labour to use (rather than to employ), it sets the relevant marginal revenue product against the wage. This general framework could explain unemployment, as well as labour hoarding, and vacancies. Elements of this have been taken up by insider-outsider analysis, but otherwise little attention has been devoted to this area. Ehrenbert, 1971, Epstein, 1982, Nickell, 1978, 1984,

Mortensen, 1973 and others have examined employment decisions when the firm faces adjustment costs, but they do not explore how these generate unemployment.

Of these ideas which lie (on the whole) in the market-clearing traditions, some (viz. the approaches concerned with customer markets and with inventories and delivery lags) are not sufficiently developed for critical appraisal; another (viz. the imperfect competition approach) deals with voluntary unemployment arising from inefficiencies due to the exercise of market power; whereas the implicit contract theories described unemployment which is voluntary when labour contracts are signed, but may be involuntary once these contracts are in operation. Only the implicit contract approach contains a new insight into the causes of involuntary unemployment and so it is this approach that will be taken up in Section IV.

III. Recent contributions in the non-market-clearing tradition

On the other hand, the investigators in the non-market-clearing tradition have endeavoured to meet the original neoclassical criticisms. Their research has proceeded along several distinct routes.

1. Adjustment costs

The rigidity of wages and prices can be rationalised on the basis of wage-price adjustment costs (i.e. costs of wage or price changes). Whenever these costs exceed the associated benefits from setting wages and prices at their market-clearing levels, markets will fail to clear without leaving agents with unexploited gains from trade. In this context, the administrative costs of price changes (e.g. writing new price tags, printing new catalogues) has commanded some attention (e.g. Barro, 1972, Sheshinski and Weiss, 1977), although it was soon noted that such costs are quite unlikely to be large in comparison to the costs of foregone production during recessions.

However, the wage-price adjustment costs may also include a wide variety of negotiation costs (e.g. employers' and workers' costs of formulating new collective bargaining agreements, the expected costs of retaliation in oligopolistic and oligopsonistic markets) and these appear to be far more substantial. Yet these latter costs have not, as yet, been incorporated in formal macroeconomic models, where their impact on unemployment can be conceptualised.

2. Union activity

In this approach (e.g. McDonald and Solow, 1981, Oswald, 1982 and Gylfason and Lindbeck, 1984) workers are assumed to belong to unions, whose objective is given by a weighted average of their employed and unemployed members' welfare. If the union is 'utilitarian', then the weights are the *number* of employed and unemployed members, respectively. Here the union is concerned with the *sum* of its members' welfares. On the other hand, if the union's objective is 'expected utility', then the weights are the *proportion* of employed and unemployed members (respectively) to the total membership. The rationale for this objective is that each union member is indifferent to risk and has an equal chance (over each period of time) to occupy the jobs at the union's disposal, in which case the weights may be interpreted as the probabilities of being employed and unemployed, respectively.

Now suppose that the union sets the wage and the firms decide how much labour to employ at that wage. The resulting wage-employment combination is that point on the labour demand curve which maximises the union's objective. In general, there will be some unemployment. If the union members are really risk-indifferent and have an equal chance at the available jobs, then this unemployment will be in each member's best interest. Otherwise there must be a conflict of interest between the union and its members and the unemployment may be 'union-voluntary' and 'membership-involuntary' (Corden, 1981).

It can be shown that the wage-employment combination above is inefficient, in the sense that the union could be made even better off without any sacrifice of firms' profits (see, for example, Leontief, 1946, McDonald and Solow, 1981). Efficient bargains involve maximising the union's objective subject to the firms' achievement of some minimum level of profit. Here, too, the upshot will generally be some unemployment, which may or may not be voluntary from the members' viewpoint.

The problem with this explanation of unemployment is that it is not complete. First, recognising that union members are usually *not* risk-indifferent and do *not* stand an equal chance at the available work, why do unions not represent their members' interests? Why is the available work not shared out equally among the members (through part-time labour arrangements or job rotation)? If unions do this (as in Snower, 1983), then involuntary unemployment would disappear (with the possible exception of the voluntary sort, due to inefficiency associated with the exercise of monopoly power).

Second, why do firms enter into agreements with the unions in the first place? Why don't they turn to non-unionised workers instead? If they could do so, the unions would lose their bargaining power and wages would fall to their market-clearing level.

This is not to say that the questions above are unanswerable, but merely that economists have not given them much attention. The second group, however, are tackled explicitly in the recent work on insider-outsider analysis (discussed below).

3. Increasing returns to scale

Weitzman, 1982 considers a market economy characterised by increasing returns and imperfect competition and aims to show how the interaction of these two elements may be responsible for involuntary unemployment.

In particular, the increasing returns apply to labour and arise out of fixed costs (i.e. the greater the firm's workforce, the lower its fixed costs per unit of output). The imperfect competition is of the Chamberlinian, 1933 monopolistic variety: each firm produces a different product and there are sufficiently many firms, so that each firm can be assured that its activity has a negligible influence on its rivals.

For simplicity, all firms are assumed to be alike. Each makes its pricing, production, and employment decisions unilaterally, maximising its profit subject to its product demand curve. When there are positive profits to be made, new firms are born; when profits are negative, some firms die. In the long run, the number of firms is such that profits are zero. Not surprisingly, it turns out that this number depends inversely on each firm's fixed costs and positively on aggregate demand (which, in turn, is inversely related to the unemployment rate).

All households are also assumed to be alike, except for their employment status. They can be either fully employed or completely unemployed (i.e. part-time work is ruled out). Their tastes are such that all firms' products are equally popular, at equal prices. The lower a product price, the higher the associated demand. Each product has substitutes, in the sense that if one product's price rises, then the demand for some other products rises as well. Every household devotes its income exclusively to the consumption of the product of its choice.

In the long-run equilibrium there are no incentives for change. Not only are profits zero, but the firms set their respective prices, production, and employment so that their marginal costs are equal to their marginal revenues, and the households find all the available jobs and buy the firms' products with the resulting purchasing power. It can be shown that each firm sets its price by a proportional mark-up depending (strangely enough) *inversely* on aggregate demand (Weitzman, 1982, p. 800). In other words, through its price-setting the firm determines the real wage (which is the inverse of the price-wage mark-up). The greater the level of aggregate demand, the greater the real wage.

Weitzman contends that the long-run equilibrium may be compatible with involuntary unemployment. An unemployed worker does not have the option of starting his own, one-man firm, because there are increasing returns to labour. Even if a group of workers should combine their efforts to take advantage of the increasing returns, the creating of a new firm would depress the market price (given the prevailing level of aggregate demand) and thereby generate negative profit.

Only through stimulation of aggregate demand— say, due to government pump-priming—can the number of firms be increased and, with it, overall purchasing power and the level of employment. The individual unemployed workers are helpless in creating jobs, but the government need not be.

4. Efficiency wages

The efficiency wage theories start from the premise that employers have less information about the productivities of their individual workers than those workers do. The inability to monitor productivity perfectly is reflected in most employment contracts, which do not provide precise specifications of productivity. In time-rate contracts, work effort is rarely (if ever) defined and prescribed accurately; in piece-rate, time-rate and other types of contracts, the exact standards according to which output is to be produced are not given either. Such practice may be explained in terms of 'bounded rationality' (e.g. Simon, 1979 and Williamson, Wachter and Harris, 1975), or the impossibility (or prohibitive expense) of observing productivity objectively (e.g. Alchian and Demsetz, 1972 and Malcolmson, 1981). In practice, worker productivity does not depend solely on abilities which can be revealed by straightforward performance tests (of the sort described by Guash and Weiss, 1980). Thus, employment contracts usually specify only those attributes of a job which can be monitored cheaply and objectively.

The next step in constructing the efficiency wage theories is to assume that firms make the wage and employment decisions unilaterally. Moreover, since their monitoring of workers' productivities is imperfect, they may use the wage offer as a screening device for productivity. The main substance of efficiency wage theories lies in providing reasons why firms may wish to do this. Different theories offer different reasons.

The economic problems arising when product quality is assessed through the product price have been studied in the context of many markets (e.g. Akerlof, 1970, Arrow, 1963), not just the labour market. They take two forms:

— adverse selection (in which product characteristics are imperfectly monitored), and
— moral hazard (in which activities of agents are imperfectly monitored).

In the same vein, the efficiency wage theories may be divided into these two categories.

In the **adverse selection approach**, the productivity of individual workers is not monitored perfectly. It come in several guises. In the '*productivity differential models*'(e.g. Malcomson, 1981) the firm cannot distinguish between high productivity and low productivity workers. Workers' skills are not entirely firm specific. Thus, the higher the firm's wage offer, the higher the quality of workers (on average) that is attracted. In the '*turn-over models*' (e.g. Stiglitz, 1974) the 'quitters' among the firm's workforce cannot be distinguished from 'stayers'. Some of the costs associated with quitting (especially training costs) are borne by the firm. The higher the wage, the more workers can be induced to stay with the firm and the lower the firm's quit-associated costs. (A moral-hazard version of this idea exists as well. Here each individual worker is less likely to leave his firm, the greater his wage relative to wages of other, comparable jobs and to the level of unemployment benefits; see also Salop, 1979).

The **moral hazard approach** supposes that workers' productivities depend on their effort on the job, which the firm cannot observe directly. This relation can take various forms. In the '*shirking models*' (e.g. Calvo and Wellisz, 1978 and Shapiro and Stiglitz, 1984), a worker who is caught 'shirking' (i.e. devoting little effort to his job) is suspended or fired. Clearly, the punishment for shirking depends (at least in part) on the difference between the firm's wage offer and the income upon suspension or expulsion. By raising the wage, the firm increases the magnitude of this punishment and thereby induces more work effort. In the

'*search models*' (Snower, 1983), effort depends inversely on the amount of on-the-job search performed by the worker. When the firm increases its wage offer, it makes search less worthwhile and thus raises worker productivity on the current job. In the '*sociological models*' (Akerlof, 1982), effort depends on whether workers believe that they are being treated 'fairly'. By offering the workers a 'gift' above the required minimum, the firm can raise group work norms above the required minimum.

In all these adverse selection and moral hazard theories, a rise in a firm's wage offer increases the average profitability of its employees. This gain to the firm must be set against the labour cost. The profit-maximising wage—at which the marginal revenue from a wage increase is equated with the associated marginal cost—may be compatible with involuntary unemployment. The unemployed may be willing to work for less pay than the incumbents, but firms who hired them on those terms would find that their marginal revenue would decline more than their marginal cost.

5. Insider-outsider theories

Whereas the efficiency wage theories place labour market power (regarding wage and employment decisions) entirely in the hands of the firms, the insider-outsider theories assume that it lies at least partially with the workers.

The latter theories' point of departure is the observation that, in general, a firm finds it costly to exchange its current, full-fledged employees (the 'insiders') for workers outside the firm (the 'outsiders'). The process of turning outsiders into insiders takes time. The length of this 'initiation period' depends on technological, administrative and legal considerations (e.g. how long it takes to train, to acquire specific legal job rights, to renegotiate the wage contract). Workers going through this process are 'entrants' to the firm.

The turnover cost generates economic rent, in that the firm would be willing to pay something to avoid a given level of turnover. The insiders are assumed to have enough bargaining power to capture some of this rent when they make their wage demands. In particular, they raise their wage above the entrant wage, but not by more than the relevant turnover cost. By implication, workers must be able to renegotiate their wage contracts periodically, for otherwise the insiders could be prevented from gaining the wage increases above.

Furthermore, entrants are in an analogous position vis-a-vis the outsiders, and thus the entrant wage is greater than the reservation wage. There are several reasons for this:

— entrants may have some rent to exploit (e.g. because the firm may already have expended advertising, screening and some training costs on them and may incur even further costs to fire them);
— entrants may have limited access to credit (e.g. for efficiency interest reasons). This could be important when insider wage is high, so that entrants' reservation wage (at which they would be indifferent between employment and unemployment over their lifetimes) turns negative;
— firms may have an incentive to set the entrant wage above their reservation wage due to efficiency wage considerations.

The upshot of this story is that the firms are quite unable to capture all the rent inherent in their employment activities. They cannot keep all of it from the insiders and, in addition, they cannot compensate for this loss by taking all the rent from the entrants.

Under these circumstances, the economy may get stuck in an equilibrium with involuntary unemployment. Here, the relative bargaining power of the firms and workers gives rise to a wage structure in which the insider wage exceeds the entrant wage which, in turn, exceeds the reservation wage. Consequently, workers prefer being employed to being unemployed. Yet there may nevertheless be unemployed workers whom no firm has an incentive to hire. The reason is that the wage differentials between insiders, entrants and outsiders do not exceed the associated turnover costs for the firm.

Clearly, the practical importance of this phenomenon hinges on the magnitude of these turnover costs. Different insider-outsider theories have focused on different forms of these costs.

First, in Lindbeck and Snower, 1984a, b and Solow, 1985, there are straightforward hiring, firing, and training costs. The hiring costs cover the expense of advertising, screening, and negotiating conditions of employment. The firing costs include severance pay and costs of negotiation, litigation and implementation of legally mandated firing procedures. The training costs cover not only the expense of running explicit training programmes, but also the diverse forms of on-the-job training. (Many of these costs—and particularly the ones for training—accrue gradually over a worker's tenure with a firm and, through the exercise of market power in an insider-outsider context, may generate wage scales within the firm.) Note that the division of these costs between the firm and its workers cannot be specified a *priori*; in the absence of the insider-outsider consideration, the firm might be able

to shift all these costs to its workers simply by reducing its wage offers. Yet if the workers have some market power, they can prevent this from happening—and this is the situation with which the insider-outsider theories are concerned.

Second, in Shaked and Sutton, 1984, the firm is assumed to conduct wage negotiations with the insiders before it can turn to the outsiders. Furthermore, the bargaining process takes time and the negotiators have positive rates of time discount. Here the firm's cost of exchanging insiders for outsiders lies in the value of time which negotiating this exchange entails.

Third, in Lindbeck and Snower, 1984c, firms have imperfect information about individual employees' effort and current effort is stochastically related to future productivity. In response, each firm offers a remuneration package containing (a) a 'time-rate' component, (b) a 'piece-rate' component (in which current productivity, and thereby past effort, is rewarded), and (c) a cut-off productivity (below which the worker is not retained). The higher the cut-off productivity (other things being equal), the greater the rate of labour turnover and thus the smaller the expected future reward for current effort. Consequently (if the substitution effect of turnover on effort exceeds the associated income effect), workers' effort declines.

The same point can be made more generally. Firms with high rates of labour turnover usually offer low job security, little opportunity for advancement, and little incentive for workers to build reputations. As a result, productivity may be low—and this is another cost of labour turnover.

Finally, Lindbeck and Snower, 1985 recognise that employees in a firm are generally able to cooperate with and harass one another and consequently their productivities and disutilities of work become interdependent. When insiders feel that their positions are threatened, they may refuse to cooperate with entrants, and the resulting productivity differential constitutes the cost of an insider-outsider exchange. Besides, the insiders may protect themselves by harassing would-be entrants. As a result, outsiders may be unwilling to replace insiders, even though the two groups of workers may have the same job characteristics.

Due to these various costs, insiders have an inherent advantage over outsiders, one they can put to use in the wage negotiations with their employers. In this context, unions have an important role to play. Through them, workers may be able to amplify the costs of hiring, firing, and training (e.g. by imposing expensive hiring and firing procedures, insisting on lengthy training periods, raising severance pay) and to expand workers' possibilities of cooperation and harassment. In addition, unions expand their members' bargaining power and create further costs by means of two potent tools: the strike and the work-to-rule (see Lindbeck and Snower, 1984a,b).

In this manner, the insider-outsider theories provide a solution to a difficulty inherent in the 'union activity approach' (above) to unemployment: why don't firms ignore unions and deal with non-unionised workers instead? Yet the proposed answer has novel implications for the determination of wages and employment. No longer do these depend solely on fulfillment of union objectives subject to a labour-demand constraint or a minimum profit constraint, but the costs which the unions can credibly impose on firms become significant as well.

Out of the above approaches to involuntary unemployment, the 'adjustment cost approach' remains rather unexplored and the 'union activity approach' is open to some questions which are partially answered by the insider-outsider theories. So they will not be examined explicitly in the next section. Rather, the 'efficiency wage theories', the 'increasing returns to scale approach', and the 'insider-outside theories' (along with the 'implicit contract theories' in the market-clearing tradition) are the salient explanations of involuntary unemployment which we now attempt to evaluate.

IV. Evaluating explanations of involuntary unemployment

We conduct our evaluation in three straightforward steps. First, we try to specify what, exactly, economists mean by the term 'involuntary unemployment'. Second, we propose several criteria which explanations of involuntary unemployment ought to fulfil. And finally we assess the contenders above with regard to these criteria.

1. What is 'involuntary unemployment'?

It is not an easy matter to nail down what politicians and journalists (not to speak of the man in the street) mean by this term. But perhaps a reasonable common understanding of it may be encapsulated as follows. It exists whenever workers are willing to work at less than the prevailing wages for jobs which they could usefully fulfill, but are unable to find such jobs. In the conventional wisdom, involuntary unemployment is fre-

quently associated with a perception of social injustice; namely, that among people of comparable labour endowments, some are not able to gain employment on the terms offered to others.

Yet this broad framework leaves much room for manoeuvre. For our purposes here, it is convenient to specify two independent definitions, each compatible with the popular conception above:

1. **Type 1** involuntary unemployment (U1): at the prevailing real wages, workers unsuccessfully seek jobs for which they are just as qualified as the current job holders.
2. **Type 2** involuntary unemployment (U2): workers unsuccessfully seek work at real wages which fall short of their potential contribution to society (given the appropriate, feasible government intervention).

The 'involuntariness' of unemployment is a private phenomenon in the first definition and a social phenomenon in the second. Under U1, the unemployed workers would prefer being employed at jobs with the prevailing characteristics to being unemployed and, if they were able to acquire such jobs, they would be equally productive (on average) as the current incumbents. Under U2, unemployment is socially inefficient in the sense that, if the unemployed workers were employed, some agents in the economy would be better off without any others becoming worse off. Furthermore, it is feasible for the government to intervene in the market mechanism so as to bring about this improvement.

Needless to say, the two definitions are logically independent in the sense that it is possible for a worker to be unemployed under one definition but not the other. As indicated below, the definitions still leave a lot of room for semantic dispute, but they will nevertheless be useful in assessing our various contending theories.

Neither of them, however, coincides with the definitions underlying the unemployment statistics of capitalist countries. Broadly speaking, people enter these statistics whenever they (a) are of 'working age', (b) are out of work and (c) submit acceptable evidence of having looked for a job in the recent past. What constitutes acceptable evidence varies from country to country (e.g. it may involve collecting unemployment benefits or merely assuring a government official that work has been sought). Of course, it is possible for a person to be included in such unemployment statistics, but nevertheless to be less qualified than the incumbents with whom he competes (thereby falling outside U1) and to demand a real wage in excess of his potential contribution to society (thereby falling outside U2). Conversely, it is also possible for someone to be U1 or U2 unemployed but not to enter the official unemployment statistics.

2. Criteria for evaluating rival explanations

In the light of these definitions and of the commonly recognised symptoms of involuntary unemployment, we now pose a number of questions which a macroeconomic theory should be able to answer in order to qualify as a successful explanation of why workers can be involuntarily unemployed in market economies:

(a) Why are involuntarily unemployed workers unwilling or unable to underbid their employed counterparts?

Since involuntarily unemployed workers would prefer holding jobs (with the prevailing job characteristics) to being unemployed, why do they not offer to work for less than the incumbents' wages? If they do so, why do firms not accept these lower wage offers?

In the market-clearing macroeconomic theories of the Old and New Classical schools, the workers above have an incentive to underbid and the firms have an incentive to accept the underbidding and thus involuntary unemployment cannot persist. Either or both of these incentives must be absent if the problem is not to be solved automatically through the process of voluntary exchange in market economies.

At first glance, it may appear that whenever workers seek jobs with the prevailing job characteristics, they must be willing to underbid. After all, they seek jobs because they strictly prefer employment at the going wages to unemployment, and thus they must also prefer employment at something less than the going wages to unemployment. Yet this does not necessarily follow. If the process of underbidding itself has an adverse influence on the prevailing job characteristics (as in the harassment version of the insider-outsider theories), then the unemployed may not be willing to underbid even though they would like to trade places with the incumbents.

Also, it may appear that whenever firms seek to maximise profits, they must be willing to accept the lowest wage bids. Yet if a drop in wages reduces not only labour costs, but productivity as well (as in the efficiency wage theories), then firms may lack this incentive.

(b) Why do employed workers accept being laid off rather than take reductions in their wages?

In practice, layoffs and hiring of previously unemployed workers play a much larger role in accounting for fluctuations in aggregate unemployment than do retirements and hiring of new entrants into the labour force. Consequently, a successful explanation of involuntary unemployment should tell us why lay-offs are not avoided through wage cuts.

In many respects, the laid-off workers are in an analogous position to the unemployed. For example, in macro models where markets clear at all times, both types of workers have an incentive to underbid the incumbents and the firms have an incentive to take advantage of their wage bids. Yet insofar as the laid-off workers have more job training than their unemployed counterparts, on average, they have an advantage in competing for jobs and thus their underbidding need not proceed as far.

On the other hand, as the implicit contract theories assume, the laid-off workers may be temporarily immobile among firms (or, at least, less mobile than the longer-term unemployed) and thus they may be unable to underbid for a limited span of time. Moreover, a worker who loses his job may find it impossible to regain it by accepting a wage cut, because the act of underbidding would induce his colleagues to harass him or to withdraw their cooperation in production (as in some insider-outsider theories). Finally, even if the laid-off workers do underbid, firms may be unresponsive for efficiency-wage reasons.

(c) Why are firms unwilling or unable to capture all the economic rent associated with employment activities (e.g. through 'long-term' wage contracts, or 'entry fees' or 'exit fees')?

This is the counterpart to questions (a) and (b). While (a) and (b) focus on the wage offers for workers, this question is concerned with those of firms. By definition, if firms would capture all rent from employment, then workers would be left indifferent between work and leisure and thus involuntary unemployment would disappear. The firms may be able to do so in various ways, of which the following are especially prominent:

'Long-term' wage contracts, whereby an entrant agrees to a particular wage trajectory covering his entire period of employment at his firm. The complete wage package offered over the whole employment period could be such that each entrant is just as well off with work (over that period) as he would be without it. (Since only the present value of wages is relevant here, it remains for the firm to decide how to divide the complete wage package among the various pay days. The firm could use this degree of freedom to influence the workers' choice concerning the length of their employment period, as in Lazear, 1981.)

'**Entry fees,**' whereby an entrant pays his firm a particular lump sum for the privilege of receiving employment. In practice, these fees need not be explicit in the wage contract; instead, they could take the form of lower wages to entrants than to senior employees. Regardless of what wages the senior employees receive, there is always an entrant wage which is sufficiently low so that the entrant is indifferent between work and leisure over the span of the prospective employment period. In that event, the entry fee is sufficiently large so that the firm captures all the rent from employing the workers.

'**Exit fees**', whereby an employee pays his firm a particular lump sum if he leaves the firm prior to retirement (for reasons other than physical incapacity). For example, entrants could post bonds which would be forfeited when they quit their firms or were fired 'with cause'. In this way the firm can extract rent from the 'quitters', but not the 'stayers', in its workforce. If no worker knows in advance whether he would be a quitter or a stayer at a particular firm (e.g. because it is impossible for him to predict his future personal circumstances), then a rise in the firm's exit fee reduces the expected reward for work relative to that of leisure. Under these conditions, it is possible for the firm to set the exit fee sufficiently high so that an entrant becomes indifferent between the prospects of employment and unemployment.

In the presence of involuntary unemployment, there appears to be a strong case for expecting firms to capture all the available rent. The reason is that the involuntarily unemployed workers often have no bargaining power (i.e. they frequently are perfect competitors for the available jobs). Thereby the firms may gain the opportunity to offer an entrant terms which leave him no better off than if he were unemployed.

(d) Why are the involuntarily unemployed workers unwilling or unable to employ themselves by starting new firms?

If workers were always able to achieve self-employment, then there could be no involuntary unemployment. Whoever could not work for someone else, would work for himself.

In principle, it is probably always possible to find *some* form of self-employment, e.g. most people have the option of going fishing, picking berries, or selling flowers from an improvised stand off the road. Yet the available forms of self-employment may not match the unemployed workers' abilities. A nuclear physicist

who, after having been laid off, prefers leisure to selling flowers, would not be classified as voluntarily unemployed.

The question remains, however, why workers are generally unable to start new firms with jobs similar to those they are seeking.

(e) Why do the involuntarily unemployed workers not form coalitions with firms and their employees in order to exploit all the relevant potential gains from trade?

Involuntary unemployment U2 is an inefficient state. Eliminating it could make some people better off without making others worse off. In effect, the involuntarily unemployed workers could 'bribe' the firms and the employed workers to create jobs and still be left better off than they were originally.

The question is why the inhabitants of market economies do not take advantage of such opportunities without government intervention. Wherein does the externality lie which makes the social benefits from coalition-formation greater than the private benefits?

On the other hand, involuntary unemployment of type U1 may be efficient, so that there are no potential gains from trade to be exploited. For example, the process of coalition-formation may involve costs (e.g. those associated with the dissemination and acquisition of information) or there may be adjustment costs providing jobs for the involuntarily unemployed. In that event, agents in market economies have no incentives to form coalitions that eliminate involuntary unemployment, nor is it socially desirable to do so.

(f) Why are the involuntarily unemployed workers unwilling or unable to engage in job sharing with their employed counterparts (e.g. through job rotation or part-time work arrangements)?

Although workers are sometimes put on short work weeks when business conditions turn adverse, lay-offs and dismissals are nevertheless common. Besides, firms rarely rotate individuals between employment and unemployment.

If workers are risk averse, they would generally prefer job sharing (and thereby receive a comparatively steady stream of income) to the prospect of being employed at some times and unemployed at others. Why do firms not satisfy this preference?

3. Evaluting the contenders

Let us now return to the contending theories of unemployment which we chose above and examine what form of involuntary unemployment they explain and how they tackle these questions.

(i) Implicit contract theories

As noted, these theories assume that firms and workers commit themselves in advance to wage-employment contracts which are contingent on some unpredictable future events. Whether such contracts can generate unemployment—voluntary or involuntary—depends on who can observe and verify these events after they have occurred.

If the events are observable and verifiable by both the firms and their workers (i.e. in the case of 'symmetric information'), then the implicit contracts create no unemployment (see, for example, Akerlof and Miyazaki, 1980 and Azariadis and Stiglitz, 1983). Perhaps the simplest way of understanding this is to consider the production and insurance components of the contracts separately. In particular, suppose that a firm first offers its workers a wage-employment contract which brings its demand for labour into equality with its available labour supply. Then it offers another contract which subsidises its workers' incomes under adverse conditions and correspondingly reduces these incomes under favourable conditions. When the two contracts are offered in conjunction, employment remains at its market-clearing levels, while wages fluctuate less than their market-clearing counterparts. In other words, unemployment is completely absent.

Now suppose that the government provides unemployment insurance which is not 'actuarily fair' in the sense that there would be excess demand for insurance under symmetric information and perfect competition. Suppose furthermore that this unemployment insurance is not wholly subsidised by the firms whose workers benefit from it. Under these conditions, the firms and their workers will take advantage of the government hand-outs by agreeing to lay offs when worker productivities are sufficiently low. Here the unemployment is 'ex ante voluntary' (since parties agreed to the contract before the state of productivity was known) and 'ex post involuntary' (in the U1, but not the U2, sense).

Yet the questions (a)–(f) are handled either through ad hoc assumptions or not at all. Recall the workers are simply assumed to be (a) unable to renegotiate their wage-employment contracts once they have been made, and (b) immobile among firms throughout the contract period. It follows immediately that the laid-off workers cannot engage in underbidding. This is an implication from rather arbitrary premises. For the same reason, the firms are unable to capture economic rent which emerges after the signing of contracts. No rationale is given for the two assumptions above, and thus there is no answer to the question why efficient coalitions cannot be formed subsequent to the contracts. The questions concerning self-employment and job-sharing remain unanswered as well.

Now suppose that the 'good' and 'bad' states of productivity (on which the implicit contract is contingent) are observable to the firm, but not to its workers. Since workers are risk averse and have limited access to capital markets, they desire stable incomes, while their labour input varies in tandem with the good and bad states. Under these circumstances, the firm has no incentive to tell the truth about the states: by announcing a good state to its workers, it can get more labour for the same wage bill. In other words, the implicit contract is not 'incentive compatible'. In order to rectify this difficulty, the contract must be changed: whenever the firm announces a good state, it must employ more labour than previously (so that in the good state the revenue from the extra output exceeds the cost of the extra labour, but in the bad state the reverse is the case).

Although this new type of contract does restore incentive compatibility, it has two unattractive features for our purposes. First, it rationalises over-employment in good states rather than unemployment in bad states and, second, it implies that workers are better off in the bad states than in the good.

If unemployment in the bad states is to be explained, then it becomes necessary to assume that the firms are more risk averse than their workers. In this case, the workers have an incentive to provide insurance to the firms.

(ii) The increasing returns approach

This approach is designed specifically to address the self-employment question (d). When there are increasing returns to labour and imperfect competition, the involuntarily unemployed workers may be unable to start new firms with jobs that are substitutes for the ones they are vainly seeking. The reason is:

— Due to increasing returns to labour, a worker operating in isolation is less productive than a group of workers operating in conjunction. Thus, a single person firm would make losses when multi-person firms are drawing even. The involuntarily unemployed are unable to find a sufficient number of comrades to operate a firm at the requisite scale. If this were the only problem that the economy faced, then the level of unemployment would always be less than the workforce of a firm of minimum efficient size. (Of course, imperfect information about where the unemployed are to be found and geographic and occupational immobilities could allow unemployment to rise above the level).

— Due to imperfect competition among sellers of goods, creation of a new firm in equilibrium involves glutting the market. If aggregate demand is given, the consumers' purchasing power is now spread more thinly among firms. If profits were initially at their equilibrium level of zero, they now turn negative. Thus, even if the unemployed are sufficiently numerous to start a firm as large as the existing ones, imperfect competition may make it unprofitable for them to do so. The crucial assumption behind this argument is that aggregate demand is given: the creation of a new firm does not generate enough purchasing power to have a significant influence on the firm's product demand.

One suspects that, in practice, increasing returns to labour are not the main reason why unemployed workers find it difficult to start new firms. Increasing returns to capital are probably even more important. But the question why the unemployed workers cannot borrow enough funds to purchase this capital is not answered within Weitzman's analysis. Credit rationing can be explained through efficiency interest theories (e.g. Stiglitz and Weiss, 1981) or loan-repayment risk (Lindbeck, 1964) rather than increasing returns *per se*.

Moreover, the process of starting firms requires organisational abilities that some workers may possess to a lesser degree than others. Perhaps on this account self-employed business people tend to be less prone to unemployment than employees in subordinate positions within organisational hierarchies.

Be that as it may, the 'increasing returns approach' offers no answers to our acid-test questions other than (d). The involuntarily unemployed workers have every incentive to underbid their employed counterparts and so we must presume that they do so in Weitzman's world. In response, Weitzman argues that if the nominal wages faced by all firms were to fall proportionately simultaneously, then each firm would have an incentive to reduce its product price proportionately as well. Thus, real wages would remain unchanged and—in the absence of a real balance effect—aggregate demand and employment would stay constant as well.

Recall that the firms in this economy set their prices by a mark-up over wages, and this mark-up depends primarily on aggregate demand. Thus, in effect, the price-setting activity of firms determines the real wages. When firms realise that all nominal wages are falling in tandem (and that their market share would shrink if they held their prices constant) their profit-maximising objectives induce them to generate real wages which are compatible with involuntary unemployment. Even so, there remains no economic rent for the firms to exploit.

The difficulty with this analysis is apparent. In practice, we do not in fact observe massive bouts of deflation whenever there is involuntary unemployment. Underbidding simply does not occur and this phenomenon deserves explanation.

Moreover, the analysis provides no reason why efficient coalitions are not formed or why there is no job sharing.

(iii) Efficiency wage theories

In a sense, these theories may be seen as complements to the increasing returns approach: they are logically compatible with it, but the questions they address are different. The efficiency wage theories are not concerned with question (d) (why there are inadequate opportunities for self-employment). Nor do they have much to say regarding question (e) (about coalition formation).

Although question (f) (about job sharing) has not received general scrutiny either, the efficiency wage theories are not incapable of addressing it. Once it is recognised that hours of work per person can play a different role in the firm's productive process from numbers of people employed, then the efficiency wage is set so that the marginal cost of a wage increase is equal to the associated marginal revenue per person and per hour. The firm chooses the size of the workforce and the length of the working day on profit-maximising grounds. Part-time work will not be offered whenever it significantly discourages high-productivity job applicants, reduces effort in current employees, or prevents the exploitation of time-scale economies. Job rotation can be analysed in the same way. However, as noted, this matter has not as yet been subject to detailed investigation and thus it is impossible to tell whether these factors can account for the limited amount of part-time work and job rotation that we find in practice.

Be that as it may, the efficiency wage theories are designed primarily to provide a response to questions (a)–(c). Workers who are involuntarily unemployed or laid-off may be willing to underbid their employed counterparts, but firms have no incentive to accept their bids, since doing so would reduce the average profitability of the workforce by more than the wage costs. The same principle explains why firms do not capture all the rent from their employees. The firms could do so if they wished, but it would not be profitable.

The weakness of this response to questions (a)–(c) is that it rests on a tenuous implicit assumption, namely, that all employees with a firm—junior and senior ones—receive the same remuneration. Thus, a firm that accepts underbidding offers a lower wage not only to its entrants, but to its existing employees as well. In response, incumbents may quit (increasing the firm's turnover costs and—if the incumbents are of above average ability—reducing the overall productivity of the workforce) or they may shirk. Yet the assumption of uniform remuneration completely side-steps the question of why firms do not capture rent through entry and exit fees or through long-term contracts with variable wages.

The turnover and shirking models are open to this criticism. By imposing entry fees, the firms can reduce not only their rates of turnover and shirking, but also their labour costs. Thus, they have an incentive to raise these fees up to the point at which the *unemployed* workers are indifferent between employment and unemployment. Consequently, involuntary unemployment disappears. Long-term wage contracts function in the same way (e.g. Lazear, 1981).

Exit fees can also reduce turnover and shirking as well as the quitters' labour costs. The firms could raise them until the *incumbents* are indifferent between employment and unemployment. Would this eliminate involuntary unemployment? Clearly, those unemployed workers who are at least as prone to quitting as the incumbents would no longer prefer work to leisure. The rest—the unemployed stayers—would be willing to underbid and, furthermore, the firms would have no incentive to accept these bids, since lower wages would not induce them to quit or shirk. (Here *low* wages act as a self-selection device for workers of high profitability.) The process of underbidding could continue until the involuntary unemployment disappears.

Of course, it may be impracticable to levy entry and exit fees because the workers who are credit-rationed may not be able to pay them. Another problem is that firms might take unfair advantage of them. In particular, firms may declare workers to be shirking (and dismiss them) in order to collect the associated fees. Since the workers cannot protect themselves against this moral-hazard problem, the dismissal decisions would cease to operate as punishment for shirking. The same sort of problem is present under long-term wage contracts. (If effort were observable by both the workers

and their firms, then the workers could protect themselves; but in that case, wages could be made contingent on effort and then there would be no need to set efficiency wages in the first place.) On the other hand, if may be true that some firms have reputations to defend and thus can be relied on not to pronounce anyone a shirker without cause.

The productivity-differential, search, and sociological models are less vulnerable to these pitfalls. Here any drop in labour remuneration from one firm relative to the other firms—regardless of whether it takes the form of entry or exit fees or long-term wage reductions—reduces the profitability of the firm's workforce, because (a) it reduces the firm's ability to attract high-productivity workers; (b) it increases the amount of on-the-job search done by the firm's current employees and thereby reduces their productivity; or (c) it leads workers to believe that they are being treated unfairly and thus reduces group work norms.

In short, the efficiency wage theories aim to show why neither the firms nor the unemployed or laid-off workers reduce their wage offers in the presence of involuntary unemployment. Some, but by no means all, of these theories fail to take account of intertemporal remuneration schemes which permit underbidding to occur without any associated fall in productivity or rise in labour costs.

(iv) Insider-outsider theories

Broadly speaking, although the insider-outsider theories address the same questions as the efficiency wage theories, the proposed answers are radically different, as are their policy implications. Whereas the efficiency wage theories ascribe the existence of involuntary unemployment to firms' profit-maximising wage decisions under asymmetric information, the insider-outsider theories ascribe it to the market power of the employees. The efficiency wage theories hang on the assumption that workers' profitability is imperfectly monitored by firms and thus the firms do not find it worthwhile to exploit all the rent from their employment activities. By contrast, the insider-outsider theories rest on the premise that it is costly to interchange employed and unemployed workers in the process of production and the current employees are able to exploit some of the resulting rent. In the former context, involuntary unemployment is reduced when firms gain more (verifiable) information about their employees' profitability; in the latter, the unemployment falls when employees lose market power (due to either a weakening of their bargaining positions or a fall in turnover costs).

The insider-outsider theories are not aimed at question (d) (about self-employment) or question (e) (about coalition formation). They do, however, offer an explanation for why job-sharing is not a pervasive response to involuntary unemployment (question (f)). Job sharing (in the form of part-time work for both the currently employed and unemployed, or job rotation among these workers) is favoured by *all* risk-averse workers only if they all face the same risk of being unemployed. Yet when there are costs of labour turnover for the employed workers to exploit, these workers find themselves with a much higher chance of retaining their jobs than the unemployed workers have in finding them. Under these circumstances, job sharing benefits the outsiders at the expense of the insiders; and the insiders use their market power to prevent this from happening. This is perhaps a more convincing answer to question (f) than one that rests on the productivity disincentives of job sharing.

One insider-outsider response to questions (a)–(c) runs as follows:

(i) underbidding is impossible because the reservation wages of the unemployed and laid-off workers do not fall below the employees' wages *minus* the appropriate turnover cost;

(ii) there is an unwillingness to underbid because the act of underbidding would raise the reservation wages of the unemployed and laid-off workers above the employees' wages; and

(iii) firms cannot capture all the rent from employment activities because the employees do not permit it.

Can firms use entry and exit fees or long-term wage contracts to eliminate the involuntary unemployment? There is a fundamental reason why this is impossible in an insider-outsider context: firms simply do not have sufficient market power to do so. As long as the employed workers get some of the rent associated with the turnover costs, firms will be unable to propose any remuneration package meagre enough to make people indifferent between work and leisure.

If we accept that employees capture some rent, then the interesting issue is not whether persistent involuntary unemployment is possible, but rather how much of this unemployment can plausibly be attributed to employees' market power. The answer depends not only on the employees' bargaining position, but also on the magnitude of the turnover costs (which determine the size of the pie to be divided between the firms and their workers).

Turning to particular insider-outsider models, the time cost of negotiating falls equally on insiders and entrants, as does the effort cost of labour turnover. On the other hand, the costs of hiring, firing, and training are generally much larger for insiders than for entrants. The same is true of the costs arising from cooperation and harassment activities. But if entrants have little rent to exploit vis-a-vis the outsiders, then firms might drive the entrants' wage down towards their reservation wage. The larger the entry fee (implicit in this activity), the smaller the problem of involuntary unemployment becomes.

However, the discrepancy between the turnover costs for insiders and entrants suggests another answer to questions (a)–(c):

The greater the insider wage (other things being equal), the smaller the entrants' reservation wage. Yet if this reservation wage is substantially negative, then entrants may find themselves unable to borrow the requisite amount to gain employment. In that event, firms are unable to extract all rent from the entrants, even though the entrants have little market power.

Here the borrowing constraints of unemployed and laid-off workers have an important role to play in explaining the magnitude of the involuntary unemployment problem.

V. Concluding remarks

From the various explanations of involuntary unemployment examined above, three appear to be particularly promising: the increasing returns approach, the efficiency wage theories, and the insider-outsider theories. The first is concerned primarily with why opportunities for self-employment are limited; the second and third show why the unemployed and laid-off workers do not underbid and why the firms do not capture all rent from their employment activities. In this sense, the two groups of explanations complement each other. However, they have yet to be brought together within a single, logically consistent framework of thought.

Although the efficiency wage and insider-outsider theories point to quite distinct sources of involuntary unemployment, they are not incompatible with one another. The insider-outsider theories may suggest why insiders receive significantly higher wages than entrants, while the efficiency wage theories may explain why entrants do not receive their reservation wages. Moreover, even when labour market power does not lie entirely with the firms, the efficiency wage theories do not become irrelevant. Rather, they can be used to determine the lower bounds for wages which are negotiated in accordance with the insider-outsider theories. In other words, the amount of involuntary unemployment may depend both on what firms are willing to give and what workers are able to get. To date, however, the interaction between the efficiency wage and insider-outsider theories remains unexplored.

References

Akerlof, G. A. (1970), 'The Market for "Lemons", Qualitative Uncertainty and the Market Mechanism', *Quarterly Journal of Economics*, **84**, 488–500.

—— (1982), 'Labour Contracts as Partial Gift Exchange', *Quarterly Journal of Economics*, **97**, 543–69.

—— and H. Miyazaki (1980), 'The Implicit Contract Theory of Unemployment Meets the Wage Bill Argument', *Review of Economic Studies*, **47**, 321–38.

Alchian, A. A. (1970), 'Information Costs, Pricing and Resource Unemployment', in E. Phelps *et al.*, 27–52.

—— and H. Demsetz (1972), 'Production, Information Costs and Economic Organisation', *American Economic Review*, **62**, 777–95.

Arrow, K. (1963), 'Uncertainty and the Welfare Economics of Medical Care', *American Economic Review*, **53**.

Azariadis, C. (1975), 'Implicit Contracts and Underemployment Equilibria', *Journal of Political Economy*, **83**, 1183–202.

—— and J. E. Stiglitz (1983), 'Implicit Contracts and Fixed Price Equilibria', *Quarterly Journal of Economics*, **98**, 1–22.

Bailey, M. N. (1974), 'Wages and Employment Under Certain Demands', *Review of Economic Studies*, **41**, 37–50.

Barro, R. J. (1972), 'A Theory of Monopolistic Price Adjustment', *Review of Economic Studies*, **39**(1), 17–26.

—— (1976), 'Rational Expectations and the Role of Monetary Policy', *Journal of Monetary Economics*, **2**(1), 1–32.

—— and H. I. Grossman (1976), *Money, Employment and Inflation*, Cambridge, Cambridge University Press.

Benassy, J. P. (1975), 'Neo-keynesian Disequilibrium in a Monetary Economy', *Review of Economic Studies*, **42**, 502–23.

Blinder, A. S. (1980), 'Inventories in the Keynesian Macro Model,' *Kyklos*.

—— (1981), 'Inventories and the Structure of Macro Models', *American Economic Review*, **71**(2), 11–16.

Brown, C., C. Gilroy and A. Kohen (1982), 'The Effect of the Minimum Wage on Employment and Unemployment', *Journal of Economic Literature*, **20**, 487–528.

Calvo, G. A., and S. Wellisz (1978), 'Supervision, Loss of Control, and the Optimum Size of the Firm', *Journal of Political Economy*, **86**, 943–52.

Carlton, D. W. (1978), 'Market Behaviour with Demand Uncertainty and Price Inflexibility', *American Economic Review*, **68**(4), 571–87

—— (1979), 'Contracts, Price Rigidity, and Market Equilibrium', *Journal of Political Economy*, **87**(5), 1034–62.

Chamberlin, E. H. (1933), *The Theory of Monopolistic Competition*, Harvard, Harvard University Press.

Corden, A. M. (1981), 'Taxation, Real Wage Rigidity and Employment', *Economic Journal*, **91**, 309–30.

Dreze, J. (1975), 'Existence of an Equilibrium under Price Rigidity and Quantity Rationing', *International Economic Review*, **16**, 301–20.

Ehrenberg, R. G. (1971), 'Heterogenous Labour, The Internal Labour Market, and the Dynamics of the Employment-Hours Decision', *Journal of Economic Theory*, 85–104.

Epstein, L. (1982), 'Competitive Dynamics in the Adjustment-Cost Model of the Firm', *Journal of Economic Theory*, 27.

Flanagan, R. J. (1973), 'The US Phillips Curve and International Unemployment Rate Differentials', *American Economic Review*, **63**, 114–31.

Friedman, M. (1968), 'The Role of Monetary Policy', *American Economic Review*, **58**, 1–17.

—— (1976), *Price Theory*, Chicago, Aldine.

Gordon, D. F. (1974), 'A Neoclassical Theory of Keynesian Unemployment', *Economic Inquiry*, **12**, 431–59.

Gordon, R. J. (1972), 'Wage-Price Controls and the Shifting Phillips Curve', *Brooking Papers on Economic Activity*, **2**, 385–421.

Gramlich, E. (1976), 'Impact of Minimum Wages on Other Wages, Employment and Family Incomes', *Brookings Papers on Economic Activity*, **2**, 409–51.

Gronau, R. (1971), 'Information and Frictional Unemployment', *American Economic Review*, **61**, 290–301.

Grossman, S., and O. Hart (1981), 'Implicit Contracts, Moral Hazard, and Unemployment', *American Economic Review*, **61**, 301–07.

—— and —— (1983), 'Implicit Contracts Under Asymmetric Information', *Quarterly Journal of Economics*, **98**, 1223–56.

Guash, J. L., and A. Weiss (1980), 'Wages as Sorting Mechanisms in Competitive Markets with Asymmetric Information: A Theory of Testing', *Review of Economic Studies*, **47**, 653–64.

Gylfason, T., and A. Lindbeck, 'Union Rivalry and Wages: An Oligopolistic Approach', *Economica*.

Hahn, F. H. (1977), 'Exercises in Conjectural Equilibria', *Scandinavian Journal of Economics*, **79**, 210–26.

Hahn, F. J. (1978), 'On Non-Walrasian Equilibria', *Review of Economic Studies*, **45**, 1–17.

Hall, R. E. (1975), 'The Rigidity of Wages and the Persistence of Unemployment', *Brookings Papers on Economic Activity*, **2**, 301–35.

Hart, O. D. (1982), 'A Model of Imperfect Competition with Keynesian Features', *Quarterly Journal of Economics*, **97**, 109–38.

—— (1983), 'Optimal Labour Contracts Under Asymmetric Information: An Introduction', *Review of Economic Studies*, **50**, 3–36.

Keynes, J. M. (1936), *The General Theory of Employment, Interest and Money*, London, Macmillan.

Layard, R., and S. Nickell (1985), 'The Causes of British Un-employment', *National Institute Economic Review*.

Lazear, E. (1981), 'Agency, Earnings Profiles, Productivity and Hours Restrictions', *American Economic Review*, 71, 606–20.

Lindbeck, A. (1963), *A Study in Monetary Analysis*, Stockholm, Almquist and Wiksell.

—— and D. J. Snower (1984a), 'Involuntary Unemployment as an Insider-Outsider Dilemma', Seminar Paper No. 282, Institute for International Economic Studies, University of Stockholm.

—— and —— (1984b), 'Strikes, Lock-Outs and Fiscal Policy', Seminar Paper No. 309, Institute for International Economic Studies, University of Stockholm.

—— and —— (1984c), 'Labour Turnover, Insider Morale and Involuntary Unemployment', Seminar Paper No. 310, Institute for International Economic Studies, University of Stockholm.

—— and —— (1985), 'Cooperation, Harassment, and Involuntary Unemployment', Seminar Paper No. 321, Institute for International Economic Studies, University of Stockholm.

Lovell, M. C. (1972), 'The Minimum Wage, Teenage Unemployment and the Business Cycle', *Western Economic Journal*, 10, 414–27.

Lucas, R. E. (1972), 'Expectations and the Neutrality of Money', *Journal of Economic Thoery* 4(2), 103–24.

—— (1973), 'Some International Evidence on Output-Inflation Trade-Offs', *American Economic Review*, 63, 326–34.

—— (1975), 'An Equilibrium Model of the Business Cycle', *Journal of Political Economy*, 83, 1113–44.

—— (1976), 'Econometric Policy Evaluation: A Critique', in *The Phillips Curve and Labour Markets*. ed. K. Brunner, supplement to the *Journal of Monetary Economics*, 1, 19–46.

—— and L. A. Rapping (1970), 'Real Wages, Employment and Inflation', in E. Phelps *et al.*, 257–305.

Malcomson, J. M., 'Unemployment and the Efficiency Wage Hypothesis', *Economic Journal*, 91, 848–66.

Malinvaud, E., *The Theory of Unemployment Reconsidered*, Oxford, Basil Blackwell.

Maurice, S. C. (1974), 'Monopsony and the Effects of an Externally Imposed Minimum Wage', *Southern Economics Journal*, 41, 283–7.

McCall, J. J. (1970), 'Economics of Information and Job Search', *Quarterly Journal of Economics*, 84, 113–26.

McCallum, B. T. (1980), 'Rational Expectations amd Macroeconomic Stabilisation Policy: An Overview', *Journal of Money, Credit and Banking*, 12(4), 716–46.

McDonald, I. M., and R. M. Solow, (1981), 'Wage Bargaining and Employment', *American Economic Review*, 71, 896–908.

Mincer, J. (1976), 'Unemployment Effects of Minimum Wages', *Journal of Political Economy*, 84, 87–105.

Moore, T. G. (1971), 'The Effect of Minimum Wages on Teenage Unemployment Rates', *Journal of Political Economy*, 79, 897–903.

Mortensen, D. T. (1970), 'Job Search, the Duration of Un-employment and the Phillips Curve', *American Economic Review*, 60, 847–61.

—— (1973), 'Generalised Costs of Adjustment and Dynamic Factor Demand Theory', *Econometrica*, 41.

Muellbauer, J., and R. Portes (1978), 'Macroeconomic Models with Quantity Rationing', *Economic Journal*, 88, 788–821.

Muth, J. F. (1961), 'Rational Expectations and the Theory of Price Movements', *Econometrica*, 29, 315–35.

Nalebuff, B. J. and J. E. Stiglitz, (1985), 'Quality and Prices', *Quarterly Journal of Economics*.

Nickell, S. (1978), 'Fixed Costs, Employment and Labour Demand over the Cycle', *Economica*.

—— (1984), 'Dynamic Models of Labour Demand', Centre for Labour Economics, Discussion Paper No. 197, London School of Economics.

Okun, A. M. (1975), 'Inflation: Its Mechanics and Welfare Cost', *Brookings Papers on Economic Activity*, 2, 351–401.

Oswald, A. J. (1982), 'The Microeconomic Theory of the Trade Union', *Economic Journal*, 92, 576–95.

Parsons, D. O. (1973), 'Quit Rates Over Time: A Search and Information Approach', *American Economic Review*, 63, 390–401.

Phelps, E. (1967), 'Phillips Curves, Expectations of Inflation, and Optimal Unemployment over Time', *Economica*, 34, 254–81.

—— (1970a), 'Money Wage Dynamics and Labour Market Equilibrium', in E. Phelps *et al.*, 124–66.

—— (1970b), 'Introduction: The New Microeconomics in Employment and Inflation Theory', in Phelps *et al.*

—— *et al.* (1979), *Microeconomic Foundation of Employment and Inflation Theory*, New York, Norton.

Ragan, J. F. (1977), 'Minimum Wages and the Youth Labour Market', *Review of Economics and Statistics*, 59, 129–36.

Salop, S. C. (1979), 'A Model of the Natural Rate of Unemployment', *American Economic Review*, 69(1), 117–25.

Sargent, T. J. (1976), 'A Classical Macroeconomic Model for the United States', *Journal of Political Economy*, 84, 207–37.

—— and N. Wallace (1975), 'Rational Expectations, the Optimal Monetary Instrument and the Optimal Money Supply Rule', *Journal of Political Economy*, 83(2), 241–54.

Shaked, A., and J. Sutton (1985), 'Involuntary Unemployment as a Perfect Equilibrium in a Bargaining Model', *Econometrica*, 52(6).

Shapiro, C., and J. E. Stiglitz (1984), 'Equilibrium Unemployment as a Worker Discipline Device', *American Economic Review*, 74(3), 433–44.

Shestinski, E., and Y. Weiss (1977), 'Inflation and Costs of Price Adjustment', *Review of Economic Studies*, 44, 287–303.

Shiller, R. J. (1978), 'Rational Expectations and the Dynamic Structure of Macroeconomic Models', *Journal of Monetary Economics*, 4, 1–44.

Simon, H. A. (1979), 'Rational Decision Making in Business Organisations', *American Economic Review*, 69, 493–513.

Siven, C. H. (1974), 'Consumption, Supply of Labour and

Search Activity in an Intertemporal Perspective', *Scandinavian Journal of Economics*, 7(1), 44–61.

Snower, D. J. (1983), 'Imperfect Competition, Underemployment and Crowding-Out,' *Oxford Economic Papers*, 35, 245–70.

—— (1983), 'Search, Flexible Wages and Involuntary Unemployment', Discussion Paper No. 132, Birkbeck College, University of London.

Solow, R. (1985), 'Insiders and Outsiders in Wage Determination', *Scandinavian Journal of Economics*, 87(2).

Stigler, G. (1946), 'The Economics of Minimum Wages Legislation', *American Economic Review*, 36, 358–65.

Stiglitz, J. E. (1974), 'Equilibrium Wage Distributions', IMMSS Technical Report No. 154, Stanford University.

—— (1976), 'The Efficiency Wage Hypothesis, Surplus Labour and the Distribution of Income in LDC's', *Oxford Economic Papers*, 28, 185–207.

—— and A. Weiss, (1981), 'Credit Rationing in Markets with Imperfect Information', *American Economic Review*, 71, 393–410.

Wachter, M. L. (1976), 'Some Problems in Wage Stabilisation', *American Economic Review*, 66, 65–71.

Weiss, A. (1980), 'Job Queues and Layoffs in Labour Markets with Flexible Wages', *Journal of Political Economy*, 88, 526–38.

Weiss, Y. (1972), 'On the Optimal Pattern of Labour Supply', *Economic Journal*, 82, 1293–1315.

Weitzman, M. L. (1982), 'Increasing Returns and the Foundations of Unemployment Theory', *Economic Journal*, 92, 787–804.

Welch, F. (1974), 'Minimum Wage Legislation in the United States', *Economic Inquiry*, 12(3), 285–318.

Williamson, O. E., M. L. Wachter and J. E. Harris (1975), 'Understanding the Employment Relation: The Analysis of Idiosyncratic Exchange', *Bell Journal of Economics*, 6, 250–78.

PART V

GROWTH AND PRODUCTIVITY

12

New approaches to economic growth

ANDREA BOLTHO

Magdalen College, Oxford

GERALD HOLTHAM

Institute for Public Policy Research

I. Introduction

Until recently, economics had little of interest to say about economic growth. Now this is changing.

So began an article devoted to the growth theme published not long ago in the London magazine *The Economist* (4 January 1992, p. 17). Several new approaches to the question of economic growth have recently emerged, in particular the expanding body of literature that goes under the name of 'endogenous growth' (see the survey by vander Ploeg and Tang in this volume).

This paper addresses four main questions:

(i) What are the major empirical issues that any theory of economic growth should try to tackle?

(ii) What interpretations have been given in the past on these issues?

(iii) What new and/or additional insights are provided by the theories of endogenous growth?

(iv) What policy implications arise from both the old and the new literature?

II. The issues

Economic growth as it is understood today, i.e. almost uninterrupted increases, decade after decade, in a country's total and *per capita* output, is a phenomenon that would seem to date from the industrial revolution.

First published in *Oxford Review of Economic Policy*, vol. 8, no. 4 (1992).

Tentative estimates of growth rates in the preceding millennium suggest that, in Europe at least, growth in this sense was virtually unknown (Maddison, 1982, p. 6). The single most important reason for a take-off at around the turn of the 18th century is both simple and widely accepted—the pace of technological progress accelerated and this allowed sustained growth in the productivities of growing inputs of capital and labour.

That simple explanation, combining innovation and accumulation, has remained with us. The most basic approach to economic growth in any macroeconomic textbook still stresses the joint importance of factor inputs and, especially, technology. The canonical model, as presented by Solow (1956), obtains the well-known result that, on conventional assumptions, an economy has a unique and stable growth path determined by the growth of the labour force and of technical progress, with the latter usually assumed to expand at a regular, if unobserved, rate.[1] This last assumption is not inevitable. Schumpeter, for instance, saw inventions, and their associated innovations, occurring in pulses which led to investment booms and creative 'gales of destruction'. But this integration of growth and trade-cycle theory has not generally been incorporated into mainstream analysis which has rather dealt with equilibrium steady states.

The mainstream model also underlay the 'growth accounting' literature (e.g. Denison, 1967), which

[1] The Harrod–Domar model, with its almost opposite result that 'an equilibrium growth rate will exist only by chance—if the natural rate is equal to the "warranted rate"' (King and Robson, 1992), was also popular, but only for a time.

attempted to quantify the role of various proximate influences on growth. The results of this literature are equally well-known—the growth of output could not nearly be accounted for by the growth of inputs. Hence the appearance of a substantial 'residual' which was attributed to technical progress (though efforts were made to decompose it into different elements, such as education, which had not been captured by the conventional measurement of factor inputs).

The major difficulty about this traditional approach, as underlined by the quotation from *The Economist* cited above, is that it provides no explanation for what causes technical progress, the most important determinant of growth. Moreover, its depiction of a stable steady state bears no resemblance to the real world. Actual profits, investment, and growth rates have exhibited long-term accelerations and decelerations, tending neither asymptotically to approach such a steady state nor even to progress smoothly along a trend provided by the technological *deus ex machina*. Nor has the world economy shown strong tendencies towards convergence in *per capita* incomes, contrary to what neoclassical growth and international trade theories would have led one to expect.

To be convincing, a theory of growth should be able to provide answers to several basic issues that arise from the economic history of the last century or so. Kaldor (1961) was probably the first to set out a list of empirical observations with which any theory had to be consistent. Among these were the purported long-run constancies of factor shares and of the capital–output ratio, neither of which seems to be borne out by the data. Factor shares have shown long-run trends in a number of countries (Kravis, 1959; Kuznets, 1959; Hill, 1979), and capital–output ratios have tended to rise in all the major economies over the last century, bar the United States (Maddison, 1991). Yet, some of Kaldor's facts are true. The following brief selection draws both from Kaldor and from other work in this area:

(i) Why have countries, or groups of countries, been able to grow for decades in succession with no apparent tendency to slow down, despite rising capital–labour ratios?

(ii) Why has convergence in *per capita* incomes across the world seemingly failed to materialize?

(iii) Why have countries, or groups of countries, generally exhibited medium- to long-term accelerations or decelerations in their growth?

First, comes the basic issue of why growth could be sustained for a century or more in the presence of rising capital–labour ratios which, according to the orthodox theory at least, should have induced falling

Table 12.1. Growth in *Per Capita* Incomes and in the Capital–Labour Ratio in Selected Countries, 1900–79 (Average Annual Percentage Changes)

	GDP *per capita*	Non-residential capital stock per employee
Germany	2.2	2.4
Italy[a]	2.5	2.8
Japan	3.1	2.6
United Kingdom	1.4	1.4
United States	1.8	1.5

Note: [a] 1900–78.
Sources: Maddison (1982, 1991).

marginal productivities of capital. Evidence on long-term growth for a sample of developed economies for which sufficient data are available is provided in Table 12.1. As was noted in a similar context: 'Output per worker and the capital–labour ratio move one-for-one with each other in the real world' (Baldwin, 1989, p. 256), contrary to a traditional theory that limits the latter's contribution to usually only 30 per cent and assigns the residual 'to conveniently unobservable technological progress' (ibid.). Dissatisfaction with this conclusion, which was basically seen as a 'confession of ignorance' (Arrow, 1962, p. 155), was a major stimulus to endogenous growth theory.

Data throwing light on the second question are presented in Table 12.2 which shows levels and growth rates of *per capita* incomes from the beginning of this century in a number of selected areas. It will be readily seen that evidence for convergence and divergence coexists in parallel. Thus, Japan and Western Europe were able to catch up on the higher income levels of the United States, slowly in the pre-Second World War period, but rapidly since 1950. The same seems to have occurred in Eastern Europe and the former Soviet Union, despite the doubtful quality of those countries' statistics.[2]

The picture for the developing world, however, is much less favourable. Before the Second World War, Asian standards of living hardly rose at all, while Latin America's process of convergence on the richer countries was minimal at best. Since the Second World War, Latin America actually regressed relative to the OECD area while Asia's convergence was far from uniform— though rapid in the East, it was slow or non-existent in

---

[2] Though the data on Eastern Europe and the Soviet Union shown in Tables 12.2 and 12.3 are Western estimates which avoid some of the known biases of the official data, the adjustments made may still be insufficient in the light of recent revelations about the extent of earlier statistical misreporting.

Table 12.2. Levels and Growth Rates of *Per Capita* Real Income (1980 Dollars and Average Annual Percentage Changes)

	Pre-Second World War		Post-Second World War	
	Level 1900	Growth 1900–38	Level 1950	Growth 1950–89
OECD Countries	1,950	1.3	3,930	3.6
United States	2,910	1.0	6,700	2.0
Europe[a]	1,830	1.2	3,090	3.2
Japan	680	1.9	1,120	6.0
Soviet Union	800	1.9	2,270	2.7
Eastern Europe	..	1.5[b]	..	3.2
Latin America[c]	640	1.5	1,380	2.5
Argentina	1,280	1.0	2,320	0.6
Brazil	440	1.7	1,070	3.0
Asia[d]	400	0.2	380	3.6
China	400	0.2	340	4.6
India	380	0	360	1.9
Africa (Sub-Saharan)	..	..	..	0.8

Notes: [a] Sample of 12 countries. [b] 1913–38 (or 1925–38 for Poland and Romania). [c] Sample of 6 countries, [d] Sample of 9 countries.
Sources: IMF, *World Economic Outlook,* May 1992; Maddison (1989); OECD, *National Accounts, 1960–1989*; World Bank, *World Development Report, 1991*, Oxford, Oxford University Press.

the Southern part of the Continent. Africa's plight is too well known to require emphasis.

A further important growth phenomenon has been the occasional presence of 'trend accelerations' or decelerations in selected countries or areas (Table 12.3). The single best-known example of such an upward spurt occurred during the so-called 'Golden Age' of the 1950s and 1960s in Japan and Europe (both East and West). In the two decades from the end of postwar reconstruction to 1973, growth rates nearly doubled in Europe and more than trebled in Japan, relative to their longer-term trends. An even more marked acceleration appears to have taken place in the Soviet Union after the Bolshevik revolution. Turning to developing countries, the East Asian NICs clearly provide a startling example of a rapid take-off in the early 1950s, as does China in the 1980s.

Conversely, a number of areas have also shown sharp growth decelerations that went well beyond periods of normal cyclical slowdown. The post-1973 near halving of growth rates in the OECD area (including, on this occasion, also the United States) is well known, as are the Eastern European and Latin American slowdowns of the 1980s.

There is the question, of course, whether such changes in trend are inherent in the growth process, in which case they should be encompassed by an adequate theory, or whether they are the result of specific, and in some sense random, events. A growth theory that tried to account for the bulk of observed fluctuations in growth rates would almost certainly be attempting too much. There must be a role for historical accidents. Nevertheless, the experience of varying growth, or even growth cycles, is so clearly widespread that a theory which predicts it has a strong appeal. Schumpeter's notion of pulses of innovation has already been mentioned. Similarly, in models like those of Arrow (1962), or in some of the more recent literature looked at below that stress the role of capital formation, growth can clearly fluctuate with the rate of investment, which is notoriously unstable in practice. This instability is, however, ignored in most models.

Table 12.3. Selected Episodes of Accelerations or Decelerations in GDP Growth (Average Annual Percentage Changes)

Accelerations			Decelerations		
Western Europe	2.5	1922–37	Western Europe	4.7	1953–73
	4.7	1953–73		2.3	1973–89
Eastern Europe	2.7	1925–37	Eastern Europe	4.8	1950–73
	4.8	1950–73		2.0	1973–89
Soviet Union	1.9	1870–1913	Soviet Union	5.0	1950–73
	6.8	1925–37		2.3	1973–89
Japan	2.8	1870–1937	Japan	9.1	1953–73
	9.1	1953–73		3.9	1973–89
NICs[a]	3.3	1900–38	Latin America	5.3	1953–79
	8.0	1953–89		1.6	1979–89
China	5.3	1956–79			
	8.9	1979–89			

Note: [a] South Korea and Taiwan only.
Sources: Bairoch (1976); IMF, *International Financial Statistics* (Yearbook); Maddison (1989, 1991).

III. Earlier explanations

The knowledge that traditional theory could not provide a convincing explanation for the 'convulsive structural, technological, and social changes' (Nordhaus and Tobin, 1972, p. 2) which have characterized the growth experience of the world economy over the last century or more was, of course, widespread well before the birth of endogenous growth models. Some economists explained this by recognizing that growth theory was equilibrium theory and much in the real world was a disequilibrium phenomenon: 'General economic growth as we have known it is not a balanced steady-state affair; ... the historical process of growth ... may best be viewed as part of a sequence of technologically induced traverses, disequilibrium transitions between successive growth paths' (Abramovitz and David, 1973, p. 429).

While older approaches were not always expressed in the formal language of more recent models, they did attempt to, and in a number of cases succeeded in, shedding light on some of the issues highlighted above. Indeed, a significant number of ideas used by endogenous growth models can be found in the earlier literature.

1. Long-term growth

The issue as to why growth has continued for well over a century in most industrialized countries at a rate that may even have accelerated slightly has, on the whole, received little attention. Applied economists have worried about growth-rate differences across countries and between time-periods, but have usually left alone the long-run sweep of history. In this area, the neoclassical paradigm, relying on exogenous technological progress, has ruled almost unopposed.

Unorthodox theories were few and far between and do not seem to have been tested empirically. A fairly complete survey is presented in Scott (1989) and work stressing 'learning by doing' (Arrow, 1962), or endogenizing human capital formation (Uzawa, 1965) and technical progress (Conlisk, 1969), is looked at in the article by van der Ploeg and Tang (this volume). Of these approaches, Arrow's contribution is one that comes near to the recent literature's preoccupations. In Arrow's 'endogenous theory of the changes in knowledge' (Arrow, 1962, p. 155), growth results from a learning process which is itself the product of experience. The latter, in turn, is a function of cumulative gross investment 'as each new machine ... put in use is capable of changing the environment in which produc-

tion takes place, so that learning is taking place with continually new stimuli' (ibid., p. 157).

The links between Arrow's model and those of more recent vintage are drawn further below. An element that seems missing from the formulation is the recognition that, in an uncertain world, optimal investment cannot be presupposed. In practice, it is likely that fixed investment will be encouraged by the experience of growth itself. A formulation that specified the latter feedback, while avoiding the *ad hoc* nature of simple accelerator models, would reinforce the endogeneity of the growth path.

2. Convergence

Research on 'why growth rates differ' has, on the other hand, a long history which predates and goes well beyond the growth-accounting exercises carried out within the confines of neoclassical theory (Denison, 1967). The idea that poorer countries should catch up on richer ones was advanced already in the 19th century to explain Continental Europe's convergence with Britain. In the 1960s one of its most persuasive advocates was Kindleberger (1967), with his well-known adaptation of the Marx–Lewis model of abundant labour supplies explaining the divergent growth experience in Western European countries. Relative backwardness and abundant labour supplies are similarly crucial in stimulating growth in the writings of Kaldor (1966) since they allow expansion in the all-important manufacturing sector and hence the reaping of static and dynamic scale economies. And scale economies also feature importantly in 'export-led' growth models, in which, for example, a favourable/unfavourable exchange rate at the outset confers lasting advantages/disadvantages as demand and supply interact in virtuous or vicious circles (Beckerman, 1962).

The accent that Kaldor put on 'Verdoorn's Law' and on economies of scale clearly foreshadows some of the new growth literature. So too do approaches that stress the technology gap between countries (Gomulka, 1971). In a variation on this theme, however, stress has also been put on research at the frontier and not merely on the catch-up of the backward economies (Fagerberg, 1987). In Schumpeterian fashion, growth differences are seen as: 'The combined result of two conflicting forces: innovation which tends to increase technological gaps, and imitation or diffusion which tends to reduce them' (ibid., p. 92).

A third set of explanations has stressed the role of policies and institutions. In this category one can find the modified physiocratic view that rapid expansion of the non-market (or public) sector retards growth because of the various financing burdens it imposes on the private economy (Bacon and Eltis, 1976). Also very influential has been the approach followed by Olson (1982), in which young societies grow more rapidly than mature ones, since they are not (yet) slowed down by the actions of 'distributional coalitions' (i.e. rent-seeking pressure groups such as oligopolies, farmers' associations, trade unions, etc.).

While the empirical testing of some of these approaches has been limited, the convergence hypothesis has been subject to numerous investigations (e.g. Gomulka, 1971; Cornwall, 1977; Baumol, 1986; Dowrick and Nguyen, 1989). The findings have suggested that convergence has occurred among industrialized countries, though possibly disturbed by the incidence of innovation in the more advanced economies acting in the opposite direction to imitation (Fagerberg, 1987). Convergence between developed and developing countries has, however, been largely absent. Indeed, some of the more recent empirical investigations suggest the presence of several 'growth clubs', e.g. rich, middle-income, and poor countries, within which convergence occurs, but between which it may not (Baumol, 1986; Dowrick and Gemmell, 1991; Chatterji, 1992).

An early explanation for widening gaps between North and South, that anticipates some aspects of the endogenous growth school, stressed that: 'Economic development is a process of circular and cumulative causation which tends to award its favours to those who are already well endowed and even to thwart the efforts of those who happen to live in regions that are lagging behind' (Prebisch, 1962, quoted in Meier, 1964, p. 345). The divergent experience of the developing world over the last decades, with some areas losing ground, but others growing rapidly, has made for a more differentiated set of explanations (some of which also foreshadow strands in the recent literature). Cumulative causation is still possible. Its reasons, however, are to be found not so much in the workings of the international trading system, as had been argued by, for instance, Nurkse (1961) or Griffin (1974), but in initial domestic conditions: 'A country's potential for rapid growth is strong not when it is backward without qualification, but rather when it is technologically backward but socially advanced' (Abramovitz, 1986, p. 388). And this 'social capability' is, in turn, importantly linked to education and hence human capital, as underlined by a recent empirical investigation that stresses the crucial role of schooling in explaining the divergent growth experience of a very large sample of countries (Barro, 1991).

3. Accelerations/decelerations

While relative backwardness and rising social capability may explain the presence or absence of convergence, they are unable to throw much light on why countries, or groups of countries, experience sudden medium-term accelerations (or slowdowns) in their growth rates. Writings in this area have drawn relatively little on formal theories of growth and have relied mainly on ad hoc explanations, tending to stress the role of economic policies or of specific disturbances to 'the institutional policy-mix . . . usually initiated by some sort of "system shock"' (Maddison, 1991, p. 85).[3]

The post-Second World War growth accelerations in Western Europe and Japan are more interesting than first appears. They did, of course, owe a good deal to the catch-up factors usually cited, but more must have been at work, given that significant gaps vis-à-vis United States productivity levels (as well as even more elastic labour supplies) had existed since the beginning of the century and had not generated similarly rapid growth before 1950.

Two possible explanations for this acceleration look at almost opposite factors. In one, the stress is on the return to unfettered market forces, resulting from the disappearance in many countries of the special interest groups of the pre-war era (Olson, 1982). In the other, the accent is on policy changes, some of which (e.g. freer international trade) made for a more liberal order (Crafts, 1992), but others (e.g. demand management) went in the direction of greater government interference in the economy (Boltho, 1982; Bombach, 1985). It is likely that the combination of greater competition and greater business confidence, instilled by the self-fulfilling belief that intervention in the economy could stabilize/promote growth, strongly contributed to the 'system shock' that raised 'animal spirits' and doubled investment ratios in the period. And shocks provide, of course, at least a proximate explanation for the slowdown that has occurred since the early 1970s.

'Animal spirits' play no role in any explanation of the Eastern European acceleration in the 1950s and 1960s (and of that of the Soviet Union already in the 1930s). Yet, the stress on economic policies remains. While the shortcomings of central planning in the

---

[3] An alternative explanation has stressed the apparent regularity of such changes in trends (Mandel, 1980; Freeman, 1983; Van Duijn, 1983). In this view, a variety of factors (of which the most plausible is that of discontinuous technical progress), has led to long-term waves, à la Kondratieff. Yet, neither the regularity nor the causal mechanisms of Kondratieff cycles seem very convincing (Chesnais, 1982; Rosenberg and Frischtak, 1983).

context of more mature societies are obvious, for less advanced countries it was an effective mechanism for mobilizing resources through forced savings and investment. This experience also provides some evidence in favour of the standard model's argument that increases in investment ratios are unable to permanently raise an economy's growth rate in the absence of technical progress. The huge investment efforts of the last few decades seem to have generated progressively smaller returns, since they occurred in a system that basically discouraged innovation. Yet, while illuminating the importance of technical progress, this explanation also shows that the latter is not fully exogenous but depends importantly on the legal and institutional framework.

Turning to developing countries, the idea that exogenous (or partly exogenous) events are crucial in explaining turning points in the growth of *per capita* incomes receives strong confirmation from the experience of forty-one countries since the mid-19th century: 'First, the turning point is almost always associated with some significant political event . . . Second, [it] is usually associated with a marked rise in exports' (Reynolds, 1983, p. 963). And the conclusion that policy changes can be crucial would seem broadly to fit the recent experience of East Asia. Despite a sharp world economic slowdown, both the NICs and China were able to maintain, or even raise, their growth tempos. In China's case this clearly reflected a gradual retreat from central planning. And in the two larger Asian NICs there was a move away from earlier policies of intervention which had contributed to rapid growth in the 1950s and 1960s (Wade, 1990). Finally, Latin America's deceleration in the 1980s owes much to the shock that came from the debt crisis, though inappropriate policies no doubt also played a role.

IV. The new literature

The theoretical literature on growth had little to say on these issues. Indeed, it had increasingly become entangled with disputes in capital theory over the legitimacy of aggregating inputs into 'factors' of production, or of aggregating production functions themselves. The theory had also become more complex with the development of multi-sectoral models or models with capital goods disaggregated either by sector or by 'vintages'. Nor could the limited data available discriminate among the many variations on the standard model. Diminishing returns in terms of general results appeared to have set in. And while practitioners, aware

of the models' limitations, searched for alternative explanations of the 'stylized facts', growth theory from the 1970s onwards went into a state of hibernation.

From this it emerged in the mid-1980s. Two elements seem to have been important in explaining the recent revival. One was the 're-discovery' of a well-known fact—convergence of *per capita* incomes across countries was either limited or absent in the world economy. Yet, if exogenous technical progress could be thought of as falling like manna from heaven, should not standards of living tend to converge, especially given the increasing mobility of capital internationally? As illustrated above, the absence of convergence had long been a datum for development economists and economic historians, but only recently have growth theorists become aware of it.

The second element was not so much a fact as dissatisfaction with existing models arising from a form of intellectual imperialism. If something called technical progress accounted for the steady-state growth rate, and for the bulk of observed growth according to the growth accountants, was it not regrettable that economists had nothing to say about what determined the rate of technical progress? Economic tools were being used to explain many decisions previously thought to be non-economic; there was an economics of crime and punishment, of marriage and parenthood; surely, the origins of the wealth of nations could not be left to be colonized by historians of science or social psychologists. Schumpeter's dictum that economic causes should be sought for economic phenomena is the motto of the endogenous growth theorists.

The latters' contribution is as yet at an early stage of development, particularly in dealing with the 'facts' that need explaining. So far the new writings have remained largely theoretical. Their main common element is to provide a broader view of what constitutes capital. This is at the heart of Scott's explanation of growth, an explanation that is not usually linked to the (largely American) endogenous growth literature, but that both predates it (in a series of earlier papers) and is virtually alone in providing empirical tests (Scott, 1989). And a broadening of the concept of capital is similarly what brings together the literature surveyed by van der Ploeg and Tang in this volume. They identify four different views which emphasize various aspects of the process: learning by doing, human capital, research and development (R & D), and public infrastructure, and also stress the potential for international spillovers in some of these areas.

As Artus (1992) has pointed out, these endogenous growth models can be further grouped into one of two broad types. Both imply that the aggregate production function exhibits increasing returns to scale. Increasing returns, although espoused by some earlier theorists, notably Kaldor, had been generally shunned as being incompatible with competitive markets and stable equilibrium.

Modern theory has escaped this dilemma by building some form of externality into the model. Individual enterprises face constant returns to scale at any time, but due to some positive production externality increasing returns are encountered at the aggregate level. In the first class of models, the elasticity of output with respect to aggregate capital is unity, implying increasing returns to capital and labour together. Though individual firms still face a conventional (constant returns) production function, their overall efficiency level is a function of the aggregate stock of capital in the economy. The principal interpretation (Romer, 1986), is that capital accumulation results in learning which cannot be internalized and emulation then raises efficiency in the economy as a whole.

This model is akin to Arrow's model of 'learning by doing' (Arrow, 1962). Arrow supposed that technical progress was a function of accumulated investment, but with an elasticity below one. The implication was that in the end growth depended on the expansion of the labour force. With a stationary population, ultimately decreasing returns to investment limited growth. The steady-state growth rate was the growth of the labour force times a factor which depended on the rate of learning by doing.

Romer broke the link between output and population growth by making the learning parameter equal to at least one. Growth can now proceed indefinitely with capital accumulation even in the presence of a stationary labour force. This type of model is simple and may well be indirectly empirically testable. A limitation is that it depends on a strong assumption about the value of the learning exponent—if this is one epsilon below one, growth falls back to the Arrow solution; if it is marginally above one, growth accelerates without bounds.

Though van der Ploeg and Tang point out that Romer has produced 'rudimentary evidence' that world growth has accelerated over the past two centuries, continuous acceleration is, of course, implausible. Yet, the two authors also point out that Hicks (1950) had dealt with this problem by positing some 'ceiling'. Locally explosive properties plus ceilings and floors lead to models with limit cycles, which have a long tradition in economics (e.g. Kondratieff waves). Indeed, on the basis of casual empiricism, cycles would seem more plausible than steady states. It is true that there is a difficulty with endogenously generated repeated cycles—surely forward-looking 'agents' would learn to anticipate them and thereby smooth them out? The force of this objection has, however, been weakened by the realization that non-linearities may lead to chaotic quasi-cycles that, though deterministic, are unpredictable. Moreover, the longer the cycle, the less likely is it to be rationally anticipated and therefore eliminated by finite-lived individuals. Perhaps, a learning exponent fractionally above one is not impossible after all.

The second class of endogenous growth models posits the existence of a specific growth factor which raises the total productivity of the other factors of production. Increasing the supply of the special growth factor diverts labour and/or capital from ordinary production. Crucially, however, not only do inputs into the growth factor enhance overall productivity, but they are themselves subject to a form of dynamic and external increasing returns to scale. Several of the specific models reviewed by van der Ploeg and Tang fall into this category.

The growth factor may consist of human capital (Lucas, 1988; Romer, 1989), or of a stock of knowledge from R & D (Romer 1990a; 1990b). In the latter case, R & D's contribution may take the form of creating new inputs (which improve productive efficiency), or new consumer goods (which raise consumer welfare). In these models, final production takes place in a perfectly competitive sector, while the R & D sector is monopolistically competitive. Firms and households maximize profits and utility in an inter-temporal framework, so that resource allocation across the two sectors is endogenized along with the growth rate. In a different model, the growth factor is represented by public infrastructure supplied by a government that raises distortionary taxes to finance it. The government, therefore, faces a trade-off and is assumed to find the optimal rate of taxation and public goods provision (Barro and Sala-i-Martin, 1990).

Most endogenous growth models have a common implication—owing to the presence of externalities, a Pareto optimum is not achieved. In the first class of models efficiency would be greater with more investment, but part of the benefit cannot be appropriated by the private investor who, therefore, ignores it in his calculations. In the second class of models, the producers of the growth factor (e.g. R & D) can charge for the contribution their output (e.g. a patent) makes, but they cannot appropriate the external benefit obtained from the patent by other growth-factor producers. For example, Romer (1990b) in effect supposes that a company engaged in R & D can patent the result of its research, but cannot prevent its own patent making it subsequently easier for other researchers to produce more patents.

Empirical implementation of these models depends on identifying a specific growth factor (e.g. public investment, human capital, or R & D), and finding measures or proxies for it. Unfortunately, the right data are not often available and certainly not in time-series form. Most empirical work has consisted of cross-country correlations between changes in labour or in total factor productivity and some particular growth factor. Such tests are evidently weak, if only because the proxies are generally inadequate—not all patents are of equal value, human capital is not well captured by the number of years spent in school, etc. A particular problem is that the tests do not encompass each other. What if more than one of the plausible growth factors is playing a role?

Results so far have thus been disappointing. In a survey of some of the empirical findings, Crafts (1992) reports that Romer himself has concluded that his 1986 model did not dominate the neoclassical one. Such evidence as we have suggests that Arrow's learning exponent is less than one. Capital's output elasticity certainly seems to exceed the share of profits in national income, but most attempts to estimate it find it substantially less than one (ibid.). More sophisticated models are much harder, perhaps fatally harder, to test. Moreover, no one has subjected to empirical tests the hypothesis that the process of generating the chosen growth factor is subject to increasing external returns to scale.

Scott's approach is similar, in that he also broadens the concept of capital, yet different in a number of other important aspects. In particular, he argues that historical observation and case studies show that the distinction between innovation and repetition is blurred. Hence, trying to find a subset of investment responsible for technical advance is a hopeless task.[4] What matters for growth is the 'learning externality' obtained from cumulative investment (defined as all past expenditures devoted to changing existing arrangements).

Scott argues that it is inappropriate to subtract scrapping from estimates of the capital stock when measuring the latter's contribution to production. With fixed costs sunk, capital is retired from production only when its marginal product is zero (or less than the wage of complementary labour). This economic obsolescence, not physical decay, is the reason for most of the scrapping that takes place. Such scrapping, therefore, subtracts nothing from potential output which is bound to be stably related to cumulated investment. Only when part of that past investment is wrongly excluded as scrapped, is it necessary to posit some process of technical progress independent of the successive acts of investment.

Some of Scott's insights are pursued by King and Robson (1992), who specify a non-linear 'technical progress function' that relates the rate of productivity growth to the rate of investment. This embodies the familiar proposition that capital accumulation generates spillovers due to 'learning by watching'.[5] The important difference in this model is that the process is non-linear. This results from the intuitively plausible (but empirically unsubstantiated) idea that the technical progress function has first increasing and then decreasing returns. Returns decline eventually because there is a limit to the rate at which ideas can be assimilated (and they do not survive to be exploited later). The S-shaped curve leads to a model with multiple equilibria, of which two are stable.

If an economy is in the vicinity of a stable equilibrium, it may be resistant to shocks. But if it has strayed or is at an unstable equilibrium, various shocks, including policy changes, can have powerful effects, moving the system away from a low growth to a high growth equilibrium or vice versa. Similarly, identical economies can be on high or low growth paths, entirely due to past history or accidents. These results are intuitively plausible and have two implications. First, they call into question simple cross-country correlations designed to test 'growth theories', particularly when the variables are averaged over short time periods. Second, they provide an explanation for how 'catch-up' can become 'overtaking'. Thus, countries such as Germany or Japan, following war-time destruction, may have moved on to a high investment path and consequently from a low to a high growth equilibrium.

Multiple equilibria also raise the possibility that even in a world with free trade, free capital movements, and international dissemination of knowledge, countries may cluster into 'growth clubs'. Chatterji (1992) examines this possibility and finds some evidence for the existence of such mutually exclusive groupings—one for the 'rich' and one for the 'poor' countries.

Overall, however, the literature on endogenous growth provides only few empirical answers to the

---

[4] Hence his lack of surprise at the difficulties that are faced in implementing 'growth factor' type models.

[5] What differentiates the model from Scott (or Romer) is that the spillover effects are immediate since output growth depends on the contemporaneous change in the capital stock rather than on cumulated past investment. In practice, little would seem to hinge on this distinction because authors of the latter type of models never specify any lags on the learning function nor do they suppose it to be stochastic. The individual firm's production function in the Romer (1986) model, for example, could be differentiated to yield a linear technical progress function.

questions posed in section II above. Differences in the strength of various growth factors can account for lack of convergence (a conclusion already anticipated in the earlier literature), and differences in past investment efforts can throw light on changes in growth trends. But the reasons for such differences are usually unexplored or left to unpredictable shocks. As for continuing long-term growth, this can occur, provided, however, that certain crucial parameters are equal to unity—something so far not confirmed by the empirical evidence.

More generally, many of the models have appealing but also arbitrary features. The King and Robson (1992) model is a good case in point. It illustrates a continuing feature of growth theory: moderate changes in assumptions can yield a model that generates very different results from the standard ones but which can also 'explain' some real world phenomena. Yet, the task of arbitrating between this model and others is almost impossible given the data sets available. Even if one model appears to have more empirical support than another, no claim for generality can be sustained. Perhaps another model would have fitted better for other times or places.

V. Policy implications

The conclusion just reached on the robustness of the existing models of endogenous growth clearly limits their usefulness for economic policy-making. In principle, of course, a number of policy implications can be drawn from them, in stark contrast to the earlier neoclassical view of the growth process which implied that policy had no long-run effects on growth. Thus, if specific growth factors can be shown to improve productivity, and given the implication in all these models of a divergence between private and social costs, a case can be made for subsidies, or other policy interventions, to raise investment, or R & D, or human capital (or, perhaps, all together).

Oddly, the authors of some of these models appear to be rather reluctant to draw the strong policy conclusions that apparently flow from their writings. One reason, no doubt, is that any government action carries dangers of its own. First, government borrowing or taxation may be required to finance new policies. Taxation is bound to be distortionary, reducing efficiency in the system. Hence, governments may face a trade-off between the beneficial effects of subsidies (or higher public investment) and the injurious effects of higher taxes. At least the new models allow many of

the live issues in public finance theory to be set in a growth context.

Second are the public-choice problems emphasized by the Buchanan school. The authors are well aware of the fact that 'the beneficial effects of government expenditures are conditional on a benevolent and efficient government' (Ehrlich, 1990, p. S8). Can the authorities be expected to make the correct choices and is this consonant with the interests of bureaucrats and politicians? And would not discretionary action lead to 'rent-seeking' behaviour? Scott, despite his stress on the growth-promoting role of capital formation, is particularly wary of the danger that pressure groups will hijack in their favour any policy that might help investment.

In addition, there is the fact that the authors of these models may be less than wholly confident that they are adequate representations of reality. After all, there are many candidates for a 'growth factor'—investment 'that changes the way of doing things' according to Scott; investment in know-how and education; investment in public infrastructure, etc. Should all of these receive preferential treatment or subsidy and, if so, how much? Given the potential costs to government action, some idea of the relative importance of the different growth factors and of their effects on growth is necessary in framing policy.

In this context, the failure of endogenous growth models unambiguously to dominate empirically their predecessors is disquieting. Moreover, tests that rigorously seek to distinguish among the different growth factors or to rank them in order of importance have yet to produce results that command general credence or assent (Levine and Renelt, 1992). The literature is largely at the stage of illustration—showing how various factors could have an influence on growth—rather than of demonstration—showing how some factors do have such an influence. Nor does the literature touch upon the institutional and legal framework which, according to Crafts (1992), is very important in explaining cross-country differences in growth rates. Governments thinking about reforms of the regulatory environment receive virtually no support from the formal theory.

The discussion of policies so far has mainly looked at the supply side and at measures designed to affect resource allocation. The development of models, such as King and Robson's, in which the equilibrium growth rate itself is path dependent, means, however, that a wider range of policies potentially affects growth. Macroeconomic instruments, for instance, by influencing the level of aggregate demand, could raise investment and thereby the growth rate. Such 'dashes for growth' had been advocated in the 1950s and 1960s, for example, in the context of various Kaldorean models of increasing returns to scale. The experience of high inflation in the 1970s has, however, discredited not only the idea of 'dashes for growth' but Keynesian demand management itself.

Nowadays economists shy away from the idea that demand changes can have any durable impact on growth. If only macroeconomic policy can avoid shocking the system, it is commonly argued, the economy would in due course settle at a 'natural' rate of unemployment. Yet, the experience of the last decade suggests that recession has had permanent effects on the rate of unemployment in many European countries. And the last five years or so have also seen a considerable investment cycle in Europe and Japan that seems to owe little to macroeconomic policy (though deregulation in financial markets has played a part).

It is plausible to think that if policy could prevent such swings it could raise the growth rate. Certainly, in models with non-linear technical progress functions, a steady rate of investment encourages learning by doing and diffusion of new ideas better than a more erratic investment path, even if on average the rate is the same. However, the best means of stabilizing demand remains a controversial issue. The endogenous growth literature, with its equilibrium models and absence of independent investment functions, has, like the neoclassical one, ignored problems of instability. Yet, maintaining an adequate level of demand and confidence so that productive investment occurs has, in practice, been a significant objective of policy in many countries.

More broadly, and surveying what governments the world over have done to promote growth, one is inevitably struck by the fact that so many of the measures taken have not only attempted to stabilize demand, but have, in fact, encouraged precisely those 'growth factors' that the new literature just reviewed has highlighted. Theory seems finally to be catching-up with practice. Education and expenditures on R & D have been subsidized almost universally. Public infrastructure has inevitably been an area in which governments have been very active. Most importantly, capital formation has generally received help and incentives throughout this century, in total disregard of neoclassical conclusions that suggested only very low returns to it. In the light of recent results showing strong positive correlations between output growth and various definitions of investment (Scott, 1989; De Long and Summers, 1991; Levine and Renelt, 1992), the policy-makers may not have been that wrong. While economists may not trust politicians, it would seem that politicians have not trusted economists either—and perhaps with at least as good a reason.

References

Abramovitz, M. (1986), 'Catching Up, Forging Ahead, and Falling Behind', *Journal of Economic History*, **46**(2), June, 385–406.

—— David, P. A. (1973), 'Reinterpreting Economic Growth: Parables and Realities', *American Economic Review*, **63**(2), May, 428–39.

Arrow, K. J. (1962), 'The Economic Implications of Learning by Doing', *Review of Economic Studies*, **29**, 155–73.

Artus, P. (1992), 'Endogenous Growth: Which are the Important Factors?', Caisse des dépôts et consignations, Document de travail, No. 25/T, September.

Bacon, R., and Eltis, W. (1976), *Britain's Economic Problem: Too Few Producers*, London, Macmillan.

Bairoch, P. (1976), 'Europe's Gross National Product: 1800–1975', *Journal of European Economic History*, **5**(2), Fall, 273–340.

Baldwin, R. (1989), 'The Growth Effects of 1992', *Economic Policy*, **9**, October, 247–81.

Barro, R. J. (1991), 'Economic Growth in a Cross Section of Countries', *Quarterly Journal of Economics*, **106**(2), May, 407–43.

—— Sala-i-Martin, X. (1990), 'Public Finance in Models of Economic Growth', NBER Working Paper, No. 3362, May.

Baumol, W. J. (1986), 'Productivity Growth, Convergence, and Welfare: What the Long-Run Data Show', *American Economic Review*, **76**(5), December, 1072–85.

Beckerman, W. (1962), 'Projecting Europe's Growth', *Economic Journal*, **72**(288), December, 912–25.

Boltho, A. (1982), 'Growth', in A. Boltho (ed.), *The European Economy: Growth and Crisis*, Oxford, Oxford University Press.

Bombach, G. (1985), *Post-war Economic Growth Revisited*, Amsterdam, North-Holland.

Chatterji, M. (1992), 'Convergence Clubs and Endogenous Growth', *Oxford Review of Economic Policy*, **8**(4), 57–69.

Chesnais, F. (1982), 'Schumpeterian Recovery and Schumpeterian Perspectives—Some Unsettled Issues and Alternative Interpretations', in H. Giersch (ed.), *Emerging Technologies*, Tübingen, J. C. B. Mohr.

Conlisk, J. (1969), 'A Neoclassical Growth Model with Endogenously Positioned Technical Change Frontier', *The Economic Journal*, **79**(314), June, 348–62.

Cornwall, J. (1977), *Modern Capitalism*, London, Martin Robertson.

Crafts, N. (1992), 'Productivity Growth Reconsidered', *Economic Policy*, **15**, October, 387–426.

De Long J. B., and Summers, L. H. (1991), 'Equipment Investment and Economic Growth', *Quarterly Journal of Economics*, **106**(2), May, 445–502.

Denison, E. (1967), *Why Growth Rates Differ*, Washington DC, The Brookings Institution.

Dowrick, S., and Nguyen, D–T. (1989), 'OECD Comparative Economic Growth 1950–85: Catch-up and Convergence', *American Economic Review*, **79**(5), December, 1010–30.

—— Gemmell, N. (1991), 'Industrialization, Catching-up and Economic Growth: A Comparative Study Across the World's Capitalist Economies', *The Economic Journal*, **101**, March, 263–75.

Ehrlich, I. (1990), 'The Problem of Development: Introduction', *Journal of Political Economy*, **98**(5), October, S1–S11.

Fagerberg, J. (1987), 'A Technology Gap Approach to why Growth Rates Differ', *Research Policy*, **16**(2–4), August, 87–99.

Freeman, C. (ed.) (1983), *Long Waves in the World Economy*, London, Butterworth.

Gomulka, S. (1971), *Inventive Activity, Diffusion, and the Stages of Economic Growth*, Institute of Economics, Aarhus.

Griffin, K. (1974), 'The International Transmission of Inequality', *World Development*, **2**(3), March, 3–15.

Hicks, J. R. (1950), *A Contribution to the Theory of the Trade Cycle*, Oxford, Clarendon Press.

Hill, T. P. (1979), *Profits and Rates of Return*, OECD, Paris.

Kaldor, N. (1961), 'Capital Accumulation and Economic Growth', in F. A. Lutz and D. C. Hague (eds.), *The Theory of Capital*, London, Macmillan.

—— (1966), *Causes of the Slow Rate of Economic Growth of the United Kingdom*, Cambridge, Cambridge University Press.

Kindleberger, C. P. (1967), *Europe's Postwar Growth—The Role of Labor Supply*, Cambridge, MA, Harvard University Press.

King, M., and Robson, M. (1992), 'Investment and Technical Progress', *Oxford Review of Economic Policy*, **8**(4), 43–56.

Kravis, I. B. (1959), 'Relative Income Shares in Fact and Theory', *American Economic Review*, **49**(5), December, 917–49.

Kuznets, S. (1959), 'Quantitative Aspects of the Economic Growth of Nations—IV. Distribution of National Income by Factor Shares', *Economic Development and Cultural Change*, **7**(3) (Part II), April, 1–100.

Levine, R., and Renelt, D. (1992), 'A Sensitivity Analysis of Cross-Country Growth Regressions', *American Economic Review*, **82**(4), September, 942–62.

Lucas, R. E. (1988), 'On the Mechanics of Economic Development', *Journal of Monetary Economics*, **22**(1), 3–42.

Maddison, A. (1982), *Phases of Capitalist Development*, Oxford, Oxford University Press.

—— (1989), *The World Economy in the 20th Century*, Paris, OECD

—— (1991), *Dynamic Forces in Capitalist Development*, Oxford, Oxford University Press.

Mandel, E. (1980), *Long Waves of Capitalist Development*, Cambridge, Cambridge University Press.

Meier, G. M. (1964), *Leading Issues in Development Economics*, New York NY, Oxford University Press.

Nordhaus, W. D., and Tobin, J. (1972), 'Is Growth Obsolete?', in NBER, *Economic Research: Retrospect and Prospect—Economic Growth*, Fiftieth Anniversary Colloquium V, New York NY, Columbia University Press.

Nurkse, R. (1961), 'International Trade Theory and Development Policy', in H. S. Ellis (ed.), *Economic Development for Latin America*, London, Macmillan.

Olson, M. (1982), *The Rise and Decline of Nations*, New Haven CT, Yale University Press.

Reynolds, L. G. (1983), 'The Spread of Economic Growth to the Third World', *Journal of Economic Literature*, **21**(3), September, 941–80.

Romer, P. (1986), 'Increasing Returns and Long-run Economic Growth', *Journal of Political Economy*, **94**(5), October, 1002–37.

—— (1989), 'Human Capital and Growth: Theory and Evidence', NBER Working Paper, No.3173, November.

—— (1990*a*), 'Capital, Labor and Productivity', *Brookings Papers on Economic Activity—Microeconomics*, 337–67.

—— (1990*b*), 'Endogenous Technological Change', *Journal of Political Economy*, **98**(5), October, S71–S102.

Rosenberg, N., and Frischtak, C. R. (1983), 'Long Waves and Economic Growth: A Critical Appraisal', *American Economic Review*, **73**(2), May, 146–51.

Scott, M. FG. (1989), *A New View of Economic Growth*, Oxford, Clarendon Press.

Solow, R. M. (1956), 'A Contribution to the Theory of Economic Growth', *Quarterly Journal of Economics*, **70**(1), February, 65–94.

Uzawa, H. (1965), 'Optimum Technical Change in an Aggregative Model of Economic Growth', *International Economic Review*, **6**(1), January, 18–31.

Van Duijn, J. J. (1983), *The Long Wave in Economic Life*, London, George Allen & Unwin.

Wade, R. (1990), *Governing the Market*, Princeton NJ, Princeton University Press.

13

The macroeconomics of growth: an international perspective

FREDERICK VAN DER PLOEG
PAUL TANG

FEE, University of Amsterdam and Tinbergen Institute[1]

I. Introduction

In the orthodox neoclassical theory, set out by Solow (1956) and others, the possibility of sustained economic growth is ascribed to an exogenous factor of production, i.e. the passage of time. This result is intimately linked to one of the properties of the neoclassical production function that is employed in this theory. This function relates output to factor inputs, the stock of accumulated physical capital goods (machinery, computers, and the like) and labour. It displays decreasing returns with respect to the use of each (reproducible) factor of production (and constant returns over all). It follows that an increase in the stock of capital goods, given the amount of labour employed, yields a less-than-proportionate increase in output. Expansion of the capital stock implies a decline in the return on a further expansion and for this reason may ultimately cease. Technical changes, however, that improve the productivity of labour and, thus, of capital, can prevent the rate of return on investment from falling. If the labour force grows at an (exogenous) rate equal to the sum of population growth and labour-augmenting technical progress, capital, output, and consumption will eventually also grow at this exogenous rate on an equilibrium growth path. Accumulation of capital is in this sense complementary to ongoing technical developments. Neoclassical theory does not provide an economic explanation for these

developments, but rather imposes a time trend on the model for the long-run rate of economic growth.

The possibility of exogenous technical progress reconciles the neoclassical theory with Kaldor's 'stylized facts' (1961): a steady growth rate of output (per worker); a more or less constant ratio between output and the capital stock; a constant return on investment; a fairly stable functional distribution of income. However, as technical progress is assumed to be exogenous in the older versions of growth theory, not much explanatory power is gained from its introduction. Furthermore, when the standard Solow model is calibrated to real data in order to explain the adjustment towards balanced growth paths, the predictions for the speed of convergence and for the national income share of capital income are, typically, too high.

Many empirical studies try to attribute the growth of output mainly to quantitative and qualitative changes in the stocks of productive factors. The residual growth in output that cannot be explained by growth in the factors of production is often referred to as the Solow residual or, in the applied literature, as total or joint factor productivity. The calculation of these residuals usually supposes perfect competition in factor and output markets, so that the contributions of the growth in capital and in labour to growth in output can be assessed by weighting them by their national income shares. Empirical studies typically find that only part of the growth in output can be accounted for in this way. The resulting Solow residuals are normally ascribed to technical progress and may be of considerable size as Table 13.1 shows.

This paper surveys the contributions that have been made in the literature to *explain* the presence of

First published in *Oxford Review of Economic Policy*, vol. 8, no. 4 (1992). This version as been updated and revised to incorporate recent developments.

[1] We thank Christopher Allsopp for his many constructive and helpful comments on an earlier version of this survey.

Table 13.1. Gross Domestic Product and Augmented Joint Factor Productivity (Annual Average Compound Growth Rate)

	1870–1913	1913–50		1950–73		1973–84	
	GDP	GDP	AJFP	GDP	AJFP	GDP	AJFP
France	1.7	1.1	0.6	5.1	3.1	2.2	0.9
Germany	2.8	1.3	0.2	5.9	3.6	1.7	1.1
Japan	2.5	2.2	0.0	9.4	4.7	3.8	0.4
Netherlands	2.1	2.4	0.5	4.7	2.4	1.6	0.1
UK	1.9	1.3	0.4	3.0	1.5	1.1	0.6
US	4.2	2.0	1.2	3.7	1.1	2.3	−0.3

Source: Maddison (1987), Tables 1 and 11*b*.
Note: The augmented joint factor productivity (AJFP) equals production growth (GDP) minus the contributions of the changes in quantity and quality of labour and of capital.

sizeable Solow residuals and at the same time to provide an understanding of what determines the long-run rate of growth of an economy. Section II discusses how a large variety of the new theories of endogenous growth builds on the classic work of Uzawa (1965) and Conlisk (1969). The assumption of decreasing returns to a narrow concept of capital is rejected in favour of constant returns to a very broad measure of capital. The long-run rate of growth then depends on a host of supply-side determinants such as learning by doing, intentional investment in human capital, research and development in the capital goods and consumption goods industries, and public infrastructure and other public goods.

Section III deals with the global distribution of welfare, focusing on the relationship between growth and development, hitherto a neglected subject in the more theoretical literature. An important issue concerns the conditions under which rates of economic growth and levels of *per capita* income will converge or diverge. The new theories of endogenous growth may be somewhat less optimistic about development and catch-up of poor with rich countries than the orthodox theory. Section IV deals briefly with the scope for growth-promoting government policy, the consequences of trade and integration for the rate of economic growth, and the political economy of growth.

Section V deals with the effects of demand-side policies on the rate of economic growth and, in particular, analyses the international spill-over effects of demand-side policies and stresses the importance of international capital mobility (see Alogoskoufis and van der Ploeg, 1991*a*,*b*). Thus this section focuses on the relation between economic growth and budgetary and monetary policy, i.e. government expenditure, public debt, and monetary growth. Once allowance is made for departures from Ricardian debt neutrality, the new theories of economic growth can explain for the first

time that high national income shares of government consumption and high ratios of government debt to national income can push up real interest rates and depress growth prospects. Section VI concludes.

II. New theories of endogenous growth

The pioneering works of Uzawa (1965) and Conlisk (1969) attempt to endogenize the rate of technical progress in the neoclassical model. Conlisk, in particular, conveys radically different conclusions from the orthodox neoclassical theory of economic growth and provides one of the first theories of endogenous growth:

In the Solow–Swan model a change in the rate of savings, in the depreciation rate or in the constant rate of unemployment will not change the equilibrium growth rate g^* (. . .); whereas g^* will indeed be affected in the model of this paper. (p. 69)

The recent explosion of the new theories of economic growth take the same route set out by Uzawa and Conlisk and amend orthodox neoclassical theory by offering an endogenous formulation of technical change.[2]

The new theories of growth abandon the assumption that production displays decreasing returns with respect to the use of capital. Instead, the definition of capital is enlarged to allow also for investment in many reproducible factors of production (such as reclamation of land through building dykes, accumulation of human capital through training, build-up of know-how through R & D, spending on infrastructure and other

[2] Scott (1989) is one of the few recent contributions to the literature on growth which acknowledge the classic work of Conlisk (1969).

public goods, etc.). It does not then seem unreasonable to assume constant, or even increasing, returns to scale with respect to this very broad measure of capital.[3]

1. Relationship to old theories of growth

The Harrod–Domar condition says that the warranted growth rate of an economy must be given by the ratio of the aggregate savings rate divided by the capital–output ratio. The natural growth rate of an economy is given by the sum of the rate of population growth and the rate of labour-augmenting technical progress. On a balanced growth path the warranted growth rate must be equal to the natural growth rate. In general, one can think of four main channels by which balanced growth can be ensured and each of these channels is associated with a particular brand of theory of economic growth (see Hahn and Matthews, 1964, and, for a more recent survey, van der Ploeg, 1984).

The first channel is through adjustments in the aggregate savings rate arising from changes in the functional distribution of income. This is a feature of the post-Keynesian models of the symbiotic contradictions of capitalism and the class struggle associated with the names of Michael Kalecki, Joan Robinson, Nicholas Kaldor, Richard Goodwin, and Luigi Pasinetti—sometimes called the UK Cambridge School. If workers save a smaller proportion of their income than capitalists, then, if income distribution can adjust in response to, for example, changes in the unemployment rate, the aggregate savings ratio can adjust and the warranted rate of economic growth can be brought into line with the natural rate of economic growth. More precisely, the aggregate savings rate and, thus, the warranted rate of growth decline when the national income share of labour increases in booms and vice versa in recessions.

By contrast, the second channel relaxes the assumption of complementarity between factors of production often used by post-Keynesians. One then arrives at balanced growth through adjustments in the capital–output ratio. This is, of course, achieved by neo-classical substitution between the factors of production and relies on factor prices being determined on competitive markets. This approach is associated with Robert Solow and others—the US Cambridge

School.[4] Not surprisingly, much of the heat in the debate between the two schools arose because of the different views about income distribution.

The third and fourth channels arrive at balanced growth through adjustments in the natural rate of growth. The third channel stresses adjustments in the population growth rate in response to economic and environmental conditions and is associated with Robert Malthus. In an open economy changes in population growth may occur through changes in international migration. The fourth channel, in contrast, does this by making the rate of technical progress endogenous and is highlighted by the new theories of endogenous growth.[5] Most of the new theories of growth are general equilibrium and neoclassical in nature.

The new theories of growth abandon the assumption of constant returns with respect to the use of (physical) capital and labour together. Instead, constant or increasing return to a broad measure of capital (including the stock of knowledge, the reclamation of land, etc.) is assumed. However, increasing returns to scale at the level of an individual firm are not compatible with perfect competition, for in this framework the condition that the sum of the factor shares has to equal unity and (supra-normal) profits are zero will be violated. This difficulty can be dealt with in two ways. First, the assumption of perfect competition can be replaced by one of imperfect competition. Second, increasing returns may be external to a firm. The last approach is mainly adopted in the literature on endogenous technical change. At least four different views on the relevant concept of capital, i.e. on the engine of economic growth, have been put forward recently.

2. Learning by doing

Romer (1986) was the first to revive the work of Arrow (1962) on learning by doing and should get the credit for making theoretical and empirical research into questions of economic growth fashionable again. In many ways one could argue that the new theories of economic growth tackle old and important questions with new tools.

---

[3] Both Scott (1989) and King and Robson (1989) postulate a technical progress function, somewhat in the spirit of Kaldor (1961), rather than a neoclassical production function. We focus on the use of neoclassical production functions to explain endogenous growth.

[4] The standard Ramsey model achieves balanced growth through adjustments in both the capital–output ratio and the aggregate savings rate. Growth in private consumption occurs when the market rate of interest exceeds the subjective rate of time preference, particularly if the elasticity of intertemporal substitution is high. The aggregate savings rate, however, only increases with the interest rate if the substitution effect dominates the income effects (i.e. if the elasticity of intertemporal substitution exceeds unity).

[5] Some determinants of the rate of (labour-augmenting) technical progress and the natural rate of grwoth, such as investment in R & D and the infrastructure, also affect aggregate savings and the warranted rate of growth.

Arrow posits the view that learning may enhance productivity (of labour) and is a product of experience. As a measure of experience he takes cumulative gross investment, for investment changes the environment and provides stimuli for learning. The effect of learning by doing, i.e. by experience, on productivity is external to an individual firm. Firms are thought not to incorporate the effect of investment on the possibilities of learning. This assumption enables Arrow to reconcile increasing returns to scale at an aggregate level with perfect competition. As learning by doing is subject to rapidly decreasing returns, the economic growth is still exogenous and determined by population growth.

The analysis of Romer resembles the work of Arrow. Romer, however, enlarges the concept of capital and considers not only the accumulation of capital goods but also investment in knowledge. As knowledge cannot be perfectly patented or hidden from other rival firms in the industry or the economy, investment in knowledge by one firm has a positive effect on the production possibilities of other firms. Contrary to Arrow, production of consumption and capital goods may display constant or increasing returns with respect to reproducible productive factors, i.e. physical capital and knowledge, at an aggregate level, but decreasing returns at a firm level. Owing to the absence of an effective patent market, the stock of knowledge is like a public good. As firms do not or cannot fully internalize the effects of their investment on the economy-wide stock of knowledge and technology, the rate of economic growth is beneath the socially optimal level. This provides a strong reason for government intervention in order to correct for the absence of patent markets. It provides a justification for why many politicians argue for public subsidies for private expenditures on R & D.

3. Intentional accumulation of knowledge

Romer (1986) is clearly a step forward in accounting for technical progress. However, the above models emphasize the side effects of investment rather than the intentional accumulation of knowledge. Later work formulates the concept of capital precisely and considers the engine of growth explicitly.

(a) Human capital

The second tack that the literature on endogenous growth has taken is based on Lucas (1988), who focused on the intentional accumulation of knowledge. Human capital can be increased by devoting time to learning, but naturally this is at the expense of time devoted to work or leisure. Human capital can be considered as an asset, so that the financial return on investment in human capital (i.e. training, education, etc.) must be compared to the return on non-human financial assets. Building on the classic work of Uzawa (1965), Lucas (1988) assumes that the accumulation of human capital is subject to constant (or increasing) returns to scale.

Lucas formulates his model for infinitely lived households. He also defends the validity of his model if households have finite lives by arguing that the stock of human capital may be transferred from older generations to younger generations. Like Romer and Arrow, Lucas presupposes that the stock of knowledge, i.e. the stock of human capital, has a positive external effect on the production of goods, though this is not a necessary assumption for a sustainable and endogenous rate of economic growth. Hence, knowledge is, according to this view, very much a public good. The most striking example of this is language. It is not much use being able to speak a particular language if the people that you have social and economic relations with do not speak the same language. Similar arguments hold for the use of computer software and many other skills.

Again, it is easy to argue that in a competitive market economy the public-good character of intentional accumulation of knowledge yields a growth rate that is lower than the socially optimal growth rate. This provides a strong rationale for government intervention in the form of publicly provided schooling (particularly at the elementary level) and training. One could also think of government subsidies to private training programmes as a way of achieving an efficient growth rate.

(b) Research and development

Third, some authors have taken research and development as central in their analysis of the engine of economic growth (for example, Romer, 1990; Grossman and Helpman, 1990a, 1991; Aghion and Howitt, 1992). The output of R & D may be seen as blueprints for new products or for a better quality of product. The initial investment in R & D is counterbalanced by a subsequent stream of profits, because the producer of a differentiated good has at least temporarily a monopolistic position arising from, for example, a patent on the blueprint for this newly developed good.

The Schumpeterian idea that innovative activities depend on the expected profitability is clearly reflected in these models. An increase in monopoly power, in other words an increase in the discounted value of future profits arising from (say) a lower price elasticity for the demand for the product that is being sold, typically implies an increase in R & D. This implication is not at odds with the common belief that competition fosters economic growth. It only emphasizes that the urge for and existence of profits are essential for firms to conduct innovation.

Consumers benefit from the production and sale of the invented goods both directly and indirectly. The reason is that, on the one hand, consumers value variety or quality while, on the other hand, productivity depends positively on the variety or on the quality of factors of production (e.g. those bought from the capital goods industry). The growth rate of the economy corresponds to the growth in the number of varieties or in the quality of consumer and capital goods.

Technological spill-overs play an important role. In the case of expanding product variety, the productivity of R & D is thought to be positively linked to the pool of public knowledge. An increase in the number of varieties decreases profits per brand, as expenditure is spread (evenly) over all varieties, but the higher number of varieties increases the productivity of R & D. As these forces tend to offset each other, investment in R & D remains profitable. In the case of rising product quality, introduction of a product on to the market contributes to the stock of knowledge because the attributes of the products can be studied and effort can be directed to improve upon the state of the art. It follows that the revenues of R & D can be partly appropriated by the firms in this sector.

4. Public infrastructure

The fourth direction in which the new theories of endogenous growth have progressed is based on the work of Barro (1990) and Barro and Sala-i-Martin (1990). They have exploited the idea that government investments in both the material infrastructure (e.g. public highways and railways) and the immaterial infrastructure (e.g. education, protection of property rights, and the like) are essential to economic growth. Effectively, the production function is extended to include government services that raise the productivity of private capital. The idea is that there is constant (or increasing) returns to scale with respect to capital of all the firms in the industry or economy together as well as the spending on public goods. Clearly, the rate of economic growth is boosted by an increase in the national income share of these types of public goods.

However, in so far as the increase in public goods must be financed by distortionary taxes, the after-tax marginal productivity of capital, the rate of interest, and the rate of economic growth are diminished. This part of the literature adds some public finance arguments by examining the optimal tax rate and provision of public goods.

5. Evaluation of various theories of endogenous growth

The recent elaborations of the orthodox neoclassical theory are not mutually exclusive and have many common features. They share the notion that technical progress is not manna from heaven, but is related to economic activity. With the exception of the fourth view, the intentional accumulation of knowledge is brought to the fore as the driving force behind economic growth. The deliberate search for new and better products or production techniques, the conscious exploration and exploitation of the environment rather than the duplication of already existing means, methods, and ideas, constitutes the basis of technical progress.

The level of technology cannot be raised drastically. Technical advancement proceeds gradually, and productivity and augmentation of productivity in the present are conditional on investment (in a broad measure of capital) in the past. The pace of economic growth depends on the intertemporal preferences of households, i.e. on the choice between consumption and saving, as is emphasized by the Ramsey (1928) model of economic growth but also by an important precursor of the new theories of growth, namely Conlisk (1969). In the Ramsey model households are thought to choose a consumption path over time. The steepness of this path, i.e. the growth rate of private consumption, depends negatively on the degree of impatience for current consumption and positively on the willingness to substitute current for future consumption and the real rate of return on savings, particularly so if the elasticity of intertemporal substitution is high.

An increase in the real rate of return induces households to save a greater fraction of their income if the substitution effect dominates the income effect. It also induces households to postpone consumption, so that the rate of growth of private consumption is increased. Many authors, however, think that the decision to save is distorted by the impossibility of collecting the full yield of the investment, as ideas cannot be kept secret and the use of ideas cannot be adequately protected by law. The development of the computer programme 'Windows' by Microsoft was inspired by the success of Macintosh. However, the external effect of investment may not be confined to investment in R & D or in education. Obviously, neither France nor the United Kingdom ever considered building the Channel Tunnel on their own. Another example of an external effect of investment in public infrastructure are the passes in Switzerland that are crucial to the efficient flow of traffic within the European Community. The downward bias in the return on investment causes the rate of private savings to be too low.

The recent neoclassical theories differ in their characterization of knowledge. Knowledge can be considered a rival or a non-rival productive factor. Means can be called non-rival if usage for one purpose does not limit usage for other purposes. Clear examples are dykes, television programmes, and the principle of the internal-combustion engine. Romer (1986, 1990) and Grossman and Helpman (1990a, 1991) represent knowledge, especially the contributions of R & D, as a non-rival productive factor. From Lucas (1988), however, can be derived the notion that knowledge in the form of human capital is a rival good. A surgeon can devote his attention to only one patient at a time.[6] Similar arguments may apply to publicly provided goods. Barro (1990) suggests that few of these goods are not subject to congestion. This argument can be illustrated by many examples, like roads or recreation areas (think of traffic jams or a coastal resort on a sunny day). It also applies to protection of property rights, as employment in the legal system is a function of the size of the population.

Non-rival technology implies that production is subject to economies of scale at an aggregate level. In the models production is at least linearly related to a broad measure of capital, so that a doubling of both the labour force and the stock of capital amount to more than a doubling of production. This outcome often incites the criticism that ongoing accumulation of capital and growth of the population, therefore, imply that the rate of economic growth accelerates rather than approaches a constant value.

The logic of this argument can be questioned. First, Romer (1986) implicitly refers to the traditional idea (see for example Hicks, 1950) that an upper boundary on the rate of economic growth may exist. Second, the argument is partial and has to be extended to include an explanation for population growth; one may

wonder whether the decrease in the population growth in the Western world is a mere coincidence or can be partially attributed in a Malthusian fashion to the increase in the standard of living. Besides, Romer (1990) provides rudimentary evidence that the rate of economic growth has risen over the last two centuries.

III. Growth and development: convergence or divergence?

Differences in growth rates between countries may be added to Kaldor's list of stylized facts. The orthodox neoclassical theory clearly predicts that the growth rates of different countries should converge in the long run. The crucial assumption for this convergence in growth rates is diminishing marginal productivity of capital. Poor countries with a dilapidated and low level of the capital stock have lots of investment opportunities and face high real interest rates, so that consumers have a strong incentive to postpone consumption and save. This is why neoclassical theory predicts that poor countries have higher growth rates than rich countries during the adjustment towards the equilibrium growth path. In fact, neoclassical theory also strongly suggests that there is a natural tendency for production per head of different countries to converge, mainly because technology is universally available and applicable.

Neoclassical theory thus has a fairly optimistic view on growth and development. However, empirically, the speed of convergence is much slower than the traditional neoclassical theory predicts. This is why Mankiw et al. (1990) include human capital as a separate factor of production into a Solow-style growth model. In this way they explain the observed too low pace of convergence and rehabilitate the main qualities of the Solow (1956) model. However, their extension of the traditional theory does not consider the possibility of capital mobility.

1. Saving, investment, and the current account

As the marginal productivity of capital in poor countries is thought to exceed that in rich countries, it is efficient for the poor countries to borrow from the rich countries on a large scale. A flow of funds from North to South could increase the speed of convergence of growth rates considerably and would be an excellent development policy. The theory suggests that, in the

---

[6] As usual, the distinction may be ambiguous in practice: a book can be read by one person at a time, but by many people over time.

absence of any restrictions on the mobility of capital (for example, irreversibility of investment), the speed of convergence should be infinite. Perfect mobility also implies that domestic saving and domestic investment should be uncorrelated. The 'golden rule' says that the optimal level of the current account deficit (being the net increase in the wealth of a nation) should be equal to the level of investment *with a market rate of return* (plus any shortfalls of the current level of production from the permanent level of production minus any discrepancies between the current level of public spending and the permanent level of public spending) (e.g. Sen, 1993).

However, empirical estimates by Feldstein and Horioka (1980) sharply contradict this prediction. Barro *et al.* (1992) claim that, in practice, the flow of capital is restrained by imperfections in the market. In particular, they assume that the collateral value of human capital is negligible in practice and the amount of debt is restricted by the collateral value of physical capital. In the case of capital mobility and an operative restriction on borrowing, the speed of convergence is faster than in the case of capital immobility but nevertheless finite. The process of convergence in a partially open economy resembles that of a closed economy. Still, capital is thought to flow during the process of adjustment from rich countries to poor countries. The imperfection of the capital market restores the link between domestic saving and domestic investment, yielding a partial explanation of the Feldstein–Horioka puzzle, as only some countries encounter restrictions on borrowing.

2. International spill-overs of investment

Theories of economic growth are usually formulated and developed for a closed economy. The predictions of the new theories of growth about convergence and development depend on the translation of these theories to the context of open and interdependent economies. Recent literature suggests that international spill-overs of investment may provide over and above the effects of capital mobility a strong reason for convergence of growth rates, although differences in levels of output and of consumption between countries may remain (see, for example, Grossman and Helpman, 1991, and Alogoskoufis and van der Ploeg, 1991*a,b*). Spill-overs of technology cause the marginal productivity of a broad measure of capital in a backward area to exceed that in an advanced area, so that the incentive to invest in the former area is higher than in the latter area. What are the territorial boundaries of the

spill-overs? Are external effects of investment confined to an area like Silicon Valley or are spillovers international? Although the external effects of investment in R & D are more likely to cross borders than those of investment in human capital, this question has to be answered empirically rather than theoretically. Note, however, that both the older and the newer theories (may) predict convergence between countries, either for levels of productivity or for rates of growth, and may be empirically hard to distinguish.

Grossman and Helpman (1991), Buiter and Kletzer (1991), and Alogoskoufis and van der Ploeg (1991*a,b*) have constructed examples in which growth rates of output differ between countries permanently. Even though international mobility of (physical) capital is perfect, differences can arise in these examples when non-tradable and reproducible inputs are used in the production of a tradable commodity. The results of Buiter and Kletzer and Grossman and Helpman rely on the assumption that international spill-overs of knowledge are absent. Buiter and Kletzer focus on the accumulation of human capital. Differences in intertemporal preferences of households or, more importantly, in public expenditure on schooling may cause countries to grow at disparate rates. Grossman and Helpman show that, under the assumption that invention and production of a variety are intrinsically related, a large country can, due to economies of scale, gain an (absolute) advantage in the research for and the development of new varieties. It follows that a large country can specialize in the conduct of R & D at the expense of innovative activity in a small country. Alogoskoufis and van der Ploeg (1991*a,b*) model international spill-overs of knowledge and find convergence of growth rates, unless the costs of adjustment for investment projects differ between countries. These adjustment costs can, in fact, be considered as a non-tradable input.

The point of international spill-overs in the production process is that there are decreasing returns to capital at a national level but constant (or increasing) returns to capital at a global level. This means that there is some scope for convergence, particularly if there is capital mobility and the importance of non-traded factors of production is not too large, while at the same time the growth rate of the global economy is endogenous.

3. Subsistence, poverty, and growth

These examples do not necessarily imply that the welfare of households in different country groups develops differently, for the households may, if capital is perfectly mobile, face the same possibilities for investing their savings. However, this has the odd implication that developing countries are best helped by an unrestricted access to the global capital market. In practice differences in levels of consumption may be persistent. Rebelo (1991), therefore, focuses attention on subsistence levels of private consumption. In poor countries resources may be devoted to subsistence consumption needs rather than to savings, and if the economy is closed or only partially open, this may mean that funds for investment are restricted. For this reason the convergence in levels of productivity or in rates of growth, as is predicted by both the traditional and recent theories of growth, may be very slow (cf. Kuznets, 1966), or, if the effect is large, poor countries may even grow (temporarily) at a slower rate than rich countries.

IV. Investment and trade-promoting policy

A remarkable feature of the recent literature on economic growth is the overwhelming support for the idea that investment has positive external effects on production possibilities. The return on investment cannot be reaped fully and the intertemporal choices by households are biased in favour of consumption and at the expense of savings. The assumption of an external effect implies an active role for the government. It may take measures to improve upon the intertemporal allocation of resources, because the outcome of decentralized decisions by the various private agents are not optimal.

At a general level an external effect may be seen to be due to inadequate definition and protection of property rights. In practice, the cost of defining and protecting these rights may be prohibitive. However, examples in which uncertainty about property rights has inhibited investment abound. One can think of the unsettled claims on land and buildings in the former East Germany or of the political instability in South America.[7] Clearly, the system of patents may be crucial to protect the research for and development of new or better products and production methods against 'cheap' imitation. After all, firms are only willing to conduct R & D if, at least temporarily, profits can be earned.

When an effective patent market does exist, however, there may be a danger of over-investment in the sense that the growth rate of a decentralized market economy may then be higher than the socially optimal growth rate. The problem may be that firms are engaged in a R & D race in which each of them tries to win a contest in which the prize at the end of the race is an infinitely lived patent (see Beath et al., 1992, for a survey of these issues). Of course, the required policy response of the government in such a situation is very different from the case in which patent markets are missing.

1. Monopoly power, public policy, and growth

Monopoly power may imply a distortion of relative prices. The government has to balance the dynamic advantage against the static disadvantage of monopoly power. The government can also change the price of future consumption relative to current consumption in other ways. Depending on the proposed engines of growth, the government may contemplate supporting R & D and/or it can direct its expenditure to schooling and invest in public infrastructure. The design of the fiscal system—the form of taxes on capital income or the fiscal method of depreciation—may also affect the rate of return on savings. The government needs to balance the sometimes inevitable distortions with the revenue of taxes. For example, a tax on capital income, on the one hand, reduces the after-tax rate of return on investment but, on the other hand, raises revenue that can be used to finance investment in public infrastructure (see Barro, 1990; Barro and Sala-i-Martin, 1990; and Alesina and Rodrik, 1991). In this case, except under the special assumption of a Cobb–Douglas production function and apart from any external effects of investment, promotion of economic growth does not necessarily imply an improvement of social welfare. In general, only in a situation where private agents do not fully internalize the benefits of investment should public policy be directed to the enhancement of economic growth. But it is important to emphasize that the external effects of investment, and so the role of the government, are theoretically assumed rather than empirically derived.

2. Trade and economic integration

Trade and economic integration can clearly affect the dynamic performance of economies. Though the nature of the advantages or disadvantages may be static,

---

[7] In fact Barro (1990) finds a negative (conditional) correlation between the number of assassinations and the number of revolutions on the one hand and the growth rate on the other hand for South American countries. Though this result nicely illustrates the text, it should not be taken too seriously.

changes in efficiency affect decisions to save and invest. As Baldwin (1989) points out, the tumult about Europe 1992 cannot be caused by the prediction of a one-time increase in productivity; rather, the excitement is based on the presumption of a (temporary) increase in growth. What are the possible effects of trade? The familiar argument emphasizes the possibility of specialization in production among countries. The pattern of inter-industry trade reflects, according to the Heckser–Ohlin theorem, the relative endowment of, for example, skilled and unskilled labour between countries. International specialization has an ambiguous effect on economic growth, for it may imply that resources are devoted less to innovative activity and more to production of goods (see Grossman and Helpman, 1990b, 1991). Furthermore, consumers may benefit from an expanded range of available products.

Compare, as an example, a closed economy with an open economy. In the open economy, the presence of intra-industry trade enriches the choice of consumers. Consumers value variety and, therefore, there is a gain due to an expansion in the range of available products. This gain is static and does not affect the dynamic performance of an economy. Intra-industry trade can, however, also influence the intertemporal choice to save and invest. Rivera-Batiz and Romer (1991a) conclude that increasing returns to scale in the production function of R & D causes free trade or economic integration to have an effect on growth. The point is that two isolated sectors do not conduct R & D as efficiently as one integrated sector.

Increasing returns can be due to specialization in the use of inputs (custom-made machines, various computer programmes, different types of fibre) or to spill-overs of knowledge. In the first case, intra-industry trade expands the available variety of inputs and enables a higher degree of specialization in the use of these inputs. It thus augments the level of productivity and, therefore, spurs the rate of economic growth. Consumers benefit now and later from the possibility of trade. In the second case, international exchange of knowledge and of goods have to be distinguished. Only through the international communication of research ideas and results can economies of scale be exploited and can the productivity of innovative activity be enhanced.

3. Increasing returns and international competition

There are several possible repercussions for trade in this case. Trade brings about an increase in the size of the

market and, in addition, increases competition in the goods market. The first effect tends to increase the reward for investment in R & D, whereas the second effect will put a downward pressure on the (temporary) profits. The result of both forces is, therefore, ambiguous. Rivera-Batiz and Romer emphasize a different mechanism. International trade of (differentiated) goods removes any redundancy in R & D, because it gives an incentive to direct efforts towards invention of new products rather than towards the imitation of already existing products. Hence, the exchange of goods induces international competition in R & D, even though the output of R & D, viz. blueprints for new or better products, is not, itself, traded. Furthermore, the distinction between the exchange of knowledge and of goods is theoretically convenient, but may not be easy to draw rigorously.

The international flow of goods and of ideas may be intimately linked; perhaps the international spill-overs of knowledge are enhanced by the trade of goods and, therefore, boost investment. In summary, trade may foster growth due to exploitation of economies of scale and to creation of an incentive to innovate rather than to imitate. These positive effects will dominate any negative effect of inter-industry specialization, as long as countries have identical relative endowments of basic inputs. A reallocation of resources at the expense of R & D may occur if dissimilar countries engage in trade.

What are the gains from the international economic integration above those from the free exchange of goods? Clearly these include the elimination of border controls and the standardization of government regulation. More generally, the removal of barriers to competition releases resources and improves overall productivity. The harmonization of VAT rates and excise duties, the open procurement of government spending, and the liberalization of financial markets, as far as trade of goods and mobility of capital are not good (short-run) substitutes, induce a reallocation of resources and imply a gain in efficiency.

4. Political economy of growth

The instruments of economic policy may not be in the hands of a benevolent and enlightened dictator, but subject to a political struggle. The abolition of subsidies on food in order to finance, for example, investment in infrastructure may promote growth but may also provoke chaos. Politicians or political parties that seek re-election and fight for votes are not solely concerned with economic growth, especially as they do not represent future generations directly. They may have to weigh the demand by voters for transfers from the rich to the poor and the distortion of taxes on (capital) income.

Persson and Tabellini (1992), Alesina and Rodrik (1991), and van Ewijk (1991) have tried formally to model the effects of political choices on the distribution of income and taxes on income in a democracy.[8] Typically, this yields the outcome that an unequal distribution of income hampers economic growth. For in a country with an unequal distribution of income the decisive (median) voter is likely to be poor and ask for a high tax on capital income to finance transfers, so that the incentive to save and, therefore, the rate of growth are low. This analysis also suggests that limiting political participation to wealthy people, which was common practice in European countries in earlier days, enables politicians or political parties to choose a low tax rate and a high growth rate. Except by van Ewijk (1991), who relates differences in income among voters to intertemporal preferences, the personal distribution of income is usually considered to be given. However, Persson and Tabellini (1992) acknowledge the possibility of an interaction between growth and income distribution, that could potentially modify their analysis. In any case there is strong empirical evidence drawn from a wide range of democracies that a fairer distribution of income and wealth induces the right political conditions for growth-promoting policies.

V. Budgetary policies and economic growth

Many politicians and political commentators argue that high levels of government consumption and government debt are bad for growth prospects. Despite this wisdom from practitioners, there have been, until recently, no theoretical textbook stories of why loose budgetary policies may damage growth prospects. Government consumption and government debt simply do not feature in the usual accounts of economic growth. Indeed, most of the new literature on endogenous growth focuses almost entirely on the effects of supply-side policies on the rate of economic growth.

Here we build on Alogoskoufis and van der Ploeg (1990, 1991b) and set out to argue that an interesting and realistic macroeconomic account of the determinants of economic growth must break with the assumption of Ricardian debt neutrality which is employed in almost all of the old and new literature on economic growth.

1. Ricardian debt neutrality and economic growth

It is curious that most of the literature on endogenous growth adopts the assumption of Ricardian debt neutrality. This is a very strong assumption. It implies that it does not matter for the level of private consumption if a government postpones taxation. In other words, private consumption is unaffected when the government finances a temporary tax-cut with short-term borrowing and a long-run increase in taxation. Private agents are assumed to be rational and simply provide for the increase in future taxation by saving. The increase in borrowing of the government is thus exactly offset by the increase in saving of the private sector. Government bonds are, then, not net wealth and private consumption is unaffected.

Under such conditions, an increase in the ratio of government debt to national income does not affect private consumption and thus does not change the resources available for investment purposes. Clearly, government borrowing does not affect the rate of economic growth when Ricardian debt neutrality prevails. An increase in the national income share of government consumption also leads under these conditions to 100 per cent crowding out of private consumption, thus leaving the resources available for investment purposes unaffected. The new theories of endogenous growth, in as far as they take on board the assumption of Ricardian debt neutrality, do not allow for an effect of government consumption on the rate of economic growth. Of course, an increase in the national income share of government spending on infrastructure does raise the rate of economic growth.

2. Breakdown of Ricardian debt neutrality

In fact, Ricardian debt neutrality may not be a very realistic assumption. It may fail for a whole range of reasons. Private agents may not be rational. When there is uncertainty about future income, a tax-cut acts like

---

[8] Usually the political choices are set in a peculiar institutional framework that has to proxy a representative democracy. Furthermore, a formal solution is subject to stringent conditions; for example, the choice has to concern a single issue and the distribution of preferences on this subject has to be single-peaked.

an insurance policy. Because the variance of after-tax income declines as the tax rate in the future goes up, private agents engage in less precautionary saving so that private consumption goes up when a postponement of taxation is announced. Private agents may be finitely lived and thus may discount the possibility of higher taxes in the future more heavily as they may not be around to have to pay them, particularly if there is no intergenerational bequest motive or, more generally, altruism. New generations may enter the economy, so that the burden of future taxation may be shared with new and stronger shoulders. Liquidity constraints may prevent private agents from borrowing to offset the saving of the government (i.e. raising taxes today and cutting taxes tomorrow).

Deaton (1992) discusses wide-ranging evidence against the hypothesis that the time-path of private consumption follows a 'random walk' in the sense that fluctuations in private consumption only depend on transitory (unanticipated) income and not on current income. There are strong reasons to believe that private consumption does depend on current disposable income and, thus, that Ricardian debt neutrality is a theoretical curiosity. It is, therefore, a challenge to see what this implies for the new theories of endogenous growth.

3. Government consumption and debt damage growth

A postponement of taxation leads, in the long run, to an increase in the ratio of government debt to national income. If Ricardian debt neutrality does not hold (as is the case in the context of a model of overlapping generations), private agents do not save all of the increase in their after-tax current income and also consume more. Owing to the rise in consumption, less resources are available for investment and the rate of economic growth falls.

An increase in the share of government consumption now induces less than 100 per cent crowding out of private consumption, because the discounted value of future taxes is less when new generations enter the economy than the current increase in taxes. It follows that there are less resources for investment and that the rate of economic growth falls.

4. Money and growth

As long as Ricardian debt neutrality holds, monetary growth does not affect the national income shares of private consumption and investment or real growth and is thus the sole determinant of inflation. Also,

government consumption leads to 100 per cent crowding out of private consumption and thus does not affect the real growth and inflation rates either. In order to have an interesting analysis of macroeconomic policy issues, Alogoskoufis and van der Ploeg (1994) show that it is crucial to depart from debt neutrality. This is achieved when there is no operational bequest motive and entry of future generations of households because then the burden of future taxes is shared with new generations. As a result, for a given stance of monetary and fiscal policy, the national income share of private consumption is higher and consequently real growth is less and inflation is higher than in an economy populated by dynasties with an operational bequest motive. An increase in monetary growth is thus not superneutral and, in addition, leads to a less than 100 per cent increase in inflation. If the increase in monetary growth is accompanied by open-market purchases of bonds rather than lump-sum subsidies, there is a larger increase in real growth and a smaller increase in inflation, so money is even less neutral. A balanced-budget increase in the national income share of government consumption leads to less than 100 per cent crowding out of private consumption, a fall in real growth, and in increase in inflation. A money-financed increase in government consumption leads to less crowding out and thus to a smaller fall in real growth and a bigger increase in inflation than a tax-financed increase. On the other hand, a bond-financed increase in government consumption leads to a bigger fall in real growth and a smaller increase in inflation than a tax-financed increase in government consumption. In general, an increase in government debt, arising from an intertemporal shift in taxation, reduces real growth and boosts inflation. If there are costs of adjustment for investment, increases in monetary growth lower real interest rates (Mundell effect) while increases in the government debt–GDP ratio raise real interest rates. Once the new theories of endogenous growth are cast within a framework of non-interconnected overlapping generation, the analysis of macroeconomic policies and the debates on the burden of government debt and the causes on inflation gain an exciting new perspective.

5. The international context

In a global economy with an integrated capital market, interest rates will converge. It follows that private agents in the various countries have the same incentive to postpone consumption and that the growth rates of the various economies also converge. In such a world there are strong links between the various economies. In particular, an expansion of government debt or government consumption in one country damages the growth prospects not only of that country but also those of all the other economies. However, those countries with loose budgetary policies will experience current account deficits and an accumulation of foreign debt, so that they end up with lower levels of private wealth and consumption than countries with tight budgetary policies.

VI. Conclusion

This survey of recent developments in the new theories of endogenous growth has set out to demonstrate that there is an important role for government policy in promoting growth prospects. In so far as there are positive external effects in the production process, the government should step in and provide subsidies to R&D, training, and schooling programmes and should also provide an optimal level of public material and immaterial infrastructure. Governments should also be careful not to have excessive national income shares of public consumption and government debt, because these hamper growth and damage the welfare of their citizens. The government may also need to promote a fair and just society in order to ensure that the political conditions are right for implementing growth-promoting and welfare-enhancing policies.

As far as issues of development and catch-up of levels and growth rates in *per capita* income are concerned, the new theories of endogenous growth are considerably more pessimistic than the orthodox theories of neoclassical growth with their strong predictions of international convergence. The global capital market simply does not work to the full advantage of developing countries and policies must be directed at improving growth in those countries, particularly as many developing countries must devote a large part of their income to subsistence consumption needs rather than to saving, investment, and growth. It is one of the achievements of the new theories of endogeneous growth that they have been able to put flesh on the arguments that have been put forward by practitioners in the field of economic development for many decades.

References

Aghion, P., and Howitt, P. (1992), 'A Model of Growth Through Creative Destruction', *Econometrica*, 2, 323–51.

Alesina, A., and Rodrik, D. (1991), 'Distributive Politics and Economic Growth', CEPR Discussion Paper, 565, London.

Alogoskoufis, G., and van der Ploeg, F. (1990), 'Endogenous Growth and Overlapping Generations', Discussion Paper 26/90, Birkbeck College, University of London.

—— (1991a), 'On Budgetary Policies, Growth, and External Deficits in an Interdependent World', *Journal of the Japanese and International Economies*, 5(4), 305–24.

—— (1991b), 'Debts, Deficits and Growth in Interdependent Economies', CEPR Discussion Paper, 533, London.

—— (1994), 'Money and Endogenous Growth', *Journal of Money, Credit and Banking*, 26(4), 771–91.

Arrow, K. J. (1962), 'The Economic Implications of Learning by Doing', *Review of Economic Studies*, 29, 155–73.

Baldwin, R. (1989), 'The Growth Effects of 1992', *Economic Policy*, 9.

Barro, R. J. (1990), 'Government Spending in a Simple Model of Endogenous Growth', *Journal of Political Economy*, 98, S103–S125.

—— and Sala-i-Martin, X. (1990), 'Public Finance in Models of Economic Growth', mimeo, Harvard University.

—— Mankiw, N. G., and Sala-i-Martin, X. (1992), 'Capital Mobility in Neo-classical Models of Growth', mimeo, CEPR.

Beath, J., Katsoulacos, Y., and Ulph, D. (1992), 'Strategic Innovation', in M. Bacharach, M. Dempster, and J. Enos (eds.), *Mathematical Models in Economics*, Oxford University Press, Oxford.

Buiter, W. H., and Kletzer, K. M. (1991), 'Persistent Differences in National Productivity Growth Rates with Common Technology and Free Capital Mobility: The Roles of Private Thrift, Public Debt, Capital Taxation and Policy towards Human Capital Formation', *Journal of the Japanese and International Economies*, 5(4), 305–24.

Conlisk, J. (1969), 'A Neoclassical Growth Model with Endogenously Positioned Technical Change Frontier', *Economic Journal*, 348–62.

Deaton, A. (1992), *Understanding Consumption*, Oxford, Oxford University Press.

Ewijk, C. van (1991), 'Distribution Effects in a Small Open Economy with Heterogeneous Agents', mimeo, Tinbergen Institute.

Feldstein, M. S., and Horioka, C. (1980), 'Domestic Savings and International Capital Flows', *Economic Journal*, 90, 314–29.

Grossman, G. M., and Helpman, E. (1990a), 'Comparative Advantage and Long-run Growth', *American Economic Review*, 80, 796–815.

—— —— (1990b), 'Trade, Innovation and Economic Growth', *American Economic Review*, Papers and Proceedings, 80, 86–91.

—— —— (1991), *Innovation and Growth in the Global Economy*, Cambridge MA, MIT Press.

Hahn, F. H., and Matthew, R. C. O. (1964), 'The Theory of Economic Growth: A Survey', *The Economic Journal*, 74, 779–902.

Hicks, J. R. (1950), *A Contribution to the Theory of the Trade Cycle*, Oxford, Oxford University Press.

Kaldor, N. (1961), 'Capital Accumulation and Economic Growth', in F. Lutz (ed.), *The Theory of Capital*, London, MacMillan.

King, M. A., and Robson, M. H. (1989), 'Endogenous Growth and the Role of History', NBER Working Paper, 3173, Cambridge MA.

Kuznets, S. (1966), *Modern Economic Growth: Rate, Structure and Spread*, New Haven, Yale University Press

Lucas, R. E. (1988), 'On the Mechanics of Economic Growth', *Journal of Monetary Economics*, **22**, 3–42.

Maddison, A. (1987), 'Growth and Slowdown in Advanced Capitalist Economies: Techniques of Quantative Assessment', *Journal of Economic Literature*, **25**, 649–98.

Mankiw, N. G., Romer, D., and Weil, D. N. (1990), 'A Contribution to the Economics of Growth', NBER Working Paper, 3541, Cambridge MA.

Persson, T., and Tabellini, G. (1992), 'Growth, Distribution and Politics', in A. Cukierman, Z. Hercowitz, and L. Leiderman (eds.), *The Political Economy of Business Cycles and Growth*, Cambridge MA, MIT Press.

Ploeg, F. van der (1984), 'Macro-Dynamic Theories of Economic Growth and Fluctuations', in F. van der Ploeg (ed.), *Mathematical Methods in Economics*, John Wiley, Chichester.

Ramsey, F. P. (1928), 'A Mathematical Theory of Saving', *Economic Journal*, **38**, 543–59.

Rebelo, S. (1991), 'Growth in Open Economies', CEPR Discussion Paper, 667, London.

Rivera-Batiz, L. A., and Romer, P. M. (1991a), 'Economic Integration and Endogenous Growth', *Quarterly Journal of Economics*, **106**, 531–56.

—— —— (1991b), 'International Trade with Endogenous Technological Change', *European Economic Review*, **35**, 971–1004.

Romer, P. M. (1986), 'Increasing Returns and Long-run Growth', *Journal of Political Economy*, **94**, 1002–37.

—— (1987), 'Growth Based on Increasing Returns due to Specialization', *American Economic Review, Papers and Proceedings*, **77**, 56–62.

—— (1990), 'Endogenous Technological Change', *Journal of Political Economy*, **98**, S71–S102.

Roubini, N., and Sala-i-Martin, X. (1991), 'Financial Development, the Trade Regime and Economic Growth', NBER Working Paper, 3876, Cambridge MA.

Scott, M. FG. (1989), *A New View of Economic Growth*, Oxford, Oxford University Press.

Sen, P. (1993), 'Saving, Investment and the Current Account', in F. van der Ploeg (ed.), *Handbook of International Macroeconomics*, Oxford, Basil Blackwell.

Solow, R. M. (1956), 'A Contribution to the Theory of Growth', *Quarterly Journal of Economics*, **70**, 65–94.

Uzawa, H. (1965), 'Optimum Technical Change in an Aggregative Model of Economic Growth', *International Economic Review*, **6**, 18–31.

14

Productivity and competitiveness

JOHN MUELLBAUER

Nuffield College, Oxford[1]

I. Introduction

The British economy is, once again, in a deep recession. As in 1980–1, it has been brought on by high interest rates and an overvalued exchange rate. Again this has been the policy response to a bout of inflation, though one caused more by domestic excess demand than was the case in 1980–1. And, unlike in 1980–1, it took place in a world context of an extended period of low inflation.

Although this article is about much more than UK manufacturing, manufacturing is a bell-wether for the whole economy because it is the key tradable sector, accounting for over 80 per cent of UK non-oil trade in 1988. It is also widely believed to be disproportionately important for wage determination despite now accounting for only around one quarter of all employment and under a third of full-time employment. Further, data availability and accuracy are better than for most of the rest of the economy. So this is the sector with which we begin. There are three key ways of measuring the international competitiveness of UK manufacturing: relative output prices, relative export prices, and relative unit labour costs, where the relativities are taken against a weighted basket of our chief competitors. Figures 14.1(a) and 14.1(b) plot data on the three measures as published by the CSO. The 1980 or 1981 high points on the graphs indicate the post-war troughs reached by competitiveness. In those years the combination of the exchange-rate appreciation resulting from the medium-term financial strategy (MTFS) and the oil revenue boom, the effects of the recent wage inflation, and the cyclical decline in labour productivity produced crisis levels of (un)-competitiveness for UK manufacturing. Ten years on, the situation is similar with the MTFS replaced by ERM membership and high interest rates. However, this time there is no boom in oil revenues.

By the end of 1990 relative export prices were only slightly below their worst annual level reached in 1980; relative producer prices were actually slightly above their worst annual level reached in 1981; relative unit labour costs were still significantly below their worst annual level reached in 1981 but deteriorating fast with the recent cyclical decline in productivity growth. However, the rapid shedding of labour in 1991 is likely to lead to an early stabilization of this measure of competitiveness.

Four years previously, in 1986, competitiveness had peaked because of the combined effects of sterling depreciation, caused in part by weak oil prices, and sustained improvements in productivity growth in manufacturing, relative to which wage inflation appeared moderate. But even at this peak, competitiveness was worse than the average for 1963–79.

Only two years ago serious commentators, e.g. Bean and Symons (1989), were still arguing about whether or not there had been a Thatcher miracle that had transformed the supply side of the economy, bringing with it a permanently better economic performance. So how did the economy get into such a difficult situation so reminiscent of 1980–1? Further, given that economic policy errors brought on this crisis, what policies are needed to extricate us from it? Specifically, can the economy recover without a sterling depreciation? Indeed, can it recover *with* one?

There are a number of forces bearing on competitiveness. They include relative productivity growth, relative wage inflation, the exchange rate, and, to some extent, relative tax regimes. The next section considers productivity growth in manufacturing, updating

First published in *Oxford Review of Economic Policy*, vol. 7, no. 3 (1991). This version has been updated and revised to incorporate recent developments.

[1] I am grateful to Rebecca Emerson for skilled research assistance, and to Tim Jenkinson for helpful comments.

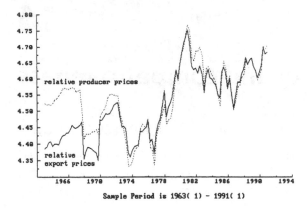

Figure 14.1(a). Two Relative Price Measures of International Competitiveness

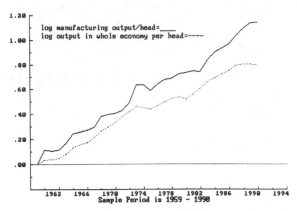

Figure 14.2. Labour Productivity in Manufacturing and in the Whole Economy

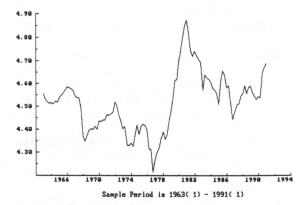

Figure 14.1(b). Relative Unit Labour Cost Measure of International Competitiveness

previous work on longer-term trends and cyclical fluctuations. Section III broadens the scope by considering productivity growth in the non-industrial sectors of the economy, primarily the service sector.

Section IV visits the econometrician's graveyard: wage determination. I investigate the reasons why wage inflation stayed relatively high throughout the 1980s, and rose again to 10 per cent in 1990.

Section V examines economic policies as they bore on the different elements of competitiveness in the 1980s. Policy errors—insufficient attention to supply-side weaknesses—and the demand overshoot of 1987–9 are dissected. The concluding section examines the sustainability of the current exchange rate and whether unit labour cost increases in the UK can be brought down fast and far enough below those of our international competitors that a depreciation can be avoided. There are policy measures which could assist the achievement of *sustained* disinflation.

II. Productivity growth in manufacturing

The introduction discussed some of the reasons for focusing on the manufacturing sector, but there are others. Historically, manufacturing in all countries has tended to experience faster rates of productivity growth. Market economies which have maintained relatively large shares of output in manufacturing have tended to have above average growth rates. Further, it has often been argued that manufacturing tends to be a leading sector in wage determination. Indeed, in the 'Scandinavian model of inflation' (Edgren *et al.*, 1973; Aukrust, 1977; Frisch, 1983, ch. 5), the above average productivity growth in manufacturing or tradables more generally, combined with productivity bargaining in that sector, imparts an inflationary bias to the whole economy. This is because wage settlements based on manufacturing productivity growth result in larger unit labour cost increases in other sectors where productivity growth is lower.

Figure 14.2 shows the evolution of output per head both in manufacturing and for the whole economy. The flattening of each curve between 1973 and 1980 is notable. Thus the speed-up in growth of output per head in UK manufacturing in the 1980s has to be seen against the slow-down beginning in 1973. Economists also like to focus on the growth of 'total factor productivity' which is the growth in output not explained by the growth of the factor inputs. When output is defined as value added, the factor inputs are labour and capital, see Muellbauer (1986) for an introduction to the concepts and literature. Growth of total factor productivity slowed in all industrial countries in the 1970s. There is still no universal agreement on the causes of this slow-down except that it was not merely a short-term cyclical phenomenon.

1. Causes of the 1970s' slow-down

One popular hypothesis relates the slow-down to the energy and raw material price increases of 1973 and, in particular, to a loss of productivity of capital after these shocks. Thus, much of the capital stock would have been designed for the low real energy and raw material prices before 1973 and so less efficient at the new real prices. This, combined with the recession induced in industrial countries, would have led to increased scrapping of buildings, plant, and machinery. The combination of such scrapping, not incorporated in official data, and loss of productivity of capital would have reduced the growth rate of output per head.

A second hypothesis, advanced by Bruno and Sachs (1982), suggesting biases in the measurement of output resulting from the input price shock, seems not to have been quantitatively very important for UK manufacturing. A third hypothesis concerns labour markets and industrial relations. It is that with reduced real income generating capacity in the industrial countries there was a sharpened conflict between reality and the aspirations for real wage growth of workers and their representatives. This conflict may have resulted in increased use of restrictive practices, strikes, and other modes of non-co-operation that showed itself in reduced productivity growth. A fourth hypothesis concerns the adjustment costs of adapting to the economic dislocation caused by the input price shocks of 1973, and the ensuing inflation to which tax and accounting systems were ill adapted. Government intervention in bailing out financially bankrupt firms may well have altered the incentive structures of managements and unions to lower the priority on cutting labour costs.

2. The 1980s' upturn

The recovery in manufacturing productivity growth from the end of 1980 is striking, even after correcting for cyclical labour utilization effects, as demonstrated by Mendis and Muellbauer (1984) and Muellbauer (1986). I have updated this research to data up to the end of 1990. The methodology consists of fitting a production function incorporating the assumption of constant returns to scale to UK manufacturing data for 1956 Q1 to 1990 Q4. Output is real value added. Various biases in the index of output are discussed in Muellbauer (1986). One of current relevance is the list

price bias. It is in the nature of this bias to vanish in the long run, but in the short run to be positively related to the rate of change of the ratio of prices of raw materials purchased by manufacturers to domestic producer prices. For example, the ratio of raw materials and energy prices paid by UK manufacturers to output prices received fell during 1990. Since list prices take longer to adjust than actual transactions prices, it is likely that prices at the end of 1990 were overstated relative to a year before and output correspondingly understated. The empirical model suggests a bias of about 1 per cent in this instance. Labour input is measured by the combination of number of workers, number of weeks worked per year (which falls because of rising holidays), and *effective* weekly hours of work. As Ball and St Cyr (1966) explain in their classic article, effective weekly hours are hard to measure and are typically different from paid-for hours. Workers are often paid for a standard work week of 38 or 39 hours without being fully utilized. As explained in Muellbauer (1984, 1986), data on overtime hours can be used to construct a utilization index which corrects labour input for variations in utilization.

The capital stock is measured by the CSO's gross capital stock series. The chief problems with it are the independence of the assumed service lives from economic conditions and the lack of direct measures of utilization. In particular, one would expect a higher rate of scrapping from 1973 because of the increases in the relative prices of raw materials and fuel, and later because of depressed demand conditions and because of Britain's steadily worsening competitive position from the beginning of 1977. Similarly, it seems plausible that in 1979–80 the further worsening of competitiveness, the effects of the second oil shock, high real interest rates, and fiscal contraction would have led to an increase in scrapping.

In the equation which is estimated, shifting time trends pick up the joint effect of changes in technology, work practices, and these measurement errors in the capital stock. In part, these measurement errors can represent under-utilization and, in part, permanent scrapping. In practice, making a clear-cut distinction between the two is hard. Vintages taken out of use temporarily may be scrapped permanently if expectations about cost and demand conditions suggest that no recovery is likely in the medium term.

There are also seasonal effects, including unusual winter temperatures, and 1972 and 1974 miners' strikes dummies. The output bias consists of four terms as in Muellbauer (1986). The estimated equations both for the sample 1956 Q1–1990 Q4 and for the data period 1956 Q1–1985 Q4 taken in Muellbauer (1986)

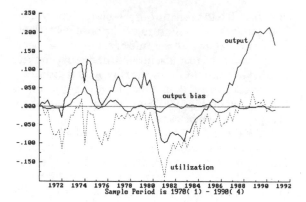

Figure 14.3. Labour Utilization, Output, and Output Bias

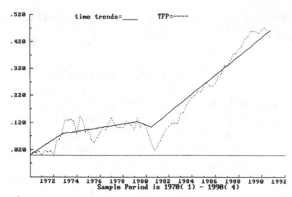

Figure 14.5. Total Factor Productivity

Figure 14.4. Utilization Measures

confirm the remarkable stability of all the parameter estimates.

Figure 14.3 plots for 1970 Q1 to 1990 Q4 output and the output bias and utilization u. This shows clearly the role of variations in the rate of utilization in accounting for variations in output and the much less important and more short-run role of the output bias.

It is interesting to compare in Figure 14.4 the increase of labour utilization with a utilization index derived, as explained in Muellbauer (1984), from the CBI capacity utilization measure. This is the proportion of firms, in the CBI Industrial Trends Survey, who answer 'yes' to whether their current level of output is 'below a satisfactory or full rate of operation'. The series move quite closely together though the CBI measure shows a smaller dip in 1980 and a much more pronounced peak in 1988–9. Even if one corrects for seasonal fluctuations in labour utilization (which the figure does not), it is clear that in 1990–1 the downturn comes sooner in the CBI measure than in labour utilization, though by 1991 Q1, the two mea-

sures coincide. There may well be a certain amount of sluggishness in the adaptation of overtime hours to economic conditions.

Let us turn to the estimates of the shifting trends. These imply the following annualized rates of 'total factor productivity growth' (defined as changes in output that cannot be explained by the other variables, chiefly weeks worked, number of workers, effective weekly hours of work, the measured capital stock, and the output bias terms):

1956 Q1 to 1959 Q3	1.8 %
1959 Q4 to 1972 Q4	2.9 %
1973 Q1 to 1979 Q2	0.7 %
1979 Q3 to 1980 Q2	−1.6 %
1980 Q3 to 1990 Q4	3.3 %

On this data set the underlying trend for 1980 Q4 to 1985 Q4 estimated for the sample ending in 1985 Q4 also comes out at 3.3 per cent.[2] Thus, the evidence suggests that the underlying trend remained steady into the second half of the 1980s.

Figure 14.5 plots total factor productivity not corrected for utilization and the output bias against the combined effect of the split time trends. These represent the combined effect of changes in technology and work practices, measurement errors in the capital stock, and improvements in labour quality. Figure 14.5 confirms that one can be seriously led astray by paying too much attention to productivity data not adjusted for utilization and output bias.

[2] Compared with the results reported in Muellbauer (1986), these results suggest an underlying trend in the 1980s somewhat more in excess of the 'golden age' from 1960 to 1972 than previously reported. The technical reason appears to lie in a slightly different way of aggregating the overtime data than previously used.

3. Causes of the 1980s' upturn

We have seen that after cyclical correction, the underlying trend beginning in late 1980 has stayed remarkably steady, at least up to the middle of 1990. This has some implications for the causes of the 1980s' upturn. An important component of any plausible package of explanations must entail the mirror image of the explanations of the 1970s' slow-down discussed above. In particular, the story about capital productivity and unobserved scrapping surely reverses after 1980. The drastic rise in the exchange rate and in interest rates in 1979–80 combined with a tight fiscal policy and a global recession led to the biggest one-year fall in manufactured output ever recorded in the UK. It seems very likely that a great deal of unrecorded scrapping took place in 1979–80. After the recovery then began in 1981, the proportionate increase in the capital stock from a lower base would have been substantially greater than recorded in the official data. This would have led to an upward bias in total factor productivity growth, i.e. in the growth of output once the growth in the capital stock and labour had been taken account of. Apart from myself, Price et al. (1990) and Darby and Wren-Lewis (1989) favour this as part of the story. In addition, Darby and Wren-Lewis have used CBI expectations data to argue that, in the late 1970s, UK manufacturing firms had consistently too optimistic expectations and hence hoarded 'too much' labour. The 1979–80 recession and the radical change in economic policy that came in with the Thatcher government in June 1979 punctured these expectations. In the great labour shake-out that occurred after the implications of the new economic policies sank in, productivity growth rose sharply.

These two hypotheses are nicely complementary in that together they can account for the steadiness of the underlying productivity trend beginning, according to our research, in the fourth quarter of 1980. Company liquidations in manufacturing continued to increase from the end of 1979 to 1982 and then remained at high levels for several more years. Even if these data lag behind the actual closure of capacity, they suggest that unobserved scrapping was far from over by 1980 Q4, implying a continued negative effect on measured total factor productivity. However, as companies compensated for years of excessive labour hoarding by bringing employment into line with their shattered and now realistic expectations, labour productivity grew unusually rapidly for a while, roughly cancelling out the negative effect of continued high rates of unobserved scrapping.

A third element is surely the 'industrial relations' hypothesis: the rise in unemployment and government legislation[3] has weakened trade unions and restored to managers the 'right to manage'. Thus workers have been less able to resist the introduction of new technology and more flexible work practices. This hypothesis could be consistent with a permanently higher rate of growth of productivity.

An associated fourth hypothesis is that the overvalued exchange rate and high interest rates, the withdrawal of subsidies and of the option of a state-financed rescue of ailing companies, and the sheer scale of the recession, put managers under unprecedented pressure to cut costs and improve work practices.

A fifth hypothesis, 'shedding of the below average', is that the improvement in the average quality of workers, management, and plant from the shedding of the less productive among them contributed temporarily to the improvement in productivity growth. There may be an element of truth in this but at least one industry study (Grant, 1985) suggests that the more dynamic and growth-oriented firms, therefore carrying bigger proportionate debt burdens, were more likely to go under in the high interest rate, high exchange rate regime of the early 1980s.

A sixth hypothesis, the 'microchip' hypothesis suggests that the 1980s experienced a wave of innovation in new technology, for example, through the spread of computer-controlled machine tools and through computer-aided design. This is, no doubt, true. But the improvement in manufacturing productivity growth internationally was by no means as uniform as this hypothesis would imply. Japan and the US also show an improved trend in manufacturing productivity growth in the 1980s. Indeed, for the US the upturn is even more striking than for the UK, though it began a little later. However, for Germany and France there is a *further* slowdown in manufacturing productivity growth in the early and mid-1980s.

A seventh hypothesis, which is part of a number of comparative analyses of international manufacturing productivity trends (Helliwell et al., 1985; Englander and Mittelstädt, 1988; and Crafts, this volume), is the catch-up hypothesis. This is that there is a tendency for levels of productivity, particularly in traded goods, to converge internationally for economies at similar stages of development and similar educational levels. The process of international competition and the diffusion of technologies, work practices, and the conscious emulation of success account for this tendency. In the 1970s, UK manufacturing slipped behind most of its

---

[3] Mayhew (1985) describes this in detail.

competitors in productivity growth. Thus, when the incentives were sharpened and union opposition weakened, there was considerable lost ground to be made up in the 1980s.

These are the main ingredients[4] of an explanation of the increase in manufacturing productivity growth in the 1980s. However, the picture is not entirely rosy. Estimating the econometric model up to 1991 Q1, the residual pattern for the last three quarters suggests some deterioration in productivity growth which is not being captured by declining utilization. This is very like what occurred between 1979 Q4 and 1980 Q4 when there was also a downward blip in productivity growth not otherwise explained by the model. The circumstantial evidence suggests that unobserved scrapping may be the explanation. The statistics on company liquidations for the last year are consistent with another major loss of capacity by UK manufacturing in the current very deep recession. Despite much discussed Japanese investment in vehicles, television, and video recorder production, this must be a reason for serious concern.

III. Productivity growth outside the industrial sector

The CSO defines the industrial or 'production' sector as manufacturing (orders 2 to 4 of the 1980 Standard Industrial Classification) plus energy extraction and water supply (order 1 of the 1980 SIC). This is 34.4 per cent of 1985 value added. The rest of the economy includes agriculture (1.9 per cent), construction (5.9 per cent), and services (57.8 per cent). Services break down into distribution, hotels and catering, and repairs (13.4 per cent), transport and communication (7.0 per cent), financial services (15.5 per cent), ownership of dwellings (5.9 per cent), public administration, national defence, and social security (7.1 per cent), education and health services (8.5 per cent), and other services (5.9 per cent).

---

[4] Among less central contributing factors, one worth mentioning is the 'subcontracting' hypothesis. This is that subcontracting and hiving off parts of the activities of manufacturing companies to the service sector increased. But because of inaccuracies in the approximations used to measure value added these did not result in decreases in measured value added corresponding to decreased direct employment. O'Mahony and Oulton (1990) take an alternative route to measuring total factor productivity by examining growth in gross output after taking account of inputs including materials and business services. Interestingly, they find productivity growth in 1982–6 to be almost the same as in 1954–73 on this basis, but significantly higher in the CSO's net value added basis adopted both in the present paper and in most others on the subject. The 'subcontracting hypothesis' could account for their finding.

The measurement of value added in the service sector is often problematic, particularly when the output is not marketed. But even when it is, there are issues of quality change. For example, computerization of insurance companies' records makes it possible to make an over-the-phone change in an insurance contract and obtain postal confirmation the next day. Twenty years ago the procedure would have been less user-friendly and would have involved much more staff time. With the advent of mobile telephones, a plumber is able to schedule his work better, reducing wasteful trips and waiting by customers. Word processors and computers have raised the output and quality of office work. The CSO attempts to take quality improvements into account, but, as CSO (1985) notes, in parts of the service sector, outputs are measured by inputs, leaving no scope for productivity change.

Smith (1989) presents some alternative output measures for financial, recreational, and catering services, and compares these with the CSO's. Smith points out that, according to the CSO, output per head in catering declined by 1.7 per cent per annum in 1971–86. Even after correcting for the changing mix of part-timers and self-employed, it declined by 1.2 per cent. This seems hardly credible and one wonders whether quality, e.g. range and quality of services, the nature of the food served, and the milieu of establishments, have not improved in compensation.

Smith makes adjustments for various types of commercial services and finds that they are not always upward revisions of the CSO figures. However, for the broadest aggregate examined, he finds growth of 45 per cent between 1980 and 1986, compared with 35 per cent on the CSO figures. And even the higher figure may not capture all of the quality improvements that have occurred. Despite these reservations about the CSO's measure of output outside the industrial sector, it is still of some interest to examine the results of a production function study on lines similar to that for the manufacturing sector in the previous section. We define an index of output at constant factor cost for the economy excluding the production sector and the ownership of dwellings. We define the gross capital stock correspondingly. Employment is defined as an index of different types of employees, male and female, and full-time and part-time, and incorporating the self-employed, who are much more important here than in the production sector. We use the following weights: 1 for a full-time male; 0.5 for a part-time male; 0.7 for a full-time female; 0.3 for a part-time female; and 1 for a self-employed person. The weight for a full-time female is somewhat more than the corresponding ratio of average earnings to reflect the likelihood that some

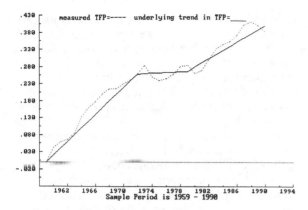

Figure 14.6. Total Factor Productivity Growth for the Non-Industrial Sector

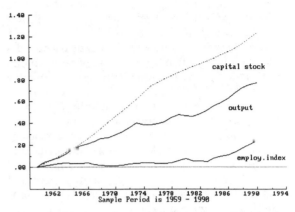

Figure 14.7. Output, Employment Index, and Capital Stock (Logs) for the Non-Industrial Sector

discrimination against women remains. Weeks of work and normal hours are taken account of as for the manufacturing sector. We have no general indicator of overtime hours or any other indicator of labour utilization outside the production sector. Since there will be some correlation in economic activity between manufacturing on the one hand, and construction plus services on the other, I have experimented with the use of the CBI capacity utilization measure and, with similar results, the measure of labour utilization in manufacturing to remove some of the cyclical variability. On annual data for 1959–90, we fit a constant returns Cobb–Douglas production function incorporating time trends shifting in 1973 and 1980 and the labour utilization measure described in the previous section.

The utilization effect is positive and very significant at the 5 per cent level. However, the residuals are serially autocorrelated suggesting, perhaps not surprisingly, that the cyclical effects are not being well modelled.

Most surprising are the shifting time trends. These suggest a 2.0 per cent rate of total factor productivity growth for 1959–72 falling to 0.1 per cent in 1973–9 and then picking up, but only to about 1.2 per cent per annum. Figure 14.6 shows total factor productivity growth and the combined effect of the shifting trends. The slow-down in productivity growth in 1989 and 1990 is marked. It owes something, no doubt, to the cyclical slow-down in service sector activity which came before the slow-down in manufacturing. If one leaves out the 1989–90 observations, measured trend total factor productivity growth in 1980–8 is restored to the pre-1973 rate.

Figure 14.7 shows more of the background: while the 1980s have seen rapid growth of output in this sector there has also been some growth in employment, especially self-employment, even taking into account the appropriate discount for part-time workers. The capital stock in this sector has grown even more spectacularly than output. One is left with the suspicion that quality improvements in service outputs in the 1980s have been significantly undercounted. If that is correct, then output and productivity growth outside the production sector and hence in the economy as a whole have been underestimated.

Insofar as one can trust the data, it appears that the 1970s' slow-down in productivity growth and the subsequent speed-up applied in the service sector as well as the manufacturing sector. This suggests that one should not look for the causes primarily in the energy and raw material-using aspects of technology but more in the other factors discussed in the previous section.

IV. Wages

When Mrs Thatcher first came to power, average earnings in the British economy were rising at around 15.5 per cent. The rate of increase peaked in 1980 at around 21 per cent, partly the result of the 1979–80 oil shock, Clegg pay awards, and the VAT increases of the previous year. Given disinflation in the world economy and rapidly rising unemployment, the rate of increase of earnings then fell in every year until it bottomed at 6 per cent in 1984. After hovering around 8 per cent for the next three years, it rose to 8.7 per cent in 1988, 9.1 per cent in 1989, and 9.7 per cent in 1990, peaking at 10 per cent around the middle of the year.

Writing an article on productivity and competitiveness for the *Oxford Review of Economic Policy* in 1986 increased my fascination with the puzzle of wage determination and led to a programme of research into some issues not previously examined. Using the Layard–Nickell (1986) framework, the models in Bover *et al.* (1989) have some fairly conventional features. Real wages paid by firms are modelled using a productivity trend (on which more below); the log male unemployment rate and its three-year rate of change capture unemployment hysteresis; trade union power is represented by union density; as in Layard and Nickell, mismatch shocks are represented by the absolute change in the proportions of employment in the industrial sector: this helps to explain why the dramatic increases in unemployment in the early 1980s were only moderately effective in bringing down wage inflation. The rise in unemployment was very concentrated in the industrial sector and in the north, and so had relatively little effect on wages in services and in the south.

However, what is new about this research is the incorporation of the housing market effects anticipated in Muellbauer (1986). Our research (Bover *et al.*,1989) examines the effects of house prices on wages. Everyone knows, of course, that higher earnings quickly drive up house prices. But it had not been realized that there is also a reverse effect with a delay of between one and three years. We find that both the average UK house price/wage ratio and the South-East/UK difference in it help to explain real wages in later years. Much controversy still surrounds the exact mechanisms by which these effects operate (see the Wages and House Prices Symposium in the *Oxford Bulletin*, 1989). The house price/earnings ratio in the South-East relative to the rest of the economy rose from 1983 to 1988, reaching unprecedented levels in 1988. Part of our story is that this helped to crowd out workers, especially manual workers, from the South-East, adding to wage pressure there. This is supported by evidence on regional migration (Muellbauer and Murphy, 1988). But the extra workers in the more depressed parts of the economy did not cause a corresponding reduction in wage pressure. This kind of asymmetry in wage setting has been much discussed ever since Keynes' General Theory. Thus overall wage pressure increased.

Another element of the story may be that the run-up in house prices in the South-East led to the large increase in south-east earnings supplements. However, because such supplements are never reduced in nominal terms, a general wage increase is the mechanism by which they narrow in proportionate terms when the increase in the South-East house price premium eventually disappears and with it the rationale for a disproportionate increase in the supplement.

There are also direct effects from house prices and mortgage costs on wage demands, since many young workers with mortgages find that 40 per cent or more of their outgoings are swallowed by mortgage interest. This links with the much discussed effects of the 'wedge' between the real wages received by workers for which unions attempt to negotiate and labour costs relative to the prices received by firms. In general, both output prices and consumer prices are relevant in wage determination. Thus, if the wage relative to output prices is to be explained, the wedge between consumer and output prices is relevant for explaining it. One of our Symposium discussants argued that, since house prices were paid by consumers and received by firms, they would wash out of the wedge. However, the national accounts output prices refer to value added in construction and exclude higher land prices implicitly paid by consumers purchasing a dwelling. Between 1981 and 1988, the price of an average building plot in England and Wales, *relative to personal disposable income*, rose from an index of 100 to one of 250, and by even more relative to output prices. It was fortunate indeed (though not a complete coincidence) that another element of the wedge, the ratio of import prices to output prices, staged a fall during the same period.[5]

Our model also tries to incorporate the recognized effects of the UK housing tenure structure on labour mobility. The lack of a free rented sector and the restrictions on the mobility of council tenants have long been seen as handicaps on the efficiency of British labour markets by writers such as Minford *et al.* (1987), and Hughes and McCormick (1981; 1987). Holmans (1987) argues that the decline of the privately rented sector has at least as much to do with tax reliefs on owner-occupation as with rent and tenure controls.

The precise results one obtains for these kinds of models are somewhat sensitive to the treatment of productivity in the wage equation. The Layard–Nickell treatment, which proxies productivity by the ratio of the gross UK capital stock to a measure of the labour force (i.e. including the unemployed) is neither theoretically nor empirically ideal. The poor quality of the capital stock data is well known and the data are sensitive to the rise of North Sea oil production, which is highly capital intensive. Layard and Nickell also ignore the rise in self-employment and in part-time work. We find broadly similar results, but a less impressive fit than

---

[5] One possibility our research did not attempt to investigate is that as unions have become less powerful, the influence of consumer prices may have declined relative to the influence of output prices. If this is so, we may have overstated the 'wedge' effects in the later 1980s.

Bover *et al.* (1989) when we replace the Layard–Nickell productivity proxy by whole economy labour productivity. Here we exclude imputed rent of housing and North Sea oil and gas from output and use an index of employment which weights full-time, part-time, and self-employed workers as in section III above. We also adjust for variations in labour utilization using the measure from section II above. This gives parameter estimates which are stable when the sample is updated from 1986 to 1990.

Let us look at some implications for understanding wage growth in the 1980s. On some conclusions there is widespread agreement between different researchers. Nickell (1987) and other researchers also find important rate of change of unemployment effects on wages. These are consistent with insider–outsider theories of wage determination. The implication is that a given demand stimulus to an economy is less inflationary if it is spread out over a longer period than crammed into a short period. The fall in unemployment began in 1986, and from the middle of 1987 to the middle of 1989 was falling by about half a percentage point per quarter. Everyone now agrees that the Treasury underestimated the growth in demand at the time. In Muellbauer and Murphy (1990), we analyse the sources of demand growth and argue that it was primarily a consumption-led boom with investment largely following on behind. For each of the years 1985–8, the growth of consumer expenditure exceeded that of personal disposable income. This was most unusual: in the previous thirty years, the reverse was always true in economic upswings. There were good reasons to expect strong growth in UK consumer expenditure in the 1980s: increased personal sector wealth; increased actual and expected productivity growth which raised actual and expected incomes; actual and expected tax cuts; all against a background of international and domestic financial liberalization which allowed consumers to borrow more. Further details of our research on the consumer boom are given in Muellbauer and Murphy (1989, 1990). Whilst we acknowledge that expectations of higher *future* incomes play a role in this story—liberalization of domestic credit was more important.

Financial liberalization gave consumers the ability to rearrange their portfolios, making it possible to access illiquid wealth more freely than before, and effectively making illiquid wealth more liquid. In our most recent consumption function, an update of Muellbauer and Murphy (1989), liquid assets and consumer debt have equal but opposite effects, which are larger than those of illiquid assets. The illiquid asset effect rises steadily during 1982–9 with financial liberalization, and is eventually about 25 per cent higher than before 1982, in an equation also including expectation effects. Combined with the large increases in the illiquid asset–income ratio in the 1980s, this accounts for a major part of the consumption boom.

V. The impact of economic policy in the 1980s on productivity and competitiveness

The channels, considered in sub-section 1 by which government may be able to affect productivity include the reform of unions and other labour institutions, changes in the incentives facing managers, the provision of education and training, incentives for investment and R & D and the (perhaps unintended) effect of macroeconomic policies that force firms into a harsher competitive climate in which survival may require the cutting of costs. Other things equal, higher productivity implies higher competitiveness but there are other ingredients in competitiveness, namely wage and other components of costs and the exchange rate, considered in sub-section 2. The possible ways in which government may be able to affect production costs include the reform of unions and other labour market institutions as above via the impact on wage cost, housing policy which may affect one of the most important elements of the cost of living and may have an impact on wage costs via labour mobility and regional labour market mismatch, and demand management. To explain the last, the setting off of domestic inflationary forces can seriously damage competitiveness at a given exchange rate, while an unstable macroeconomic environment increases adjustment costs for firms and the risk premia built into the cost of capital. The exchange rate itself will be influenced by interest rate policy and the decision to belong or not to belong to the Exchange Rate Mechanism and at what rate.

1. Productivity

In section III, I examined various hypotheses for the strong recovery of manufacturing productivity growth in the 1980s. The sterling shock of 1979–80, high interest rates, 20 per cent wage inflation in 1980, tight fiscal policy, a world recession, and the government's refusal to bail out loss-making companies imposed an unprecedentedly ferocious cost-cutting discipline on UK firms in the tradables sector. All commentators agree that this was a significant factor in the recovery of productivity growth. However, it was accompanied by, eventually, a loss of two million jobs in manufacturing and the shedding of much capacity. High productivity growth is all very well, but British economic performance would have been better if it had been achieved on a larger base. Manufacturing output at the beginning of 1991 is not much above the 1973–4 peak.

Regarding the early 1980s, Brittan (1989) has remarked:

The government's great mistake, especially given its gradualist approach to union reform, was to underrate the perversities of the British labour market . . . The result was that the government underestimated both the job losses in tackling inflation solely from the demand side and the underlying forces making for high unemployment.

The weakening of union power (Acts of 1980, 1982, and 1984) withdrew some legal immunities, e.g. for secondary strikes and picketing, made union officials more accountable, eroded the closed shop, and put union funds at risk of seizure in cases of breaches of the law. Many would agree with Brittan (1989) that this contributed to a blitz on restrictive practices. Metcalf (1989), indeed, finds that productivity growth was greater for unionized plants than for non-unionized plants in the 1980s—presumably because of an initially greater incidence of restrictive practices. Gregg and Machin (1990) confirm Metcalf's finding and, further, that within unionized plants higher productivity growth is linked to the removal of union privileges such as closed shop agreements.

In the long run, education and training of workers and management should have a critical effect on productivity growth and on competitiveness, including that of the 'non-price' variety. However, one should beware of expecting too much in the short run from increased investment in human capital and of monocausal explanations of productivity growth. Steedman (1990) documents the great increase in investment in human capital in France in the 1970s and 1980s by comparison with the UK. Yet, for 1980–7, British productivity growth exceeded that in France. More recently, French performance has improved and this is likely to be sustained.

The UK government's record on education and training is, at best, patchy. As pointed out in the *Oxford Review of Economic Policy* issue on this theme (Vol. 4 No. 3), training policy for much of the 1980s was, to a great extent, a short-term palliative to keep the unemployed off the streets. It is hard to fault Mayhew's thumbnail sketch of the overall situation (1991). On training, two overriding long-term problems remain. One concerns the poor quality, lack of external assessment, and comparability of many of the qualifications which the new National Council for Vocational Qualifications is designing. The other is that, despite the new, locally based Training and Enterprise Councils, little has been done to increase the financial incentives for training, either for individuals or companies. Where training has a general element, such incentives are needed to compensate for the external diseconomies of 'free-riding' firms who poach trained workers from others.

Finally, private investment in training has, like investment in plant, been one of the major casualties of the current recession induced by government policy to combat inflation and the balance of payments deficit. The consequences will be with us for many years.

2. Wages, other costs, and the exchange rate

It was argued above that union power declined in the 1980s. This is reflected in measures such as union density and the prevalence of closed shop agreements. The inclusion of these measures in wage equations, typically results in significant positive coefficients with the conclusion that policy had some success in restraining wage inflation via its impact on union power. However, there is another measure of the trade union influence on wages, the union/non-union wage differentials, which as Oswald (1991) documents, remained steady in the 1980s and on the face of it, implies that policy on trade unions had no effect on wages. There are at least two reasons for doubting that the conventionally measured union/non-union wage differential correctly captures union power. The first is that, among manual workers, union members tend to exhibit higher skills, greater seniority and lower turnover. Hence part of the reason for a maintained union/non-union wage differential may lie in the rise in the relative wages of skilled workers compared with unskilled workers. The second reason may lie in sample selection: if trade unionists are more likely to have lost their jobs in the 1980s than non-union members with otherwise similar characteristics, surviving trade union members will, on average, have more favourable relative productivity levels than before unemployment rose. Union/non-union wage differentials are typcially unable to control for these productivity characteristics of individual workers and misattribute to union membership what is partly the effect of these characteristics.

While government policy on unions has had some beneficial effects on wage inflation, there is more doubt about the government's attempts to foster decentralized and localized pay bargaining. At a time of disinflation, co-ordination could bring wage settlements down faster.

As noted in section IV, the UK housing market has hindered labour mobility. The 1988 Housing Act and the BES subsidies for providers of new rented accommodation were a very belated attempt to breathe life into the private rented sector which would encourage labour mobility and a more flexible labour market. The cut in building social housing to about 25 per cent of its 1979 volume and sales of council homes have not been beneficial to labour mobility. Hughes and McCormick (1991) find evidence that the cuts in social housing disproportionately affected migration to the South-East where the housing need was the greatest.

The house price boom, with its initial concentration in the South-East, was inflationary because it contributed to the much stronger than expected growth in demand, because it had direct effects on the cost of living, and because it reduced labour mobility and raised regional mismatch in the labour market.

There has been some controversy over the causes of the unexpectedly rapid growth of demand after 1986. Many commentators, of whom Brittan (1989) is one, lay part of the blame on the 'excessive' devaluation of 1985–6 followed by the shadowing of the German mark by sterling until 1988 at 'too low' an exchange rate. However, as Figure 14.1 reveals, UK international competitiveness in 1986 was still substantially worse than the average levels of 1963–79. UK oil revenues would shrink in the years ahead and production of other tradables needed to be expanded to fill the gap. Moreover, with unemployment at three million, capacity utilization in manufacturing at relatively low levels, and both world and domestic inflation subdued, there was scarcely a better window of opportunity for a successful devaluation.

Given the demand boost from improved competitiveness and from lower energy prices, it was clear in 1986–7 that a fiscal stimulus should be avoided though many in the economic establishment continued to press for reflation for most of the next two years, and could not believe the data when unemployment began to fall in the autumn of 1986. Fiscal policy was excessively expansionary in 1987 and 1988, and monetary policy became very loose between the October 1987 stock market crash and the summer of 1988 when the overheating in the economy at last became apparent to everyone. I discussed the main

causes of this overheating in section IV. Few now disagree with the proposition that financial liberalization played a significant role and that the housing market was overstimulated.

The process of financial liberalization in the UK was sparked by a series of policy decisions. An important pre-condition was the abolition of exchange controls in 1979. One milestone was the removal of the 'corset' on bank lending in 1980–1 which led to the major entry of the clearing banks into the mortgage market: by 1982 they provided 36 per cent of net mortgage advances. Hire purchase controls on consumer durable purchases were removed in 1982. The removal of administrative guidance on building society lending allowed them to access money markets and the 1986 Building Society Act allowed building societies to provide general loans backed by housing collateral. It then became possible for even the retired to access the wealth tied up in their houses much more freely than before. For example, in the late 1980s a retired person could borrow up to 25 per cent of the value of their home and never pay interest, the accumulated interest and debt simply being repaid from the estate at death. Further, a new breed of financial institutions, usually backed by foreign assets, took a rapidly increasing share of the mortgage market beginning around 1985.

Thus, the greatest policy error of all, as I have argued ever since autumn 1986, was not to compensate for the radical reform of the financial system by sweeping away the main fiscal privileges for owner-occupiers. Financial deregulation *alone* was a classic instance of something economists familiar with the 'general theory of the second best' have warned of for decades: in an economy with several important distortions, removing one, while leaving the others in place, can make the overall situation worse than doing nothing. The fiscal privileges distorted the tax system and the housing market, especially given tight land-use planning policies in the areas where growth was greatest, so feeding into house and land prices. Until Mrs Thatcher was forced out of office in November 1990, it was hardly even possible to discuss reform. Indeed, the flagship of her post-1987 policies, the abolition of domestic rates and their replacement by the poll tax, made matters very significantly worse by further raising the temperature of the speculative fever in housing. It is much to Mr Lawson's credit that he strongly opposed this policy, though without the ultimate threat of resignation, and that he resisted the Prime Minister's annual proposals to raise the £30,000 tax relief ceiling. The position of the Prime Minister was so dominant and the imperviousness of her mind to economic logic on this score so well known, that the policy advice from this pen was treated until 1989 with a kind of embarrassed pity by the key advisers to Lawson and Thatcher.

There were other problems. None of the quarterly econometric models incorporated physical wealth, though most included financial wealth. Why this was so remains a mystery, especially given Pattersons's (1984) annual consumption function incorporating housing wealth. The simple explanation is that the CSO does not routinely produce quarterly estimates of personal sector physical wealth and the modellers could not be bothered to undertake the quarterly interpolations using house price indices. At any rate, after the October 1987 stock market crash, all predicted a much bigger slow-down of demand than would have occurred with a comprehensive definition of wealth and than did in fact occur. A further problem was the well-known deterioration in the quality of official statistics, in part the result of deregulation and of the structural changes in the economy, and in part the result of cuts in the Government Statistical Service. Finally, there was too much confidence in the 'Thatcher miracle' by which tax cuts and other improvements in incentives and institutions had supposedly revolutionized the performance of the UK economy.

It was not until the March 1988 Budget that the first policy move was made on the fiscal privileges of owner occupiers. This was to limit mortgage tax relief to one relief with a £30,000 ceiling per property instead of one per borrower. However, Mr Lawson committed the serious blunder of announcing that the policy would only come into effect on August 1, 1988.[6] Instead of dampening the speculative fever in the housing market, it therefore fuelled a final speculative spurt which drew in many young purchasers, particularly in the South, who subsequently found themselves bearing large capital losses and, in all too many cases, eventually lost their homes. When the demand overshoot signalled by a soaring balance of payments deficit and higher growth and inflation figures at last become obvious, base rates were raised from $7^{1}/_{2}$ per cent in May 1988 in 6 separate moves to reach 15 per cent in October 1989 where they remained until October 1990. Unfortunately, in the UK, higher interest rates feed straight into the mortgage cost component of the Retail Price Index to which many other prices are indexed. It was not always so: until 1974 the UK followed the practice of Germany and the US in imputing changes in home-

owners' housing costs by parallel movements in market rents. However, mainly as a consequence of the cumulative effects of rent controls, the rental market had become too small by 1974 to sustain such imputations. The structure of the UK housing market therefore makes interest rate rises a peculiarly unsatisfactory weapon against inflation.

Tight monetary policy applied too late and ERM entry at an over-valued exchange rate in October 1990 took the brunt of the response to the inflationary domestic demand overshoot. Fiscal policy in the 1980s had, if anything, become destabilizing: faster growth generated higher tax revenue and reduced income support payments and the lower PSBR was then used to justify tax cuts that further fuelled growth. Unstable growth, inflation and particularly interest rates raise adjustment costs of firms and create a climate of uncertainty that must raise the risk premium in the cost of capital and must impede investment in capital, R & D and training. Together with the resort to an overvalued exchange rate to squeeze inflation when its causes were domestic, such policies reduce the capacity of the economy relative to what it woud otherwise have been and this either causes inflation or reduced economic growth in later periods.

VI. Prospects

1. Productivity

As argued in section III, manufacturing productivity growth, after correcting for cyclical effects, is due for a temporary set-back because of the shedding of capacity. Official measures of the capital stock do not incorporate such capacity losses. While there remains a substantial productivity gap between the UK and its chief competitors, there are forces that will continue to close the gap. But these forces are likely to meet increasing resistance from the underlying disadvantage experienced by the UK in terms of education and training of its workforce. Nevertheless, the growth of output per head is soon likely to recover as the shedding of labour proceeds.

In the service sector a sharp cyclical deterioration in productivity growth was already under way in 1989 and continued in 1990. Given the fall in demand for services and tougher domestic competitive conditions, a sharp recovery in productivity growth in services has probably already begun.

There is little scope for policy to have a short-run effect on productivity growth except through cyclical

[6] This was on the adivice of Inland Revenue who said that mortgage lenders would otherwise have difficulties in reprogramming their computers in time.

demand variation. But in the longer term the weaknesses in vocational training mentioned in section V need to be corrected, and the quality of schooling for the mass of pupils needs to be raised, see Prais and Wagner (1985) for a comparative analysis of UK deficiencies.

2. Wages

The econometric work discussed in section IV has fascinating implications for wage disinflation over the next two or three years. The rise in unemployment will plainly continue well into 1992. A major difference from 1980–3 is that the rise in unemployment is much more uniformly distributed over the economy. Indeed, the recession began in the South-East, which, because of its higher indebtedness, was more exposed to the rise in interest rates. And the shake-out of jobs and in bankruptcies in the service sector continues apace. This means that the rise in unemployment will be much more effective in reducing wage pressure. As far as the housing market is concerned, the house price to wage ratio peaked in 1989 and has fallen sharply since, though it is still above its historical average in 1991. However, regional house-price differentials are back at historical average levels, thus removing a significant impediment to regional mobility. Our estimates suggest that most of the benefits of these changes have still to feed into wages. Furthermore, there are signs of a revival in the private rented market. The 1988 Housing Act has improved the contractual framework for renting. The Business Expansion Scheme subsidies have helped to compensate for the tax advantages of owner-occupation and given a moderate stimulus to the supply of rented accommodation. The large losses made by those who bought houses in recent years, and the wave of mortgage repossessions, have dented the image of owner-occupation. Further policy changes are needed: depreciation allowances for tax should be extended to landlords generally, and the fiscal privileges of owner-occupiers contracted further. For example, the next Budget could permanently restrict mortgage interest tax relief to eight years for all owner-occupiers, or reduce the tax relief ceiling from £30,000 to £25,000, or both. Such policies would help expand the private rented sector, improving labour mobility and the functioning of labour markets. Construction of social housing should be greatly expanded and land-use planning more concerned with development and less merely with restriction. All of this would help the recovery in housing markets, due in 1993, to go into more dwellings rather than into higher prices. Such

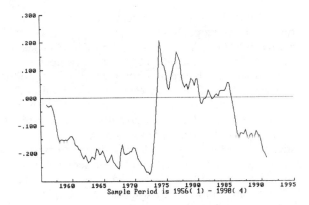

Figure 14.8. Log Input Prices/Output Prices for Manufacturing

policies would sustain disinflation through the 'wedge' between consumer and producer prices discussed in section IV.

The fiscal position of any government in the next few years will be a tight one. The pressure to raise revenue will be great and governments will be well advised to minimize the impact on the wedge. It is desirable at the margin to shift more taxation on to rentiers, particularly those who derive their returns from the most immovable asset of all—land. A land tax, far from distorting resource allocation, has net benefits for the efficiency of resource allocation.

3. The exchange rate

How sustainable is the current exchange rate of sterling within the EMS at around 2.95 DM? In Muellbauer and Murphy (1990) we have given a comprehensive account of the underlying balance of payments position. Despite considerable doubts that the UK's net foreign asset position is as strong as recorded in the Pink Book, the asset position is probably fairly healthy. Increasing European integration will, in any case, reduce the importance of a balance of payments constraint on policy. There are other reasons why part of the answer to the above question is: the current level of sterling is less unsustainable than the 1980–1 exchange rate. Figures 14.1(a) and 14.1(b) suggest that international competitiveness is not quite as bad as the worst point of 1980–1, particularly in unit labour costs. Manufacturing also faces substantially lower real raw material prices than in 1980–1, see Figure 14.8. If a new world recession can be avoided, the current manufacturing recession will be less deep than in 1980–1.

However, while wage inflation is likely to fall to

remarkable low levels in the next two years (leading many commentators to see it entirely and mistakenly as a result of the exchange-rate constraint) it would take years of wage inflation below those of our competitors to make a noticeable dent in our international lack of competitiveness. Given that real interest rates will have to remain high to sustain the current level of sterling and that fiscal policy is under pressure, it is hard to see where growth will originate in the next year or so. In such circumstances of continuing rising unemployment, rising spare capacity, and much reduced skills shortages, an exchange-rate realignment will emerge as a more and more attractive option, especially as part of a general EMS realignment. Provided domestic disinflationary policies are in place, a modest adjustment will turn out to be relatively free of inflationary risk.

VII. Postscript

This postscript has two parts. The first comments on the aftermath of the September 1992 devaluation which I argued in 1991 would 'emerge as a more and more attractive option' and one 'relatively free of inflationary risk'. The second raises some basic questions over the measurement of manufacturing output and the extent of a Thatcher 'productivity miracle'. These suggest that my article was too sanguine about the pickup in manufacturing productivity growth in the 1980s and probably too pessimistic about the slowdown after 1973.

1. 'Golden Wednesday' and its aftermath

On Wednesday September 16, 1992, the UK left the Exchange Rate Mechanism and Sterling stabilized at 2.45DM in the last 3 months of 1992, and at 2.48DM both in 1993 and 1994, compared with 2.88 in the first 8 months of 1992. 2.45DM represented a devaluation of 15 per cent against the German mark, though only 13 per cent on the trade weighted exchange rate index, given other currency realignments. It was Black Wednesday for the reputation of the policy makers though the Chancellor, Norman Lamont, later confessed to singing in his bath when the immediate trauma was over.

In the Summer of 1992, it had become increasingly clear that the government's commitment to the ERM band centred on 2.95DM was incredible. There were four arguments.[7] The first centred on the evidence from indices of international competitiveness, reviewed above, for the over-valuation of Sterling. Church (1992) estimated the 'Fundamental Equilibrium Exchange Rate' (FEER), suggesting an over-valuation of 15–18 per cent. The FEER aims to compute the exchange rate consistent with macroeconomic equilibrium. It is a comprehensive measure taking into account competitiveness, trade performance, oil, capital flows and economic activity.

The second argument centred on the loss of long-term capacity in the production of tradable goods which had occurred in the UK as a result of competitiveness worse since 1979 than in the previous 15 years. International comparisons suggest a highly significant relationship between competitiveness and subsequent growth in manufacturing output and hence capacity. For example, by contrast with the UK, Italy experienced a roughly constant competitiveness index between 1977 and 1988 and growth of manufactured output of 37 per cent in 1979–89 in contrast with the UK's 11 per cent (which, as we shall see below, may in fact have been zero or negative) over the same period. This capacity effect is ignored in most econometric models, which thus underestimate the long-run output effects of *appropriate* devaluations and overestimate their inflation effects.

The third argument hinged on trade performance: despite weak domestic demands stemming from high real interest rates, falling income and wealth and the debt trap faced by many households, the UK current account deficit in 1992 was running at £9.5bn per annum and at a £13bn visible trade deficit. Moreover, the more positive conclusion some commentators drew from the apparent stability of the UK's share of world exports manufactures was likely to be illusory given the increasing use of imported components in UK exports which thus overstated the net value added of exports.

Fourthly, 5 years of trade deficits had shrunk Britain's international wealth and so its ability to finance further deficits.

Given these fundamentals, it would have taken sharp rises in interest rates to counter speculative pressure against Sterling. But given the already depressed state of the domestic economy and the debt overhang for households and in the commercial property sector and hence the fragile state of the financial system[8], it was

---

[7] Which I put in an article published in the *Financial Times* On Monday September 14, 1992.

[8] It later emerged that the Bank of England provided support of the order of £4b in the early 1990's to shore up financial institutions.

clear that sharply higher interest rates could not be sustained.

The outcome was not only inevitable but beneficial. Devaluation and departure from the ERM made it possible to bring down base rates, bringing relief to in-debted households, and gave manufacturing produc-tion and exports a much needed boost. The inflationary consequences were mild, in contrast to what most fore-casters said at the time. For example, UK consumer prices rose 7.4 per cent between 1992Q3 and July 1993, only 2.1 per cent more than in France over the same period, where growth and unemployment both fared worse. The policy makers scrambled to compensate for the loss of anti-inflationary credibility they had endured. One of the outcomes was an increase in the effective autonomy of the Bank of England and a clari-fication of its anti-inflationary objectives. Another was a rise in taxation to repair the large hole in the public finances. Given the large stimulus to the economy from the lower exchange rate and lower interest rates, this was sound not only as stabilization policy but helped to make the devaluation effective by ensuring that resources were free to move into the tradable sector. In the 1991 Budget, mortgage interest tax relief had already been restricted to the standard rate of income tax. In March 1993, it was announced that tax relief would be restricted to 20 per cent from April 1994 and this fell to 15 per cent from April 1995.

In two other respects policy was less supportive of long-term objectives of growth without inflation. First, the one year rise announced in November 1992 in de-preciation allowances to stimulate investment, was far too brief in duration to provide a significant offset for the battering which business confidence and the belief by business in the stability of the macroeconomic en-vironment had received as a result of the policy errors of the preceding 6 years. As a result, fixed capital for-mation has been disappointingly weak in the 1993–5 recovery. Second, as far as education and training, im-portant for long-run economic performance, are con-cerned, there has been no sign of improved finance and little move on the reforms still needed, though the in-tegration of education and training under a single able Minister was a welcome development.

2. Doubts about the Thatcher productivity miracle

Recent research by Francis and Stoneman (1995) and Greenhalgh and Gregory (1995) has profound impli-cations for the assessment of the Thatcher years. They suggest that, correctly measured, real value added in UK manufacturing and hence productivity grew of the order of 1 per cent per annum more slowly over the 1980s than reported by the CSO. Moreover, the logic of this critique, which has the same basis as that of Bruno and Sachs (1982), suggests that between 1973 and 1979 the official figures may have understated the growth of output and therefore of productivity. Esti-mates by Gavin Cameron suggest an underestimate in this period of between half and three quarters of one per cent per annum. The implication is that total fac-tor productivity growth in the 1980s was less than a percentage point per annum greater than in the 1973 to 1979 period and was probably below the rates achieved between 1959 and 1973. The 1980s were there-fore far from the Thatcher productivity miracle, the belief in which itself contributed to the policy errors of the 1980's.

Why then was faith in the CSO figures, shared by my 1991 article, misplaced? To understand the problem, consider how the growth rate of value added *should* be measured. For UK manufacturing as a whole we can define 'final' gross output as gross output minus the sum of outputs of the various sec-tors of manufacturing used as inputs by other sectors of manufacturing. Suppose that final gross output is given by a constant returns function of capital, labour, raw materials, and service inputs, see Oulton and O'Mahony (1994), p. 119. Then the growth rate of value added is defined as the growth rate of final gross output minus the cost share weighted growth rates of raw material and service inputs divided by one minus the sum of the cost shares of these last inputs. The growth rates are of constant price 'volume' indices. The algebra, which also shows the connection with total factor productivity growth or the Solow residual R, is as follows:

$$\Delta\ln Q = w_k\Delta\ln K + w_L\Delta\ln L + w_M\Delta\ln M + w_S\Delta\ln S + R$$

where Q = final gross output, K = capital, L = labour, M = raw materials including imported intermediate manufactures and S = services and the w's are the re-spective cost shares. Then the growth rate of value added is defined as

$$(\Delta\ln Q - w_M\Delta\ln M - w_S\Delta\ln S)/(1 - w_M - w_S)$$

which measures the contribution of capital and labour plus the Solow residual to the growth of final gross out-put.

To measure (1) requires estimates of the volume indices M and S or estimates of price indices for mater-ials and services purchased by the manufacturing sec-tor so that the current price figures reported in the

Censuses of Production can be correctly deflated. The technical term for this deflation is 'double inflation'. However, the CSO does not go in for these complications and instead approximates the growth rate of value added by averaging the growth rates of gross outputs in the different sectors weighting these by the shares in value added[9] of the different sectors.

When volumes of inputs of raw materials, imported semi manufactures, and services rise more rapidly than the volume of final gross output, the CSO's method will *overstate* the growth of value added. An increase in the 'roundaboutness' of production with UK manufacturing firms buying more manufactured intermediate goods from other UK manufacturers would also result in such an overstatement. In the 1980s, real raw material prices have been largely on a downward trend, particularly after 1985. The figures for imports of semi-manufactures show more rapid growth than does gross manufacturing output. There has also been a trend towards the outsourcing of services so that volumes of bought in services have also risen more rapidly than gross manufacturing output. Whether 'roundaboutness' has also risen is less clear. The other factors, however, conspire to lead to the CSO overestimating the growth of value added in manufacturing in the 1980s.

Stoneman and Francis (1995), using annual Census of Production data, estimate the bias between 1979 and 1989 at 1.2 per annum, with the biggest overestimates of output growth occurring in 1985 and 1986 when input prices fell relative to output prices as shown in Figure 8 of my 1991 article. Greenhalgh and Gregory (1995) use the 1979, 1985 and 1990 input-output tables which, in some respects, should give cleaner estimates of value added. They split manufacturing into 'high tech' (including pharmaceuticals, mechanical engineering, electronic equipment, motor vehicles and aerospace) and 'other'. They find that high tech value added was 8.6 per cent lower in 1990 than in 1979 while gross output was 15.4 per cent higher. In other manufacturing value added fell 9.1 per cent while gross ouput fell 6.7 per cent. For total manufacturing, real value added grew about 1.2 per cent per annum more slowly than gross output, a similar bias to the one reported by Stoneman and Francis.

It seems likely that the massive rise in the ratio of input to output prices in manufacturing between 1972 and 1974 would have led to slower growth in raw material inputs and therefore to have led to the official figures underestimating the growth of real value added (or overestimating its decline). In the longer run, the trends to increased outsourcing of services and the increased use of components manufactured abroad, which were probably already in train in the 1970s, would have led to a partial off-set of this bias.

The result is that the official figures probably overestimate by large margins the improvement in manufacturing productivity growth in the 1980s compared with the 1970s. They also probably overstate by large margins, as Stoneman and Francis (1995) argue, the relative performance of UK productivity growth by comparison, for example, with France and the US whose figures are not similarly biased. Finally, they imply that the cost in real output loss of the policies pursued by the Thatcher government in 1979–84 was much greater and longer in duration than implied by the official figures.

References

Aukrust, O. (1977), 'Inflation in the Open Economy: a Norwegian Model', in L. B. Krause and W. S. Salant (eds.),*Worldwide Inflation*, Washington DC, Brookings, 107–53.

Ball, R. J., and St Cyr, E. B. A. (1966), 'Short-Term Employment Functions in British Manufacturing Industry', *Review of Economic Studies*, **33**, 179–207.

Bean, C., and Symons, J. (1989), 'Ten Years of Mrs T', *NBER Macroeconomics Annual 1989*, 13–72 (with discussion).

Bover, O., Muellbauer, J., and Murphy, A. (1989), 'Housing, Wages, and UK Labour Markets', in *Wages and House Price Symposium, Oxford Bulletin of Economics and Statistics*, **51**, 97–162 (with discussion).

Brittan, S. (1989), *The Thatcher Government's Economic Policy*, the Esmee Fairbairn Lecture 1989, University of Lancaster.

Bruno, M., and Sachs, J. (1982), 'Input Price Shocks and the Slowdown in Economic Growth: the Case of UK Manufacturing', *Review of Economic Studies*, **49**, 679–706.

Central Statistical Office (1985), *United Kingdom National Accounts: Sources and Methods*, third edn., London, HMSO, ch. 5.

Church, K. B. (1992), 'Properties of the Fundamental Equilibrium Exchange Rate in Models of the UK Economy', *National Institute of Economic Review*, August, 62–70.

Darby, K., and Wren-Lewis, S. (1989), 'Trends in Manufacturing Labour Productivity', National Institute of Economic and Social Research, discussion paper.

Edgren, G., Faxen, K. O., and Odhner, C. E. (1973), *Wage Formation and the Economy*, London, Allen and Unwin.

Englander, A. S., and Mittelstädt, A. (1988), 'Total Factor Productivity: Macroeconomic and Structural Aspects of the Slowdown', *OECD Economic Studies*, **10**, 7–56.

Frisch, H. (1983), *Theories of Inflation*, Cambridge, Cambridge University Press.

---

[9] Value added here is the gross output in current prices minus the cost of inputs, excluding capital and labour, in current prices.

Grant, R. M. (1985), 'Capacity Adjustment and Restructuring in the UK Cutlery Industry 1974–84', NEDO working paper 21.

Gregg, P., and Machin, S. (1990), 'Unions, the Demise of the Closed Shop, and Wage Growth in the 1980s', National Institute of Economic and Social Research, discussion paper no. 195.

Greenhalgh, C., and Gregory, M. (1995), 'International Trade, De-industrialisation and Labour Demand—an Input-Output Study for the UK 1979–90', mimeo, September.

Helliwell, J. F., Sturm, P. G., and Salou, G. (1985), 'International Comparisons of the Sources of Productivity Slowdown, 1973–82', *European Economic Review*, **28**, 157–91.

Holmans, A. E. (1987), *Housing Policy in Britain: A History*, London, Croom Helm.

Hughes, G. A., and McCormick, B. (1981), 'Do Council Housing Policies Reduce Migration Between Regions?', *Economic Journal*, **91**, 919–39.

—— (1987), 'Housing Markets, Unemployment, and Labour Market Flexibility in the UK', *European Economic Review*, **31**, 615–41.

—— (1991), 'The Decline of the Rental Sector and the Regional Mobility of Labour', unpublished manuscript.

Layard, R., and Nickell, S. (1986), 'Unemployment in Britain', *Economica*, **53**, Supplement, S121–70.

Mayhew, K. (1985), 'Reforming the Labour Market', *Oxford Review of Economic Policy*, **1**(2), 60–79.

—— (1991), 'The Assessment: the UK Labour Market in the 1980s', *Oxford Review of Economic Policy*, **7**(1).

Metcalf, D. (1989), 'Water Notes Dry Up', *British Journal of Industrial Relations*, **27**, 1–31.

Mendis, L., and Muellbauer, J. (1984), 'British Manufacturing Productivity 1955–1983: Measurement Problems, Oil Shocks, and Thatcher Effects', Centre for Economic Policy Research, discussion paper no. 32.

Minford, P., Peel. M., and Ashton, P. (1987), *The Housing Morass*, London, IEA, Hobart paperback.

Muellbauer, J. (1984), 'Aggregate Production Functions and Productivity Measurement: a New Look', Centre for Economic Policy Research, discussion paper no. 34.

—— (1986), 'The Assessment: Productivity and Competitiveness in British Manufacturing', *Oxford Review of Economic Policy*, **2**(3), 1–25.

—— and Murphy, A. (1988), 'House Prices and Migration: Economic and Investment Implications', Shearson, Lehman and Hutton Securities research report, December.

—— —— (1989), 'Why has UK Personal Saving Collapsed?', Credit Suisse First Boston research report, July.

—— —— (1990), 'Is the UK Balance of Payments Sustainable?', *Economic Policy*, October, 348–95 (with discussion).

Nickell, S. (1987), 'Why is Wage Inflation in Britain so High?', *Oxford Bulletin of Economics and Statistics*, **49**, 103–28.

O'Mahony, M. and Oulton, N. (1990), 'Growth of Multi-Factor Productivity in British Industry, 1954–86', National Institute of Economic and Social Research, discussion paper no. 182.

Oswald, A. (1991), 'Pay-Selling, Self-Employment and the Unions', *Oxford Review of Economic Policy*, **7**(3), 31–40.

Oulton, N., and M. O'Mahony (1994), *Productivity and Growth*, Cambridge: Cambridge University Press.

Patterson, K. D. (1984), 'Net Liquid Assets and Net Illiquid Assets in the Consumption Function: Some Evidence for the UK', *Economics Letters*, **14**, 389–95.

Prais, S. J., and Wagner, K. (1985), 'Schooling Standards in England and Germany: Some Summary Comparisons Bearing on Economic Performance', *National Institute Economic Review*, no. 112.

Price, S., Meen, G., and Dhar, S. (1990), 'Employment, Capital Scrapping and Capacity Utilization: Some Results for UK Manufacturing', Oxford Economic Forecasting, discussion paper no. 1.

Smith, A. D. (1989), 'New Measures of British Service Outputs', *National Institute Economic Review*, no. 131, May, 75–88.

Steedman, H. (1990), 'Improvements in Workforce Qualifications, Britain and France 1979–88', *National Institute Economic Review*, no. 133.

Stoneman, P., and N. Francis (1995), 'Double Deflation and the Measurement of Output and Productivity in UK Manufacturing 1979–89', *International Journal of the Economics of Business*, **1**, 423–37.

15

Reversing relative economic decline? The 1980s in historical perspective

NICK CRAFTS

London School of Economics

I. Introduction

During the 1980s the Conservatives under Mrs Thatcher sought to reverse a century or more of relative economic decline in the UK. They were elected at the end of a period of particularly poor British economic performance and adopted policy reforms which not only represented a radical departure from the basic beliefs of previous post-war governments but also from pre-war policy trends.

Among the key elements of this new policy stance were the following.

(a) Supply-side policy moved towards increasing pressure for cost reductions and away from the pursuit of allocative efficiency. Government regulatory failures rather than market failures were targeted for policy action, subsidies were reduced, and privatization given a high priority. This amounted to a rejection of the central thrust of post-war industrial policies.

(b) The government finally abandoned counter-cyclical policy in the form of demand management as practised by the Keynesian policy-makers of the 1950s and 1960s. The Thatcherites, if not the Wets, accepted rapidly rising unemployment as the consequence both of seeking to establish a credible anti-inflationary policy and of reducing the productivity gap between the UK and competitor countries.

(c) There was a heightened emphasis on efficiency relative to equity in the resolution of policy conflicts. Thus, priority was given to cutting personal taxes rather than expanding welfare spending and to faster growth

and restructuring rather than income redistribution or support for lame ducks.

In all these respects the government was proceeding contrary to the wishes of trade union leaders and can be seen as rejecting corporatist approaches to economic policy.

Certainly the government saw itself as seeking to escape from the trade unions' veto on economic reform (Holmes, 1985) and many of the changes of the 1980s would have been regarded as inconceivable by informed opinion in the 1960s and 1970s—for example, the reduction of the top marginal income tax rate to 40 per cent, the privatization of the major public utilities and many council houses, the indexing of benefits to prices rather than wages, and the decimation of the NUM. Similarly, the government's re-election in 1983 and 1987 notwithstanding the very high levels of unemployment in those years flew in the face of the conventional wisdom previously accepted by all governments.

This review of policy developments poses a number of linked questions which this paper addresses. First, what underlay relatively slow British economic growth through to the 1970s and why was it not corrected by government? Second, how was the government able to break out of the constraints imposed by the political economy of the previous thirty-five years? Third, and most important, does the new policy stance promise permanently to reverse relative economic decline?

Economists' answers to these questions have been influenced by recent developments in the analysis of growth and productivity. I propose to draw on several of these ideas in the remaining sections of the paper

First published in *Oxford Review of Economic Policy*, vol. 7. no. 3 (1991).

and a brief introduction to them is appropriate at this point.

It is important to recognize that long-run economic growth among OECD countries has been characterized by processes of catch-up and convergence (Abramovitz, 1986; Dowrick and Nguyen, 1989). In general, the existence of a large productivity gap has meant the potential for faster growth while catching-up the leader country, and allowance needs to be made for this in judging relative growth performance over time. As a more developed country, Britain had less scope than continental Europe for catch-up in the 1950s, but by the 1980s this had been reversed.

Countries differ, however, in their ability to take full advantage of the possibilities of technology transfer in part because of technology policy (Ergas, 1987) and in part because of weaknesses in human capital or industrial relations (Prais, 1981). Recent work on productivity change in Britain both by historians (Lewchuk, 1987) and economists (Bean and Symons, 1989; Machin and Wadhwani, 1989) has stressed the importance of bargaining structures and bargaining power between management and workers in productivity outcomes and the realization of the gains from technological change. Recognition of this draws attention to a number of impacts of government policy on productivity, for example, through demand management and competition policy, which are ignored in standard textbook analysis.

Attention has also been drawn to the political economy of growth, particularly by the work of Olson (1982). He emphasized the danger that interest groups would operate in long-established democracies to slow down growth through rent and vote-seeking behaviour which is inimical to the efficient allocation and redeployment of resources. This is an interesting line of argument in the British case where many of the economy's failings were long-appreciated but not remedied by successive governments.

Finally, the so-called 'new growth theory' associated particularly with Romer (1986) has developed new neoclassical models of growth whose predictions contrast with those of the familiar Solow (1970) model. Thus, through the absence of decreasing returns to (a broad concept of) capital, persistent divergences in growth rates between countries are possible, and shocks (say to the investment rate or to the capital to output ratio) can have permanent effects on growth. If the world works as these new models predict, the growth effects from the Thatcherite reforms could continue strongly through the 1990s rather than petering out relatively quickly.

II. Economic performance in the 1980s in historical perspective

This section examines economic performance in the 1980s in terms of the standard economic policy goals—growth, unemployment, inflation, and income distribution—in comparison with that of earlier decades. Although the central focus of this paper is relative economic decline, it is important to keep this whole set of variables in view since there are trade-offs to be made and, at the heart of the 1980s revolution in British economic policy, there was a revised weighting of objectives, even if this was somewhat less full-blown than the political rhetoric seemed to suggest.

Table 15.1 summarizes recent trends in productivity growth between years which are business cycle peaks. Labour productivity is based on real output per person employed and total factor productivity takes account of the use of both labour and capital inputs. In theory this is a superior measure of productivity change but in practice in recent years measurement of capital inputs may well have been particularly subject to error. In any event, for present purposes the picture is broadly similar in both cases.

Three features of Table 15.1 stand out. First, in the 1960s and 1970s the UK is close to the bottom of the league table. Second, since the 1960s there has been a pervasive fall in productivity growth which has lasted too long to be readily explained by short-run demand shocks. Third, in the 1980s the relative performance of the UK improves sharply to near the top of the league, although in absolute terms productivity growth did not regain the level of the 1960s.

Table 15.2 relates to productivity levels rather than growth and here measurement errors may be much more serious. In particular, it is necessary to confront the index number problem of what prices to use in valuing output in different countries given that it is known that comparisons at prevailing exchange rates are often seriously misleading. The estimates in Table 15.2 attempt to compare real levels of productivity by estimating purchasing power parity exchange rates, in the case of manufacturing by different methods. Despite doubts over the details, some broad trends are probably quite reliable and are useful for the understanding of comparative economic growth outcomes.

Britain's long-run relative economic decline shows up clearly in these data. Before the Second World War there emerged a major productivity gap with the United States while the European overtaking of the UK is revealed as a new phenomenon of the 1960s and

Table 15.1. Productivity Growth in the Business Sector of OECD Countries (% per year)

	Labour productivity			Total factor productivity		
	1960–73	1973–9	1979–88	1960–73	1973–9	1979–88
Australia	3.2	2.0	1.1	2.9	1.2	1.0
Austria	5.8	3.3	1.8	3.4	1.4	0.7
Belgium	5.0	2.8	2.1	3.7	1.5	1.1
Canada	2.8	1.5	1.5	2.0	0.7	0.3
Denmark	4.3	2.6	1.5	2.8	1.2	0.8
Finland	5.0	3.4	3.2	3.4	1.7	2.3
France	5.4	3.0	2.4	3.9	1.7	1.5
Germany	4.6	3.4	1.9	2.7	2.0	0.7
Greece	8.8	3.4	0.2	5.8	1.5	−0.7
Italy	6.3	3.0	1.6	4.6	2.2	1.0
Japan	9.4	3.2	3.1	6.4	1.8	1.8
Netherlands	4.9	3.3	1.5	3.1	2.0	0.6
New Zealand	1.8	−1.5	1.4	1.0	−2.2	0.6
Norway	4.1	0.1	2.0	3.6	−0.4	1.4
Spain	6.1	3.8	3.4	4.2	1.7	2.1
Sweden	3.9	1.4	1.6	2.5	0.3	0.9
Switzerland	3.2	0.7	0.9	1.6	−0.9	0.2
UK	3.5	1.5	2.6	2.2	0.5	1.9
USA	2.8	0.6	1.6	1.8	0.1	0.7

Source: Kendrick (1990).

Table 15.2. Output Per Hour Worked (UK = 100)

(*a*) Real GDP/hour worked

	France	Germany	Japan	USA
1913	60	65	23	135
1929	70	65	30	158
1938	82	73	35	154
1950	70	54	24	185
1960	84	84	33	188
1973	105	100	62	156
1979	114	113	70	150
1986	119	105	68	133

Source: Maddison (1989).

(*b*) Manufacturing output/hour worked

Basis:	Industry of Origin				Expenditure			
	France	Germany	Japan	USA	France	Germany	Japan	USA
1911	60	85		172				
1929	61	76		235				
1937	56[a]	76		197				
1950	64	74		247				
1960	81	105		242	103	118	50	242
1973	113	133	125[b]	220	134	145	105	212
1979	142	162		226	166	173	133	214
1988	126	137	175[c]	207	147[d]	146[d]	139[d]	195

Notes: [a] 1938, [b] 1975, [c] 1985, [d] 1987.
Sources: Based on Broadberry (1990); Hooper and Larin (1989); Szirmai and Pilat (1990); and Van Ark (1990).

Table 15.3. World Trade Shares and De-industrialization (%)

(*a*) Shares of world trade in manufactures (%)

	1950	1960	1970	1979	1990
France	9.9	9.6	8.7	10.5	9.7
Germany	7.3	19.3	19.8	20.9	20.2
Japan	3.4	6.9	11.7	13.7	15.9
UK	25.5	16.5	10.8	9.1	8.6
USA	27.3	21.6	18.6	16.0	16.0

Sources: Brown and Sheriff (1979) and CSO (1991).

(*b*) Deployment of the labour force (%)

	France	Germany	Japan	UK	USA
1950					
Agriculture	27.4	23.2	41.0	4.9	11.9
Industry	37.0	44.4	24.2	49.4	35.9
Services	35.6	32.4	34.8	45.7	52.2
1970					
Agriculture	13.9	8.6	17.4	3.2	4.5
Industry	39.7	48.5	35.7	44.8	34.4
Services	46.4	42.9	46.9	52.0	61.1
1990					
Agriculture	6.1	3.4	7.2	2.1	2.8
Industry	30.0	39.7	34.1	28.8	26.2
Services	63.9	56.9	58.7	69.1	70.9

Sources: Bairoch (1968) and OECD (1991); figures for France and Japan are for 1954 and 1955 respectively not 1950.

Table 15.4. Inflation and Unemployment: a Long-Run Picture (%)

	Adjusted unemployment[a]	Inflation[b]	Misery Index
1925–9	5.3	−1.0	4.3
1930–7	8.1	−0.4	7.7
1952–64	1.4	3.6	5.0
1965–73	1.9	5.9	7.8
1974–9	3.7	16.2	19.9
1980–8	9.3	7.2	16.5

Source: Derived from Feinstein (1972) and Central Statistical Office (1990); Misery Index is the sum of the unemployment and inflation rates.
Notes: [a] The series has been adjusted to allow for changes in definition to approximate to a 1988 basis throughout. These data are only provided by the CSO back to 1971; 1946–70 figures are original estimates x 0.765 based on comparison of 1970s old and new official series and interwar data x 0.478 to allow also for restricted coverage of national insurance figures, as proposed by Metcalf *et al.* (1980).
[b] Measured by the GDP deflator.

1970s. The major weaknesses in British productivity performance appear to be in manufacturing rather than services. Some evidence of at least a temporary reversal of relative decline since 1979 can also be seen, particularly in manufacturing. A dramatic change in relative manufacturing productivity with Japan has taken place in the last twenty years which has not been pegged back in the 1980s. Britain's relatively poor productivity performance in manufacturing has been accompanied by a much discussed de-industrialization of the labour force and a loss of world market share in manufactures. These trends are reported in Table 15.3, where perhaps the most striking contrasts are those of the last thirty years between the UK and Germany. Table 15.3 shows that de-industrialization has been particularly pronounced in the UK. The main losses in the British share of manufactured trade came in the 1950s and 1960s, and in the 1980s there was relatively little change.

In Table 15.4 performance in terms of the key short-term macroeconomic variables is reported for various business cycles. The data for unemployment are adjusted to allow for the changes in the series underlying the raw data which are normally reported. In the light of the many changes in the method of counting in recent years this is essential if a long-run perspective is to be obtained, even though the adjustments employed are necessarily somewhat crude. Table 15.4 indicates that, when the data are made comparable, the 1980s witnessed the worst unemployment of any period in the table including the 1930s. Not so surprisingly, the 1970s stand out as a period of exceptionally high inflation while the 1980s saw inflation at a much higher level than in the early post-war period. The much used, albeit arbitrarily weighted, Misery Index has been very high in both the last two cycles and again on this measure the 1980s compare very unfavourably with the 1930s.

In section I it was noted that the Conservatives in the 1980s adopted new priorities with regard to taxes and welfare spending. Table 15.5 attempts to reflect this shift and also provides some information on changes in the distribution of income and wealth. Personal income taxation rates were indeed reduced considerably during the 1980s, particularly for top rate payers, and reached levels not seen since the 1930s. This philosophy of taxation contrasts greatly with that of the early post-war period, although overall tax burdens were not reduced in the 1980s as indirect and local taxes have been increased.

Table 15.5 also reports unemployment benefits relative to wages for a single man. Here the level of benefits fell relative to wages in the 1980s, reversing the trend of the 1950s and 1960s and by the end of the 1980s falling back below the level of the 1930s. Nevertheless, the magnitude even of this change, which represented the sharpest reversal of previous welfare policy, should not

Table 15.5. Fifty Years of Taxes, Benefits, and Changes in Income and Wealth Distributions

(*a*) Personal income tax rates (%)

	Standard rate	Top rate
1938	25	62.5
1949	45	97.5
1961	38.75	88.75
1973	38.75	88.75
1979	33	83
1989	25	40

Source: Reports of the Inland Revenue Commissioners; rates in force at 1 January.

(*b*) Benefit/wage rate ratio

1938	0.29	1973	0.31
1951	0.19	1979	0.30
1961	0.24	1988	0.25

Sources: Metcalf *et al.* (1980) and Atkinson and Micklewright (1989).

(*c*) Distribution of income (%)

	Original income			Gross income		
	Top 20%	Bottom 20%	Gini	Top 20%	Bottom 20%	Gini
1938	50		43	50		46
1949				47		41
1961			40	45	5.3	40
1973	44		42	42	6.2	37
1979	45	0.5	45	42	6.3	35
1986	51	0.3	52	44	6.1	40

	Disposable income			Final income		
	Top 20%	Bottom 20%	Gini	Top 20%	Bottom 20%	Gini
1938	46		43	41–3		
1949	42		36			
1961	42	6.5	33	42		33
1973	39	6.8	33	39		32
1979	39	6.5	33	38	7.1	32
1986	42	6.2	36	42	6.3	36

Sources: Barna (1945), Lydall (1959), Central Statistical Office (1988), Royal Commission (1979).
Notes: Original income is income from value added; gross income also includes transfer payments in cash; disposable income is gross income less income tax and national insurance contributions; final income is income after allowing for all taxes and transfers.

(*d*) Distribution of wealth in Great Britain (%)

	Top 1%	Top 5%	Top 10%
1938	55	77	85
1950	47	74	
1960	34	60	72
1970	30	54	69
1980	20	43	60
1988	17	38	53

Sources: Atkinson *et al.* (1989) and Central Statistical Office (1990).

be exaggerated and at the end of the 1980s the benefit/wage rate ratio was still above that of the 1950s.

The early post-war years were characterized by much public and political sympathy for the notion of a more egalitarian society, whereas in the 1980s the pendulum has swung back somewhat against this ideal. In this context it is interesting to consider the trends in the distribution of income and wealth shown in Table 15.5. The picture which emerges is a striking one. The available data suggest a distinct reduction in inequality of incomes in the late 1940s compared with the pre-war scene and on these measures this seems to have been accentuated somewhat through to the 1970s. The 1980s has seen a sharp reversal of this trend, notably in original income where the share of the top 20 per cent now appears to be slightly higher than in 1938. Both the revival of profits and the rise in unemployment in the 1980s can be presumed to have played a part in this turn-around. Declines in the inequality of the wealth distribution appear to have continued steadily throughout the last fifty years, prompted by the rise in 'popular wealth' in the form of wider home ownership.

Nevertheless, it is important to recognize that the social security safety net lessened the impact on final income and that the welfare state survived relatively little changed through 1987, the main difference being that the increase in its importance relative to other areas in the economy stopped (Le Grand, 1990). Moreover, opinion polls continued to register solid support for welfarist policies; for example, in 1989 56 per cent declared themselves in favour of higher taxes to fund higher government expenditure on health, education, and welfare benefits, against only 3 per cent supporting the opposite (Taylor-Gooby, 1990).

III. The Thatcher reforms in the context of earlier experience

It is generally accepted that the Conservative strategy with respect to supply-side policy represented a major shift of direction from anything seen since 1945. It is less well known, but none the less true, that this policy stance was also in marked contrast to that pursued before 1939 (Broadberry and Crafts, 1990a). In this section the Thatcher reforms are considered in the context both of the evidence about obstacles to faster growth in the earlier post-war period and also the apparent failure of policies designed to remedy supply-side failures and to increase productive potential.

In retrospect, post-war Britain's relatively slow growth compared with France or Germany appears to be due primarily to poor productivity performance

rather than low investment in physical capital. This conclusion emerges from growth accounting, whether at the level of the economy as a whole (Maddison, 1987) or simply for manufacturing (Panic, 1976), is confirmed by recent econometric work (Crafts, 1991; Englander and Mittelstädt, 1988) and is suggested most starkly by the observation that in 1958–72 the increase in net output per unit of investment in Germany was 1.9 times the level in the UK whereas the German investment rate was only 25 per cent higher (CBI, 1977).

The proximate reasons for this weak productivity performance are widely agreed—even if their relative importance is difficult to quantify accurately—and were apparent to well-informed observers from at least the late 1940s on, as the reports of the Anglo-American Productivity Council demonstrate. The chief of these persistent weaknesses relative to other leading economies lay in poor quality of management, unfortunate structure and conduct of industrial relations, weak research and development by the business sector, and inadequate training of workers (Alford, 1988; Batstone, 1988; Prais, 1981; Pratten and Atkinson, 1976; Sheldrake and Vickerstaff, 1987).

Post-war governments were more interventionist than their pre-war counterparts. A major overhaul of policy took place in the 1960s when slow growth became a matter of great public concern and debate. Major features of the development of supply-side policy include the following. (For a fuller account, see Kirby, 1991, and Morris and Stout, 1985.)

(a) There was a strong emphasis on subsidizing investment from the 1950s through a variety of schemes starting with investment allowances and building up to selective assistance under the Industry Act of 1972. There was frequently a very strong element of regional policy in support for investment.

(b) Public ownership was seen by some as an appropriate solution to market failures. There were substantial extensions of nationalization in the 1940s (including coal, electricity, gas, railways) and in the 1970s (Rolls-Royce, British Leyland, British Aerospace, British Shipbuilders). In the 1950s and 1960s nationalized industries accounted for almost 20 per cent of investment in the economy.

(c) There was a gradual revival of competition policy against the background of an economy full of anti-competitive practices inherited from the years of depression and war. The Monopolies Commission in 1948 and the Restrictive Practices Act in 1956 are early landmarks followed by the abolition of resale price maintenance in 1964, the extension of the Monopolies Commission remit to cover mergers in 1965, and the Fair Trading Act in 1973.

(*d*) The policy reappraisals of the 1960s emphasized the pursuit of economies of scale and merger promotion via the Industrial Reorganization Corporation (1966), the establishment of Industrial Training Boards (1964), the expansion of higher education and a short-lived experiment with indicative planning.

(*e*) With the exception of the ill-fated Industrial Relations Act of 1971, the over-riding approach was to continue the tradition of voluntarism in industrial relations rather than to attempt reform via legislation.

Assessments of these policy initiatives have rightly tended to be critical. In some cases the problems were those of inexperience or straightforward poor policy design. To some extent this seems to be the case in the running of nationalized industries which was given far too little thought initially (Chester, 1975), and in technology policy which concentrated unduly on aiming to produce radical innovations and too little on facilitating the diffusion of improvements (Ergas, 1987). Policy was based on an exaggerated belief in the importance of promoting investment (Driver, 1983), the effectiveness of mergers as a means of raising efficiency (Meeks, 1977), and achieving an Americanization of industrial structure (Broadberry and Crafts, 1990*b*), and too little on establishing a framework within which X-inefficiency would be minimized.

It is also important, however, to recognize, as the political economy of growth literature of the Olson (1982) tradition suggests, that in many cases political or interest-group constraints dictated choices and that the economy seemed to a serious extent to be locked in to suboptimal policies. A particularly important feature of the post-war world was the very strong commitment by all parties to full employment and the continual worry that this would lead to wage inflation. In these circumstances it was natural to seek co-operative arrangements with organized interest groups in the form of corporatist agreements or implicit social contracts (Flanagan *et al.*, 1983) and to give priority to avoiding job losses.

The new bargaining models of productivity (Bean and Symons, 1989; Machin and Wadwhani, 1989) predict that this would be conducive to poor productivity performance, particularly in an economy characterized by a historical legacy of multi-unionism. Thus, such strategies undermined the credibility of or even eliminated policy options which would have raised productivity through reducing trade union bargaining power or changing incentive structures, and also operated to slow down exit of inefficient producers.

Similar problems had occurred in the 1930s when government attempts to rationalize industries such as coal and steel were undermined by fears of job losses (Supple, 1987; Tolliday, 1987) and when the macro-economic policy responses to the shock of the depression focused on the need to raise prices to restore profitability and lower real wages via devaluation, protectionism, and encouragement of cartels (Booth, 1987) and led to much reduced pressures for industrial efficiency.

Thus it is not surprising to find Silberston's assessment of the industrial policies of the 1960s and 1970s as 'directed at helping old industries to survive rather than encouraging new products and new technology' (1981, p. 49) or the condemnation of the regulatory arrangements and ministerial manipulation of decisions in the nationalized sector (Vickers and Yarrow, 1988). Similarly, key policy reforms such as those in competition policy in the early post-war period (Mercer, 1989) or in training policy in the 1960s (Vickerstaff, 1985) were emasculated—in the former case to keep business support for the export drive and in the latter to avoid trade union opposition.

It is useful to consider Conservative supply-side policy since 1979 against this background. The government's programme has involved tax reform, privatization, deregulation, harsh treatment of lame ducks, trade union reform, and a general distrust of public sector interventions to correct market failure. Trade union inputs into policy-making ceased. The new policy stance affected both the framework in which investment decisions and bargains between management and workers were made as well as having direct effects on the use of capital and labour.

The thrust of policy was to attack restrictive practices and overmanning rather than to aim for investment-led growth. The reforms were carried out in the context initially of a stern counter-inflationary policy which implied in the short term a severe profits squeeze and a major contraction of manufacturing activity, a steep rise in unemployment, and a decline in trade union strength. This prompts the question of how such a radical shift in policy was possible, given that to previous governments (including notably the Heath government of 1970–4) it would have appeared politically suicidal.

An answer to this question so soon after the event can only be very provisional but an argument along the following lines may be plausible. First, the unsatisfactory economic performance of the 1970s discredited the Keynesian, corporatist policy stance in many quarters and left the Labour Party initially in severe difficulty and then with the splintering off of the SDP. Second, the government successfully gambled on being able to place the blame for rising unemployment elsewhere, given the unpopularity of trade unions

Table 15.6. Regression Estimates of Bonus from Catch-Up and Reconstruction and of Residual Growth (% per year)

| | 1950–60 | | 1960–73 | | 1979–88 | |
	Bonus	Residual	Bonus	Residual	Bonus	Residual
Australia	1.2	−0.43	0.5	−1.58	0.4	−1.28
Austria	2.8	0.98	1.3	−0.02	0.5	−0.92
Belgium	2.1	−0.66	1.1	0.31	0.4	0.80
Canada	0.2	0.96	0.3	0.33	0.3	0.05
Denmark	1.9	−1.41	1.1	0.09	0.6	−0.92
Finland	2.1	−0.72	1.4	0.16	0.7	0.73
France	2.1	0.19	1.0	0.58	0.3	1.10
Germany	2.9	1.97	1.0	−0.34	0.3	0.09
Italy	2.4	0.16	1.2	0.83	0.4	0.07
Japan	4.0	0.37	2.2	2.17	0.9	0.16
Netherlands	1.9	−0.16	0.8	−0.33	0.2	0.58
Norway	1.8	−1.30	0.9	−0.93	0.4	0.03
Sweden	1.3	0.05	0.8	0.33	0.3	−0.30
Switzerland	1.1	0.10	0.7	−1.04	0.5	−0.57
UK	1.6	−0.80	0.8	−0.63	0.5	0.84
USA	0.0	0.69	0.0	0.07	0.0	0.26

Source: Derived from regressions reported in Crafts (1991).
Note: The catch-up and reconstruction bonus is calculated relative to the USA in each period. The residuals for 1950–60 and 1960–73 come directly from estimation; those for 1979–88 are found by comparing predicted and actual growth using the estimated coefficients.

which had developed by the end of the 1970s and the general tendency for European unemployment to rise (Holmes, 1985, p. 83). Third, the composition and behaviour of the electorate had changed enough to make unemployment much less of a vote loser in any case— by the mid-1980s it seemed to have much less impact than interest rates and inflation (Sanders, 1991). Fourth, this was a period of strong prime-ministerial government when the doubts of the risk-averse majority of the Conservative Party and cabinet, who had not originally anticipated the unemployment consequences of Thatcherite policies, were swept aside.

IV. The Thatcher experiment in retrospect

In Tables 15.1 and 15.2 it was shown that the UK experienced a distinct improvement in relative productivity performance in the 1980s. In order to assess this in its historical context it is important to allow for differential scope for productivity growth both over time and between countries, as suggested by the work of Abramovitz (1986) and Dowrick and Nguyen (1989) noted in section I. In particular, we would like to measure productivity growth normalized for catch-up potential. Table 15.6 provides estimates of this kind. In the table 'bonus' refers to the predicted gain from this factor based on an analysis of a cross-section of countries. 'Residual' refers to the extent to which growth exceeded or fell short of that predicted on the basis of allowing not only for catch-up but also for growth of factor inputs.

Two important points emerge from Table 15.6. First, allowing for catch-up is not sufficient to explain slower British growth in the 1950s and 1960s—the economy is seen to have had an inferior performance even allowing for this factor. Second, there seems to have been an impressive turn-around in the 1980s with a strong positive residual, even though our rise in the relative growth league was aided by the exhaustion of fast catch-up elsewhere. Thus, there is a prima-facie case for suggesting at least a temporary pay-off to the Thatcher Experiment in terms of reversing relative economic decline.

There is, indeed, evidence to suggest that in several ways the new policy stance led to improved productivity. Econometric studies have suggested that the productivity surge in British manufacturing of the early 1980s can be quite well explained along the lines of a model in which adverse employment shocks and declining product market power of firms lead to new bargaining equilibria in which overmanning is reduced and/or work effort is increased (Bean and Symons, 1989; Machin and Wadhwani, 1989; Metcalf, 1989).

This is plausible in the light of the evidence in comparative studies of productivity and industrial relations structures in post-war British industry from the Anglo-American Productivity Council onwards. Thus productivity gains were obtained which had been precluded by the earlier post-war policy configuration. Moreover, it seems that the early 1980s' recession did not have a similar impact in France or Germany given their different history and institutions so that relative productivity performance was enhanced (Crafts, 1991).

Similarly, the government had some success in improving productivity in the public enterprise sector, notably after defeating major strikes in steel (1980) and coal (1984/5), even though both industries remained in public ownership with no immediate prospect of privatization. In coal-mining labour productivity in a sample of matched pits in 1985 was 21 per cent higher than in 1983 as management exploited a strong position to cut manpower and eliminate restrictive practices (Glyn, 1988). In British Steel total factor productivity grew at 12.9 per cent per year over 1979–88 (Bishop and Kay, 1988). A calculation based on data in Bishop and Kay shows that the major nationalized industries of 1979 accounted for about 17 per cent of labour productivity growth in 1979–88 despite accounting for only 5.7 per cent of 1979 employment.

The Conservatives under Mrs Thatcher in moving rapidly away from corporatism and accepting unemployment as a concomitant of modernization of the economy followed a strategy which intensified competitive pressures on many managers. An average effective tariff rate of 8.7 per cent in 1972 had fallen to 1.2 per cent in 1986 (Ennew *et al.*, 1990) (although non-tariff barriers to trade remained high in a few sectors, e.g. cars and textiles) compared with 8.3 per cent in Germany in the same year (Klodt, 1988). The increase in import penetration was pronounced—the share of imports in home demand for manufactures was 16.6 per cent in 1970 but 35.2 per cent by 1987 (Hewer, 1980, p. 208; Central Statistical Office, 1989, p. 219). Subsidies which had averaged 2.7 per cent of GDP in 1975–9 had fallen to 1.7 per cent by 1985–8 (Ford and Snyker, 1990)—spending by the DTI on regional and investment subsidies was cut by 69 per cent between 1979 and 1987 (Shepherd, 1987).

The reform of personal taxation, on which the government placed so much emphasis during the 1980s, amounted to a considerable change, as Table 15.5 reported. Economists remain sceptical, however, that tax changes had any major role in the productivity revival but it must be admitted that the evidence is of poor quality. Any sizeable effect can be expected to have come from the lowering of top tax rates. Study of the 1979/80 reduction in the top marginal rate from 83 per cent to 60 per cent suggests they may have had a modest positive impact on tax receipts —Dilnot and Kell found an upper bound effect of £1.2 billion (3 per cent) (1988, p. 90). The Conservatives' reliance on this route to increasing economic growth remains largely an act of faith rather than a policy for which large gains have been empirically demonstrated.

While convergence on the income and productivity level of the leading country will require elimination of relatively high levels of X-inefficiency this will not be sufficient; it is also important that policies and institutions are conducive to realizing the possibilities of technology transfer. In this regard it is doubtful that the UK improved its relative standing in the 1980s.

Assessments of the government's policies towards research and development during the 1980s have generally been quite critical, as have reviews of the UK's relative technological capabilities (Patel and Pavitt, 1987; Stoneman and Vickers, 1988). Based on managerial skills, educational standards, established in-house expertise, and commitment to particular technologies against a background of capital market institutions less prone to short termism, Pavitt and Patel (1988, p. 52) concluded that the UK was a comparatively weak performer with a myopic system which contrasted unfavourably with the dynamic systems in countries such as Germany and Sweden. This comment would apply equally to the pre-Thatcher period (Saul, 1979). Greenhalgh (1990) emphasizes the impact of this in explaining weak British trade performance in manufacturing. Table 15.7 shows that the share of UK GDP spent on R & D remained at the same level while in the average of the other countries there was a rise of 0.43 percentage points.

An important, continuing reason for differences in private sector investment in R & D between Britain and Germany may be found in the capital market. This is for two reasons. First, both in the 1970s and still recently the relative cost of capital for long-term R & D projects has been about 60 per cent higher in the UK than in Germany (McCauley and Zimmer, 1989). Second, there is a high level of hostile take-over activity in the UK but not in Germany. The British arrangements might be thought more likely to correct managerial failure, but less likely to encourage managers loath to lose the benefits of firm-specific investments through a change in ownership to forego current earnings for long-term R & D (Franks and Mayer, 1990). In practice, the effectiveness of German banks as monitors of firm performance (Cable, 1985) and the ineffectiveness of the British takeover mechanism in singling out poor

Table 15.7. Education and Training and R & D: Some Comparisons

(*a*) Enrolment rates for 15–19 year age group in 1986–7 (%)

Australia	49.3[a]	Italy	na
Austria	na	Japan	71.0[a]
Belgium	78.6	Netherlands	72.8
Canada	72.7	Norway	72.8
Denmark	76.0	Sweden	64.5
Finland	72.0	Switzerland	77.6
France	73.4	UK	54.3
Germany	70.2	USA	78.8

Source: OECD (1989, Chart VB).

(*b*) Share of labour force with intermediate vocational qualifications (%)

	1979	1988
France	32	40
Germany	61	64
UK	23	26[b]

Sources: NEDO (1983); Steedman (1990).

(*c*) Numbers qualifying in engineering, c.1985 (thousands)

	PhD	MSc[c]	BSc	Technicians	Craftsmen
France	0.3	6	15	35	92
Germany	1.0	4	21	44	120
UK	0.7	2	14	29	35

Source: Prais (1989).

(*d*) Expenditure on R & D as a share of GDP (%)

	1970	1980	1988
Australia	1.23	0.98	1.19
Austria	0.61	1.10	1.32
Belgium	1.31	1.36	1.61
Canada	1.36	1.15	1.32
Denmark	0.96	1.04	1.43
Finland	0.78	1.07	1.76
France	1.91	1.84	2.31
Germany	2.06	2.41	2.83
Italy	0.88	0.86	1.34
Japan	1.85	2.18	2.92
Netherlands	2.01	1.89	2.30
Norway	1.10	1.27	1.81
Sweden	na	na	2.99
Switzerland	2.25	2.31	2.88
UK	2.18	2.24	2.27
USA	2.65	2.39	2.86

Sources: Englander and Mittelstädt (1988), and OECD (1990).
Notes: [a] Not including Third Level enrolments. [b] On a comparable basis including only examined qualifications would be 19. [c] Includes 'enhanced' bachelor degrees.

performance and leading to efficiency gains post merger (Meeks, 1977; Singh, 1975) may make the German arrangements unambiguously superior.

Similarly, Table 15.7 suggests continuing weaknesses in relative education and training standards in the UK, despite increased efforts in this direction—real expenditure by employers on training appears to have roughly trebled since 1970/1 (OECD, 1977, p. 352; Department of Employment, 1989). Thus in 1987 a third of all workers in the UK had never received any training and a fifth of all companies did no training, although the proportion of employees in job-related training had risen from 9.1 per cent in 1984 to 13.3 per cent in 1987 (Department of Employment, 1989). In Germany, however, the proportion aged 19–64 participating in vocational training rose from 10 per cent in 1979 to 18 per cent in 1988 when seven times as many passes at Meister level were recorded as in Britain (Wagner, 1991).

The productivity surge of the 1980s appears to have owed little to greater investment in plant, people, or research and development but to have come primarily from more efficient use of existing factors of production and in effect to have been based on a shake-out of inefficiencies which had accumulated in the earlier decades. Should this be thought of, then, as a one-off catch-up whose effects on growth rates will peter out with the prospect of a further relative economic decline in the 1990s? This question is especially important for the assessment of the costs and benefits of the change in policy stance.

The change in fashion among growth theorists signalled above in section I makes the answer less clear-cut than might have seemed to be the case until recently. While traditional neoclassical growth theory would have said yes, in the new growth theory the answer is maybe, as is nicely set out by Baldwin (1989) in his discussion of the effects of 1992. This approach considers a production function of the type

$$Y = AK^{a+b}L^{1-a} \tag{1}$$

which is a conventional Cobb–Douglas modified to include the term *b* which captures external effects, spillovers, etc. of capital formation. The literature has contemplated the possibility that $a + b = 1$ in which case there are no diminishing returns to capital and a positive productivity shock which lowered the capital to output ratio would permanently increase growth of capital and of output. On balance, however, the evidence appears to be in favour of $b > 0$ but $a + b$ equals, say, 0.5 rather than 1 (Crafts, 1991) and it seems likely that the traditional view is still valid.

Nevertheless, even though it is likely that there was

a strong once-and-for-all element to the gains from the new get-tough policies, there are grounds for believing that there may be some long-run growth effects from the Thatcher shock, if the bargaining power of workers over manning levels remains weak. Prais (1981) found in a series of case studies that bargaining practices in the UK tended in the 1970s to retard and dilute the gains from the introduction of new technology. The gains from the flow of future technological improvements may be greater as a result of the changed conduct of British industrial relations.

The chief benefits of the Thatcher reforms were the realization of a once-and-for-all catch-up that had been foregone under the earlier policy regime. On balance, it seems likely that we are not yet in a position to anticipate the complete elimination of the productivity gap with the most advanced countries and it seems quite possible that relative decline (albeit with UK GDP growth at 2–3 per cent) will re-emerge in the 1990s because of weaknesses in training and technology where market failures imply the need for solutions for which the present government has a distaste.

The costs of the Thatcher Shock involved a large rise in unemployment and also, through mismatch, the NAIRU (Nickell, 1990) which contributed to the very high Misery Index for the 1980s. This may have been hard to avoid. It would also seem possible to argue that there was an avoidable increase in income inequality which could have been mitigated by different tax and benefits policies without seriously jeopardizing the productivity improvements which probably did not depend on personal tax cuts. Whether the shock was worth it remains very much an open question.

So, too, does the political sustainability of Thatcherite supply-side policy, even assuming continued Conservative government, given the obvious appeal of the middle ground to Mr Major in the context of a reformed Labour Party and the destruction by recession of the notion of a 'Thatcher Miracle'. Even at the peak of Thatcherism the onslaught on the welfare state was quite limited, as was noted in section II. Recent surveys suggest that much of the electorate remains stubbornly 'social democratic' in its values and that policies focused on further trade union reform, privatization, and income tax cuts are not necessarily election winners (Rentoul, 1990). The change in rhetoric and the softening of policy stance in areas like child benefits are already apparent.

On the other hand, a return to the degree of featherbedding of inefficiency so characteristic of earlier post-war Britain seems unlikely even under a Labour administration, given that to a significant extent we can expect the bargaining framework in which firms and unions operate to be set by European commitments which imply continued competitive pressure and quite possibly a further period of high unemployment. Perhaps under either party the greater acceptance of market mechanisms characteristic of the 1980s will largely survive but tempered somewhat more by realization that market failures require well-designed policy interventions.

V. Conclusions

The main conclusions of the paper are easily summarized.

(a)   There was a distinct improvement in productivity performance in the 1980s which led to a reversal of relative economic decline.

(b)   The productivity revival was based on a successful attack on the inefficiency characteristic of post-war Britain which had previously been precluded by the pursuit of corporatist solutions to economic problems.

(c)   Other factors contributing to relative economic decline before 1979 still remained at the end of the 1980s, in particular weaknesses in technology and human capital.

(d)   It is therefore quite possible that relative economic decline will resume in the 1990s since a large part of the impact of the reduction of inefficiency would have only a short-term effect on the growth rate.

(e)   There were costs, some of which were avoidable, as well as benefits from the Thatcher reforms and the net effect on economic welfare remains arguable.

(f)   A partial retreat from Thatcherism, which should perhaps be seen as the product of unusual circumstances, seems likely.

References

Abramovitz, M. (1986), 'Catching Up, Forging Ahead, and Falling Behind', *Journal of Economic History*, **46**, 385–406.

Alford, B. W. E. (1988), *British Economic Performance, 1945–1975*, London, Macmillan.

Atkinson, A. B., Gordon, J. P. F., and Harrison, A. (1989), 'Trends in the Shares of Top Wealth Holders in Britain, 1923–1981', *Oxford Bulletin of Economics and Statistics*, **51**, 315–32.

—— Micklewright, J. (1989), 'Turning the Screw: Benefits for the Unemployed, 1979–1988', in A. B. Atkinson, *Poverty and Social Security*, London, Harvester Wheatsheaf.

Bairoch, P. (1968), *La Population Active et Sa Structure*, Brussels, OECD.

Baldwin, R. (1989), 'The Growth Effects of 1992', *Economic Policy*, 9, 248–81.

Barna, T. (1945), *Redistribution of Incomes through Public Finance in 1937*, London, Oxford University Press.

Batstone, E. (1988), *The Reform of Workplace Industrial. Relations*, Oxford, Clarendon Press.

Bean, C., and Symons, J. (1989), '*Ten Years of Mrs T.*', CEPR Discussion Paper 316.

Bishop, M., and Kay, J. (1988), *Does Privatization Work? Lessons from the UK*, London, London Business School.

Booth, A. (1987), 'Britain in the 1930s: A Managed Economy', *Economic History Review*, 40, 499–522.

Broadberry, S. N. (1990), 'Comparative Levels of Productivity in Manufacturing, 1870–1990', mimeo, University of Warwick.

—— Crafts, N. F. R. (1990a), 'The Implications of British Macroeconomic Policy in the 1930s for Long Run Growth Performance', *Rivista di Storia Economica*, 7, 1–19.

—— —— (1990b), 'Britain's Productivity Gap in the 1930s: Some Neglected Factors', Warwick Economics Discussion Paper no. 364.

Brown, C. J. F., and Sheriff, T. D. (1979), 'De-industrialisation: A Background Paper', in F. T. Blackaby (ed.), *De-industrialisation*, London, Heinemann.

Cable, J. R. (1985), 'Capital Market Information and Industrial Performance: The Role of West German Banks', *Economic Journal*, 95, 118–32.

Central Statistical Office (1988), *Economic Trends Annual Supplement*, London, HMSO.

—— (1989), *Annual Abstract of Statistics*, London, HMSO.

—— (1990), *Economic Trends Annual Supplement*, London, HMSO.

—— (1991), *Monthly Review of External Trade Statistics*, London, HMSO.

Chester, N. (1975), *The Nationalisation of British Industry, 1945–1951*, London, HMSO.

Confederation of British Industry (1977), *Britain Means Business*, London.

Crafts, N. F. R. (1991), 'Productivity Growth Reconsidered', paper presented to Economic Policy Panel, London.

Department of Employment (1989), *Training in Britain. A Study of Funding, Activity and Attitudes*, London, HMSO.

Dilnot, A. W., and Kell, M. (1988), 'Top Rate Tax Cuts and Incentives: Some Empirical Evidence', *Fiscal Studies*, 9, 70–92.

Dowrick, S., and Nguyen, D-T. (1989), 'OECD Comparative Economic Growth 1950–85: Catch-Up and Convergence', *American Economic Review*, 79, 1010–30.

Driver, C. (1983), 'Import Substitution and the Work of the Sector Working Parties', *Applied Economics*, 15, 165–76.

Englander, A. S., and Mittelstädt, A. (1988), 'Total Factor Productivity: Macroeconomic and Structural Aspects of the Slowdown', *OECD Economic Studies*, 10, 7–56.

Ennew, C., Greenaway, D., and Reed, G. (1990), 'Further Evidence on Effective Tariffs and Effective Protection in the UK', *Oxford Bulletin of Economics and Statistics*, 52, 69–78.

Ergas, H. (1987), 'Does Technology Policy Matter?', in B. R. Guile and H. Brooks (eds.), *Technology and Global Industry*, Washington, National Academy Press.

Feinstein, C. H. (1972), *National Income, Expenditure and Output of the United Kingdom, 1855–1965*, Cambridge, Cambridge University Press.

Flanagan, R., Soskice, D. W., and Ulman, L. (1983), *Unionism, Economic Stabilization and Incomes Policies: The European Experience*, Washington, Brookings.

Ford, R., and Snyker, W. (1990), 'Industrial Subsidies in the OECD Economies', OECD Department of Economics and Statistics Working Paper 74.

Franks, J., and Mayer, C. (1990), 'Capital Markets and Corporate Control: A Study of France, Germany and the UK', *Economic Policy*, 10, 191–231.

Glyn, A. (1988), 'Colliery Results and Closures after the 1984–85 Coal Dispute', *Oxford Bulletin of Economics and Statistics*, 50, 161–73.

Greenhalgh, C. (1990), 'Innovation and Trade Performance', *Economic Journal*, 100, Supplement, 105–18.

Hewer, A. (1980), 'Manufacturing Industry in the 1970s', *Economic Trends*, 320, 97–109.

Holmes, M. (1985), *The First Thatcher Government, 1979–1983*, Brighton, Wheatsheaf.

Hooper, P., and Larin, K. A. (1989), 'International Comparisons of Labor Costs in Manufacturing', *Review of Income and Wealth*, 35, 335–55.

Kendrick, J. W. (1990), 'International Comparison of Productivity Trends and Levels', George Washington University Discussion Paper 90–2.

Kirby, M. W. (1991), 'Supply-Side Management', in N. F. R. Crafts and N. W. C. Woodward (eds.), *The British Economy Since 1945*, Oxford, Clarendon Press.

Klodt, H. (1988), 'Industrial Policy and Repressed Structural Change in West Germany', Kiel Institute of World Economics Working Paper 322.

Le Grand, J. (1990), 'The State of Welfare', in J. Hills (ed.), *The State of Welfare*, Oxford, Clarendon Press.

Lewchuk, W. (1987), *American Technology and the British Vehicle Industry*, Cambridge, Cambridge University Press.

Lydall, H. (1959), 'The Long Term Trend in the Size Distribution of Income', *Journal of the Royal Statistical Society*, Series A, 122, 1–37.

McCauley, R. N., and Zimmer, S. A. (1989), 'Explaining International Differences in the Cost of Capital', *Federal Reserve Bank of New York Quarterly Review*, 14, 7–28.

Machin, S., and Wadhwani, S. (1989), 'The Effects of Unions on Organizational Change, Investment and Employment: Evidence from WIRS', Centre for Labour Economics, London School of Economics, Discussion Paper no. 355.

Maddison, A. (1987), 'Growth and Slowdown in Advanced Capitalist Countries: Techniques of Quantitative Assessment', *Journal of Economic Literature*, 24, 649–98.

—— (1989), *The World Economy in the Twentieth Century*, Paris, OECD.

Meeks, G. (1977), *Disappointing Marriage*, Cambridge, Cambridge University Press.

Mercer, H. (1989), 'The Evolution of British Government Policy towards Competition in Private Industry, 1940–1956', Ph.D. Thesis, London School of Economics.

Metcalf, D. (1989), 'Water Notes Dry Up: The Impact of the Donovan Reform Proposals and Thatcherism at Work on Labour Productivity in British Manufacturing Industry', *British Journal of Industrial Relations*, 27, 1–31.

—— Nickell, S., and Floros, N. (1980), 'Still Searching for an Explanation of Unemployment in Interwar Britain', Centre for Labour Economics, London School of Economics Discussion Paper no. 71.

Morris, D. J., and Stout, D. K. (1985), 'Industrial Policy', in D. J. Morris (ed.), *The Economic System in the UK*, Oxford, Oxford University Press.

National Economic Development Office (1983), *Education and Industry*, London.

Nickell, S. (1990), 'Inflation and the UK Labour Market', *Oxford Review of Economic Policy*, 6(4), 26–35.

OECD (1977), *Learning Opportunities for Adults, Vol. IV Participation in Adult Education*, Paris.

—— (1989), *Education in OECD Countries, 1986/7*, Paris.

—— (1990), *Main Science and Technology Indicators*, Paris.

—— (1991), *Labour Force Statistics*, Paris.

Olson, M. (1982), *The Rise and Decline of Nations*, New Haven, Yale University Press.

Panic, M. (1976), *UK and West German Manufacturing Industry, 1954–1972: A Comparison of Performance and Structure*, London, NEDC.

Patel, P., and Pavitt, K. (1987), 'The Elements of British Technological Competitiveness', *National Institute Economic Review*, 122, 72–83.

Pavitt, K., and Patel, P. (1988), 'The International Distribution and Determinants of Technological Activities', *Oxford Review of Economic Policy*, 4(4), 35–55.

Prais, S. J. (1981), *Productivity and Industrial Structure*, Cambridge, Cambridge University Press.

—— (1989), 'Qualified Manpower in Engineering', *National Institute Economic Review*, 127, 76–83.

Pratten, C. F., and Atkinson, A. G. (1976), 'The Use of Manpower in British Industry', *Department of Employment Gazette*, 84, 571–6.

Rentoul, J. (1990), 'Individualism', in R. Jowell, S. Witherspoon, and L. Brook (eds.), *British Social Attitudes*, 7th Report, Aldershot, Gower.

Romer, P. (1986), 'Increasing Returns and Long Run Growth', *Journal of Political Economy*, 94, 1002–37.

Royal Commission on the Distribution of Income and Wealth (1979), *Report*, London, HMSO.

Sanders, D. (1991), 'Government Popularity and the Next General Election', *Political Quarterly*, 62, 235–61.

Saul, S. B. (1979), 'Research and Development in British Industry from the End of the Nineteenth Century to the 1960s', in T. C. Smout (ed.), *The Search for Wealth and Stability*, London, Macmillan.

Sheldrake, J., and Vickerstaff, S. (1987), *The History of Industrial Training in Britain*, Aldershot, Avebury.

Shepherd, J. (1987), 'Industrial Support Policies', *National Institute Economic Review*, 122, 59–71.

Silberston, A. (1981), 'Industrial Policies in Britain, 1960–1980', in C. Carter (ed.), *Industrial Policy and Innovation*, London, Heinemann.

Singh, A. (1975). 'Takeovers, Natural Selection and the Theory of the Firm: Evidence from the Postwar UK Experience', *Economic Journal*, 85, 497–515.

Solow, R. (1970), *Growth Theory: An Exposition*, Oxford, Oxford University Press.

Steedman, H. (1990), 'Improvement in Workforce Qualifications: Britain and France, 1979–88', *National Institute Economic Review*, 133, 50–61.

Stoneman, P., and Vickers, J. (1988), 'The Assessment: The Economics of Technology Policy', *Oxford Review of Economic Policy*, 4(4), i–xvi.

Supple, B. E. (1987), *History of the British Coal Industry, Vol. 4, 1913–1946*, Oxford, Oxford University Press.

Szirmai, A., and Pilat, D. (1990), 'Comparisons of Purchasing Power, Real Output and Labour Productivity in Manufacturing in Japan, South Korea and the USA, 1975–85', *Review of Income and Wealth*, 36, 1–31.

Taylor-Gooby, P. (1990), 'Social Welfare: the Unkindest Cuts', in R. Jowell, S. Witherspoon, and L. Brook (eds.), *British Social Attitudes*, 7th Report, Aldershot, Gower.

Tolliday, S. (1987), *Business, Banking and Politics: The Case of British Steel, 1918–1939*, Cambridge, Mass., Harvard University Press.

Van Ark, B. (1990), 'Comparative Levels of Manufacturing Productivity in Postwar Europe: Measurement and Comparisons', *Oxford Bulletin of Economics and Statistics*, 52, 343–74.

Vickers, J., and Yarrow, G. (1988), *Privatization: An Economic Analysis*, Cambridge, Mass., MIT Press.

Vickerstaff, S. (1985), 'Industrial Training in Britain: The Dilemmas of a Corporatist Policy', in A. Cawson (ed.), *Organised Interests and the State*, London, SAGE Publications.

Wagner, K. (1991), 'Training Efforts and Industrial Efficiency in West Germany', in J. Stevens and R. Mackay (eds.), *Training and Competitiveness*, London, Kogan Page.

Index